P9-DIA-728

Fundamental Constants

Quantity	Symbol	Approximate Value	Current Best Value†
Speed of light in vacuum	c	3.00×10^8 m/s	2.99792458×10^8 m/s
Gravitational constant	G	6.67×10^{-11} N·m²/kg²	$6.6742(10) \times 10^{-11}$ N·m²/kg²
Avogadro's number	N_A	6.02×10^{23} mol^{-1}	$6.0221415(10) \times 10^{23}$ mol^{-1}
Gas constant	R	8.314 J/mol·K = 1.99 cal/mol·K = 0.0821 L·atm/mol·K	8.314472(15) J/mol·K
Boltzmann's constant	k	1.38×10^{-23} J/K	$1.3806505(24) \times 10^{-23}$ J/K
Charge on electron	e	1.60×10^{-19} C	$1.60217653(14) \times 10^{-19}$ C
Stefan-Boltzmann constant	σ	5.67×10^{-8} W/m²·K⁴	$5.670400(40) \times 10^{-8}$ W/m²·K⁴
Permittivity of free space	$\epsilon_0 = (1/c^2\mu_0)$	8.85×10^{-12} C²/N·m²	$8.854187817\ldots \times 10^{-12}$ C²/N·m²
Permeability of free space	μ_0	$4\pi \times 10^{-7}$ T·m/A	$1.2566370614\ldots \times 10^{-6}$ T·m/A
Planck's constant	h	6.63×10^{-34} J·s	$6.6260693(11) \times 10^{-34}$ J·s
Electron rest mass	m_e	9.11×10^{-31} kg = 0.000549 u = 0.511 MeV/c^2	$9.1093826(16) \times 10^{-31}$ kg = $5.4857990945(24) \times 10^{-4}$ u
Proton rest mass	m_p	1.6726×10^{-27} kg = 1.00728 u = 938.3 MeV/c^2	$1.67262171(29) \times 10^{-27}$ kg = 1.00727646688(13) u
Neutron rest mass	m_n	1.6749×10^{-27} kg = 1.008665 u = 939.6 MeV/c^2	$1.67492728(29) \times 10^{-27}$ kg = 1.00866491560(55) u
Atomic mass unit (1 u)		1.6605×10^{-27} kg = 931.5 MeV/c^2	$1.66053886(28) \times 10^{-27}$ kg = 931.494043(80) MeV/c^2

† CODATA (12/03), Peter J. Mohr and Barry N. Taylor, National Institute of Standards and Technology. Numbers in parentheses indicate one-standard-deviation experimental uncertainties in final digits. Values without parentheses are exact (i.e., defined quantities).

Other Useful Data

Joule equivalent (1 cal)	4.186 J
Absolute zero (0 K)	−273.15°C
Acceleration due to gravity at Earth's surface (avg.)	9.80 m/s² (= g)
Speed of sound in air (20°C)	343 m/s
Density of air (dry)	1.29 kg/m³
Earth: Mass	5.98×10^{24} kg
Radius (mean)	6.38×10^3 km
Moon: Mass	7.35×10^{22} kg
Radius (mean)	1.74×10^3 km
Sun: Mass	1.99×10^{30} kg
Radius (mean)	6.96×10^5 km
Earth–Sun distance (mean)	149.6×10^6 km
Earth–Moon distance (mean)	384×10^3 km

The Greek Alphabet

Alpha	Α	α	Nu	Ν	ν
Beta	Β	β	Xi	Ξ	ξ
Gamma	Γ	γ	Omicron	Ο	o
Delta	Δ	δ	Pi	Π	π
Epsilon	Ε	ε	Rho	Ρ	ρ
Zeta	Ζ	ζ	Sigma	Σ	σ
Eta	Η	η	Tau	Τ	τ
Theta	Θ	θ	Upsilon	Υ	υ
Iota	Ι	ι	Phi	Φ	ϕ, φ
Kappa	Κ	κ	Chi	Χ	χ
Lambda	Λ	λ	Psi	Ψ	ψ
Mu	Μ	μ	Omega	Ω	ω

Values of Some Numbers

$\pi = 3.1415927$	$\sqrt{2} = 1.4142136$	$\ln 2 = 0.6931472$	$\log_{10} e = 0.4342945$
$e = 2.7182818$	$\sqrt{3} = 1.7320508$	$\ln 10 = 2.3025851$	1 rad = 57.2957795°

Mathematical Signs and Symbols

$\propto$	is proportional to	$\leq$	is less than or equal to
$=$	is equal to	$\geq$	is greater than or equal to
$\approx$	is approximately equal to	Σ	sum of
$\neq$	is not equal to	$\bar{x}$	average value of x
$>$	is greater than	Δx	change in x
$\gg$	is much greater than	$\Delta x \to 0$	Δx approaches zero
$<$	is less than	$n!$	$n(n-1)(n-2)\ldots(1)$
$\ll$	is much less than		

Properties of Water

Density (4°C)	1.000 kg/m³
Heat of fusion (0°C)	333 kJ/kg (80 kcal/kg)
Heat of vaporization (100°C)	2260 kJ/kg (539 kcal/kg)
Specific heat (15°C)	4186 J/kg·C° (1.00 kcal/kg·C°)
Index of refraction	1.33

Unit Conversions (Equivalents)

Length

1 in. = 2.54 cm
1 cm = 0.3937 in.
1 ft = 30.48 cm
1 m = 39.37 in. = 3.281 ft
1 mi = 5280 ft = 1.609 km
1 km = 0.6214 mi
1 nautical mile (U.S.) = 1.151 mi = 6076 ft = 1.852 km
1 fermi = 1 femtometer (fm) = 10^{-15} m
1 angstrom (Å) = 10^{-10} m = 0.1 nm
1 light-year (ly) = 9.461×10^{15} m
1 parsec = 3.26 ly = 3.09×10^{16} m

Volume

1 liter (L) = 1000 mL = 1000 cm^3 = 1.0×10^{-3} m^3 = 1.057 qt (U.S.) = 61.02 $in.^3$
1 gal (U.S.) = 4 qt (U.S.) = 231 $in.^3$ = 3.785 L = 0.8327 gal (British)
1 quart (U.S.) = 2 pints (U.S.) = 946 mL
1 pint (British) = 1.20 pints (U.S.) = 568 mL
1 m^3 = 35.31 ft^3

Speed

1 mi/h = 1.467 ft/s = 1.609 km/h = 0.447 m/s
1 km/h = 0.278 m/s = 0.621 mi/h
1 ft/s = 0.305 m/s = 0.682 mi/h
1 m/s = 3.281 ft/s = 3.600 km/h = 2.237 mi/h
1 knot = 1.151 mi/h = 0.5144 m/s

Angle

1 radian (rad) = 57.30° = 57°18′
1° = 0.01745 rad
1 rev/min (rpm) = 0.1047 rad/s

Time

1 day = 8.64×10^4 s
1 year = 3.156×10^7 s

Mass

1 atomic mass unit (u) = 1.6605×10^{-27} kg
1 kg = 0.0685 slug
[1 kg has a weight of 2.20 lb where $g = 9.80\ m/s^2$.]

Force

1 lb = 4.45 N
1 N = 10^5 dyne = 0.225 lb

Energy and Work

1 J = 10^7 ergs = 0.738 ft·lb
1 ft·lb = 1.36 J = 1.29×10^{-3} Btu = 3.24×10^{-4} kcal
1 kcal = 4.186×10^3 J = 3.97 Btu
1 eV = 1.602×10^{-19} J
1 kWh = 3.60×10^6 J = 860 kcal

Power

1 W = 1 J/s = 0.738 ft·lb/s = 3.42 Btu/h
1 hp = 550 ft·lb/s = 746 W

Pressure

1 atm = 1.013 bar = 1.013×10^5 N/m^2
= 14.7 $lb/in.^2$ = 760 torr
1 $lb/in.^2$ = 6.90×10^3 N/m^2
1 Pa = 1 N/m^2 = 1.45×10^{-4} $lb/in.^2$

SI Derived Units and Their Abbreviations

Quantity	Unit	Abbreviation	In Terms of Base Units†
Force	newton	N	$kg \cdot m/s^2$
Energy and work	joule	J	$kg \cdot m^2/s^2$
Power	watt	W	$kg \cdot m^2/s^3$
Pressure	pascal	Pa	$kg/(m \cdot s^2)$
Frequency	hertz	Hz	s^{-1}
Electric charge	coulomb	C	$A \cdot s$
Electric potential	volt	V	$kg \cdot m^2/(A \cdot s^3)$
Electric resistance	ohm	Ω	$kg \cdot m^2/(A^2 \cdot s^3)$
Capacitance	farad	F	$A^2 \cdot s^4/(kg \cdot m^2)$
Magnetic field	tesla	T	$kg/(A \cdot s^2)$
Magnetic flux	weber	Wb	$kg \cdot m^2/(A \cdot s^2)$
Inductance	henry	H	$kg \cdot m^2/(s^2 \cdot A^2)$

† kg = kilogram (mass), m = meter (length), s = second (time), A = ampere (electric current).

Metric (SI) Multipliers

Prefix	Abbreviation	Value
yotta	Y	10^{24}
zeta	Z	10^{21}
exa	E	10^{18}
peta	P	10^{15}
tera	T	10^{12}
giga	G	10^9
mega	M	10^6
kilo	k	10^3
hecto	h	10^2
deka	da	10^1
deci	d	10^{-1}
centi	c	10^{-2}
milli	m	10^{-3}
micro	μ	10^{-6}
nano	n	10^{-9}
pico	p	10^{-12}
femto	f	10^{-15}
atto	a	10^{-18}
zepto	z	10^{-21}
yocto	y	10^{-24}

PHYSICS

PRINCIPLES WITH APPLICATIONS

PHYSICS

PRINCIPLES WITH APPLICATIONS

SIXTH EDITION

Volume 2

DOUGLAS C. GIANCOLI

PEARSON
Prentice Hall

Upper Saddle River, New Jersey 07458

Library of Congress Cataloging-in-Publication Data

Giancoli, Douglas C.
Physics : principles with applications / Douglas C. Giancoli.-- 6th ed.
p. cm.
Includes index.
ISBN 0-13-035257-8 (vol.2 pbk. : alk. paper) — ISBN 0-13-060620-0 (full book vol 1 & 2. casebound. : alk. paper) —
ISBN 0-13-035256-X (vol 1 reprint. pbk.. : alk. paper) — ISBN 0-13-184661-2 (Nasta edition : alk. paper) —
ISBN 0-13-191183-X (International edition : alk. paper)

1. Physics. I. Title.

QC23.G399 2005
530—dc22

2004017226

Editor-in-Chief, Science: John Challice
Senior Acquisitions Editor: Erik Fahlgren
Senior Development Editor: Karen Karlin
Vice President of Production and Manufacturing: David Riccardi
Executive Managing Editor: Kathleen Schiaparelli
Senior Production Editor: Susan Fisher
Production Editor: Chirag Thakkar
Manufacturing Manager: Trudy Pisciotti
Manufacturing Buyer: Alan Fischer
Managing Editor, Audio and Visual Assets: Patricia Burns
AV Project Managers: Adam Velthaus and Connie Long
Assistant Managing Editor, Science Media: Nicole Bush
Associate Editor: Christian Botting
Media Editor: Michael J. Richards
Director of Creative Services: Paul Belfanti
Advertising and Promotions Manager: Elise Schneider
Creative Director: Carole Anson
Art Director: Maureen Eide
Illustration: Artworks
Marketing Manager: Mark Pfaltzgraff
Editor-in-Chief of Development: Carol Trueheart
Director, Image Research Center: Melinda Reo
Photo Research: Mary Teresa Giancoli and Jerry Marshall
Manager, Rights and Permissions: Cynthia Vincenti
Copy Editor: Jocelyn Phillips
Indexer: Steele/Katigbak
Editorial Assistant: Andrew Sobel
Composition: Emilcomp srl / Prepare Inc.

Cover Photo: K2, the world's second highest summit, 8611 m (see pp. 10–11) as seen from Concordia. It is said to be the most difficult of the world's highest peaks (over 8000 m—see Table 1–6 and Example 1–3), and was first climbed by Lino Lacedelli and Achille Compagnoni in 1954. (Art Wolfe/Getty Images, Inc.)

PEARSON Prentice Hall
Published by Pearson Education, Inc.
Pearson Prentice Hall
Pearson Education, Inc.
Upper Saddle River, NJ 07458

Printed in the United States of America
10 9 8 7 6 5 4 3 2 1

ISBN 0-13-035257-8

Pearson Education Ltd., *London*
Pearson Education Australia Pty., Limited, *Sydney*
Pearson Education Singapore, Pte. Ltd.
Pearson Education North Asia Ltd., *Hong Kong*
Pearson Education Canada, Ltd., *Toronto*
Pearson Educación de Mexico, S.A. de C.V.
Pearson Education—Japan, *Tokyo*
Pearson Education Malaysia, Pte. Ltd.

CONTENTS

Contents of Volume 2

31 Nuclear Energy; Effects and Uses of Radiation 863

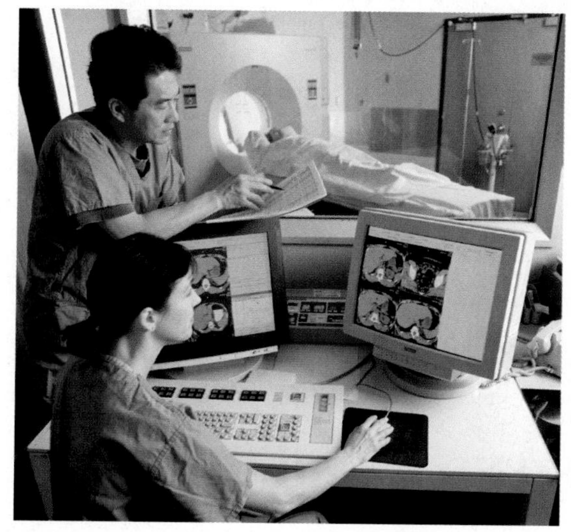

32 Elementary Particles 889

33 Astrophysics and Cosmology 914

Appendices

APPLICATIONS TO BIOLOGY AND MEDICINE

APPLICATIONS TO OTHER FIELDS AND EVERYDAY LIFE

PROBLEM SOLVING BOXES

NEW ▶ ***Marginal notes: Caution.*** Margin notes, in blue, point out main topics acting as a sort of outline and as an aid to find topics in review. They also point out applications and problem-solving hints. A new type, labeled CAUTION, points out possible misunderstandings discussed in the adjacent text.

Deletions. To keep the book from being too long, and also to reduce the burden on students in more advanced topics, many topics have been shortened or streamlined, and a few dropped.

New Physics Topics and Major Revisions

Here is a list of major changes or additions, but there are many others:

- Symmetry used more, including for solving Problems
- NEW ▶ Dimensional analysis, optional (Ch. 1)
- More graphs in kinematics (Ch. 2)
- Engine efficiency (Chs. 6, 15)
- Work-energy principle, and conservation of energy: new subsection (Ch. 6); carried through in thermodynamics (Ch. 15) and electricity (Ch. 17)
- NEW ▶ Force on tennis ball by racket (Ch. 7)
- NEW ▶ Airplane wings, curve balls, sailboats, and other applications of Bernoulli's principle: improved and clarified with new material (Ch. 10)
- Distinguish wave interference in space and in time (beats) (Ch. 11)
- Doppler shift for light (Ch. 12 now, as well as Ch. 33)
- NEW ▶ Giant star radius (Ch. 14)
- First law of thermodynamics rewritten and extended, connected better to work-energy principle and energy conservation (Ch. 15)
- Energy resources shortened (Ch. 15)
- NEW ▶ SEER rating (Ch. 15)
- NEW ▶ Separation of charge in nonconductors (Ch. 16)
- NEW ▶ Gauss's law, optional (Ch. 16)
- NEW ▶ Photocopiers and computer printers (Ch. 16)
- Electric force and field directions emphasized more (Chs. 16, 17)
- Electric potential related better to work, more detail (Ch. 17)
- NEW ▶ Dielectric effect on capacitor with and without connection to voltage plus other details (Ch. 17)
- NEW ▶ Parallel-plate capacitor derivation, optional (Ch. 17)
- NEW ▶ Electric hazards, grounding, safety, current interrupters: expanded with much new material (Chs. 17, 18, 19 especially, 20, 21)
- NEW ▶ Electric current, misconceptions discussed in Chapter 18
- Superconductivity updated (Ch. 18)
- Terminal voltage and emf reorganized, with more detail (Ch. 19)
- Magnetic materials shortened (Ch. 20)
- NEW ▶ Right-hand rules summarized in a Table (Ch. 20)
- Faraday's and Lenz's laws expanded (Ch. 21)
- AC circuits shortened (Ch. 21), displacement current downplayed (Ch. 22)
- NEW ▶ Radiation pressure and momentum of EM waves (Ch. 22)
- NEW ▶ Where to see yourself in a mirror; where you can actually *see* a lens image (Ch. 23)
- NEW ▶ Liquid crystal displays (LCD) (Ch. 24)
- NEW ▶ Physics behind digital cameras and CCD (Ch. 25)
- NEW ▶ Seeing under water (Ch. 25)
- Relativistic mass redone (Ch. 26)
- NEW ▶ Revolutionary results in cosmology: flatness and age of universe, WMAP, SDSS, dark matter, and dark energy (Ch. 33)
- NEW ▶ Specific heats of gases, equipartition of energy (Appendix)

Problem Solving, with New and Improved Approaches

Being able to solve problems is a valuable technique in general. Solving problems is also an effective way to understand the physics more deeply. Here are some of the ways this book uses to help students become effective problem solvers.

Problem Solving Boxes, about 20 of them, are found throughout the book (there is a list on p. xiii.). Each one outlines a step-by-step approach to solving problems in general, or specifically for the material being covered. The best students may find these "boxes" unnecessary (they can skip them), but many students may find it helpful to be reminded of the general approach and of steps they can take to get started. The general Problem Solving Box in Section 4–9 is placed there, after students have had some experience wrestling with problems, so they may be motivated to read it with close attention. Section 4–9 can be covered earlier if desired. Problem Solving Boxes are not intended to be a prescription, but rather a guide. Hence they sometimes follow the Examples to serve as a summary for future use.

Problem Solving Sections (such as Sections 2–6, 3–6, 4–7, 6–7, 8–6, and 13–8) are intended to provide extra drill in areas where solving problems is especially important.

Examples: Worked-out Examples, each with a title for easy reference, fall into four categories:

(1) The majority are regular worked-out Examples that serve as "practice problems." New ones have been added, a few old ones have been dropped, and many have been reworked to provide greater clarity, more math steps, more of "why we do it this way," and with the new Approach paragraph more discussion of the reasoning and approach. The aim is to "think aloud" with the students, leading them to develop insight. The level of the worked-out Examples for most topics increases gradually, with the more complicated ones being on a par with the most difficult Problems at the end of each Chapter. Many Examples provide relevant applications to various fields and to everyday life.

(2) ***Step-by-step Examples:*** After many of the Problem Solving Boxes, the next Example is done step-by-step following the steps of the preceding Box, just to show students how the Box can be used. Such solutions are long and can be redundant, so only one of each type is done in this manner. ◀ NEW

(3) ***Estimating Examples,*** roughly 10% of the total, are intended to develop the skills for making order-of-magnitude estimates, even when the data are scarce, and even when you might never have guessed that any result was possible at all. See, for example, Section 1–7, Examples 1–6 to 1–9.

(4) ***Conceptual Examples:*** Each is a brief Socratic question intended to stimulate student response before reading the Response given.

APPROACH paragraph: Worked-out numerical Examples now all have a short introductory paragraph before the Solution, outlining an approach and the steps we can take to solve the given problem. ◀ NEW

NOTE: Many Examples now have a brief "note" after the Solution, sometimes remarking on the Solution itself, sometimes mentioning an application, sometimes giving an alternate approach to solving the problem. These new Note paragraphs let the student know the Solution is finished, and now we mention a related issue(s). ◀ NEW

Additional Examples: Some physics subjects require many different worked-out Examples to clarify the issues. But so many Examples in a row can be overwhelming to some students. In those places, a subhead "Additional Example(s)" is meant to suggest to students that they could skip these in a first reading. When students include them during a second reading of the Chapter, they can give power to solve a greater range of Problems. ◀ NEW

Exercises within the text, after an Example or a derivation, which give students a chance to see if they have understood enough to answer a simple question or do a simple calculation. Answers are given at the bottom of the last page of each Chapter. ◀ NEW

Problems at the end of each Chapter have been increased in quality and quantity. Some old ones have been replaced or rewritten to make them clearer, and/or have had their numerical values changed. Each Chapter contains a large group of Problems arranged by Section and graded according to (approximate) difficulty: level I Problems are simple, designed to give students confidence; level II are "normal" Problems, providing more of a challenge and often the combination of two different concepts; level III are the most complex and are intended as "extra credit" Problems that will challenge even superior students. The arrangement by Section number is to help the instructors choose which material they want to emphasize, and means that those Problems depend on material up to and including that Section: earlier material may also be relied upon.
General Problems are unranked and grouped together at the end of each Chapter, accounting for perhaps 30% of all Problems. These are not necessarily more difficult, but they may be more likely to call on material from earlier Chapters. They are useful for instructors who want to give students a few Problems without the clue as to what Section must be referred to or how hard they are.
Questions, also at the end of each Chapter, are conceptual. They help students to use and apply the principles and concepts, and thus deepen their understanding (or let them know they need to study more).

Assigning Problems

I suggest that instructors assign a significant number of the level I and level II Problems, as well as a small number of General Problems, and reserve level III Problems only as "extra credit" to stimulate the best students. Although most level I problems may seem easy, they help to build self-confidence—an important part of learning, especially in physics. Answers to odd-numbered Problems are given in the back of the book.

Organization

The general outline of this new edition retains a traditional order of topics: mechanics (Chapters 1 to 9); fluids, vibrations, waves, and sound (Chapters 10 to 12); kinetic theory and thermodynamics (Chapters 13 to 15); electricity and magnetism (Chapters 16 to 22); light (Chapters 23 to 25); and modern physics (Chapters 26 to 33). Nearly all topics customarily taught in introductory physics courses are included here.

The tradition of beginning with mechanics is sensible because it was developed first, historically, and because so much else in physics depends on it. Within mechanics, there are various ways to order topics, and this book allows for considerable flexibility. I prefer to cover statics after dynamics, partly because many students have trouble with the concept of force without motion. Furthermore, statics is a special case of dynamics—we study statics so that we can prevent structures from becoming dynamic (falling down). Nonetheless, statics (Chapter 9) could be covered earlier after a brief introduction to vectors. Another option is light, which I have placed after electricity and magnetism and EM waves. But light could be treated immediately after waves (Chapter 11). Special relativity (Chapter 26) could be treated along with mechanics, if desired—say, after Chapter 7.

Not every Chapter need be given equal weight. Whereas Chapter 4 or Chapter 21 might require $1\frac{1}{2}$ to 2 weeks of coverage, Chapter 12 or 22 may need only $\frac{1}{2}$ week or less. Because Chapter 11 covers standing waves, Chapter 12 could be left to the students to read on their own if little class time is available.

The book contains more material than can be covered in most one-year courses. Yet there is great flexibility in choice of topics. Sections marked with a star (*) are considered optional. They contain slightly more advanced physics material (perhaps material not usually covered in typical courses) and/or interesting applications. They contain no material needed in later Chapters, except perhaps in later optional Sections. Not all unstarred Sections must be covered; there remains considerable flexibility in the choice of material. For a brief course, all optional material could be dropped, as well as major parts of Chapters 10, 12, 19, 22, 28, 29, 32, and 33, and perhaps selected parts of Chapters 7, 8, 9, 15, 21, 24, 25, and 31. Topics not covered in class can be a resource to students for later study.

New Applications

Relevant applications of physics to biology and medicine, as well as to architecture, other fields, and everyday life, have always been a strong feature of this book, and continue to be. Applications are interesting in themselves, plus they answer the students' question, "Why must I study physics?" New applications have been added. Here are a few of the new ones (see list after Table of Contents, pages xii and xiii).

◀ ALL ARE NEW

- Digital cameras, charge coupled devices (CCD) (Ch. 25)
- Liquid Crystal Displays (LCD) (Ch. 24)
- Electric safety, hazards, and various types of current interrupters and circuit breakers (Chs. 17, 18, 19, 20, 21)
- Photocopy machines (Ch. 16)
- Inkjet and Laser printers (Ch. 16)
- World's tallest peaks (unit conversion, Ch. 1)
- Airport metal detectors (Ch. 21)
- Capacitor uses (Ch. 17)
- Underwater vision (Ch. 25)
- SEER rating (Ch. 15)
- Curve ball (Ch. 10)
- Jump starting a car (Ch. 19)
- *RC* circuits in pacemakers, turn signals, wipers (Ch. 19)
- Digital voltmeters (Ch. 19)

Thanks

Over 50 physics professors provided input and direct feedback on every aspect of the text: organization, content, figures, and suggestions for new Examples and Problems. The reviewers for this sixth edition are listed below. I owe each of them a debt of gratitude:

Zaven Altounian (McGill University)
David Amadio (Cypress Falls Senior High School)
Andrew Bacher (Indiana University)
Rama Bansil (Boston University)
Mitchell C. Begelman (University of Colorado)
Cornelius Bennhold (George Washington University)
Mike Berger (Indiana University)
George W. Brandenburg (Harvard University)
Robert Coakley (University of Southern Maine)
Renee D. Diehl (Penn State University)
Kathryn Dimiduk (University of New Mexico)
Leroy W. Dubeck (Temple University)
Andrew Duffy (Boston University)
John J. Dykla (Loyola University Chicago)
John Essick (Reed College)
David Faust (Mt. Hood Community College)
Gerald Feldman (George Washington University)
Frank A. Ferrone (Drexel University)
Alex Filippenko (University of California, Berkeley)
Richard Firestone (Lawrence Berkeley Lab)
Theodore Gotis (Oakton Community College)
J. Erik Hendrickson (University of Wisconsin, Eau Claire)
Laurent Hodges (Iowa State University)
Brian Houser (Eastern Washington University)
Brad Johnson (Western Washington University)
Randall S. Jones (Loyola College of Maryland)
Joseph A. Keane (St. Thomas Aquinas College)
Arthur Kosowsky (Rutgers University)
Amitabh Lath (Rutgers University)
Paul L. Lee (California State University, Northridge)
Jerome R. Long (Virginia Tech)
Mark Lucas (Ohio University)
Dan MacIsaac (Northern Arizona University)
William W. McNairy (Duke University)
Laszlo Mihaly (SUNY Stony Brook)
Peter J. Mohr (NIST)
Lisa K. Morris (Washington State University)
Paul Morris (Abilene Christian University)
Hon-Kie Ng (Florida State University)
Mark Oreglia (University of Chicago)
Lyman Page (Princeton University)
Bruce Partridge (Haverford College)
R. Daryl Pedigo (University of Washington)
Robert Pelcovits (Brown University)
Alan Pepper (Campbell School, Adelaide, Australia)
Kevin T. Pitts (University of Illinois)
Steven Pollock (University of Colorado, Boulder)
W. Steve Quon (Ventura College)
Michele Rallis (Ohio State University)
James J. Rhyne (University of Missouri, Columbia)
Paul L. Richards (University of California, Berkeley)
Dennis Rioux (University of Wisconsin, Oshkosh)
Robert Ross (University of Detroit, Mercy)
Roy S. Rubins (University of Texas, Arlington)
Wolfgang Rueckner (Harvard University Extension)
Randall J. Scalise (Southern Methodist University)
Arthur G. Schmidt (Northwestern University)
Cindy Schwarz (Vassar College)

Bartlett M. Sheinberg (Houston Community College)
J. L. Shinpaugh (East Carolina University)
Ross L. Spencer (Brigham Young University)
Mark Sprague (East Carolina University)
Michael G. Strauss (University of Oklahoma)
Chun Fu Su (Mississippi State University)
Ronald G. Taback (Youngstown State University)
Leo H. Takahashi (Pennsylvania State University, Beaver)
Raymond C. Turner (Clemson University)
Robert C. Webb (Texas A&M University)
Arthur Wiggins (Oakland Community College)
Stanley Wojcicki (Stanford University)
Edward L. Wright (University of California, Los Angeles)
Andrzej Zieminski (Indiana University)

I am grateful also to those other physicist reviewers of earlier editions:

David B. Aaron (South Dakota State University)
Narahari Achar (Memphis State University)
William T. Achor (Western Maryland College)
Arthur Alt (College of Great Falls)
John Anderson (University of Pittsburgh)
Subhash Antani (Edgewood College)
Atam P. Arya (West Virginia University)
Sirus Aryainejad (Eastern Illinois University)
Charles R. Bacon (Ferris State University)
Arthur Ballato (Brookhaven National Laboratory)
David E. Bannon (Chemeketa Community College)
Gene Barnes (California State University, Sacramento)
Isaac Bass
Jacob Becher (Old Dominion University)
Paul A. Bender (Washington State University)
Michael S. Berger (Indiana University)
Donald E. Bowen (Stephen F. Austin University)
Joseph Boyle (Miami-Dade Community College)
Peter Brancazio (Brooklyn College, CUNY)
Michael E. Browne (University of Idaho)
Michael Broyles (Collin County Community College)
Anthony Buffa (California Polytechnic State University)
David Bushnell (Northern Illinois University)
Neal M. Cason (University of Notre Dame)
H. R. Chandrasekhar (University of Missouri)
Ram D. Chaudhari (SUNY, Oswego)
K. Kelvin Cheng (Texas Tech University)
Lowell O. Christensen (American River College)
Mark W. Plano Clark (Doane College)
Irvine G. Clator (UNC, Wilmington)
Albert C. Claus (Loyola University of Chicago)
Scott Cohen (Portland State University)
Lawrence Coleman (University of California, Davis)
Lattie Collins (East Tennessee State University)
Sally Daniels (Oakland University)
Jack E. Denson (Mississippi State University)
Waren Deshotels (Marquette University)
Eric Dietz (California State University, Chico)
Frank Drake (University of California, Santa Cruz)
Paul Draper (University of Texas, Arlington)
Miles J. Dresser (Washington State University)
Ryan Droste (The College of Charleston)
F. Eugene Dunnam (University of Florida)
Len Feuerhelm (Oklahoma Christian University)
Donald Foster (Wichita State University)
Gregory E. Francis (Montana State University)
Philip Gash (California State University, Chico)
J. David Gavenda (University of Texas, Austin)
Simon George (California State University, Long Beach)
James Gerhart (University of Washington)
Bernard Gerstman (Florida International University)
Charles Glashausser (Rutgers University)
Grant W. Hart (Brigham Young University)
Hershel J. Hausman (Ohio State University)
Melissa Hill (Marquette University)
Mark Hillery (Hunter College)
Hans Hochheimer (Colorado State University)
Joseph M. Hoffman (Frostburg State University)
Peter Hoffman-Pinther (University of Houston, Downtown)
Alex Holloway (University of Nebraska, Omaha)
Fred W. Inman (Mankato State University)
M. Azad Islan (SUNY, Potsdam)
James P. Jacobs (University of Montana)
Larry D. Johnson (Northeast Louisiana University)
Gordon Jones (Mississippi State University)
Rex Joyner (Indiana Institute of Technology)
Sina David Kaviani (El Camino College)
Kirby W. Kemper (Florida State University)
Sanford Kern (Colorado State University)
James E. Kettler (Ohio University, Eastern Campus)
James R. Kirk (Edinboro University of Pennsylvania)
Alok Kuman (SUNY, Oswego)
Sung Kyu Kim (Macalester College)
Amer Lahamer (Berea College)
Clement Y. Lam (North Harris College)
David Lamp (Texas Tech University)
Peter Landry (McGill University)
Michael Lieber (University of Arkansas)
Bryan H. Long (Columbia State College)
Michael C. LoPresto (Henry Ford Community College)
James Madsen (University of Wisconsin, River Falls)
Ponn Mahes (Winthrop University)
Robert H. March (University of Wisconsin, Madison)
David Markowitz (University of Connecticut)
Daniel J. McLaughlin (University of Hartford)
E. R. Menzel (Texas Tech University)
Robert Messina
David Mills (College of the Redwoods)
George K. Miner (University of Dayton)
Victor Montemeyer (Middle Tennessee State University)
Marina Morrow (Lansing Community College)
Ed Nelson (University of Iowa)
Dennis Nemeschansky (USC)
Gregor Novak (Indiana University/Purdue University)
Roy J. Peterson (University of Colorado, Boulder)
Frederick M. Phelps (Central Michigan University)
Brian L. Pickering (Laney College)
T. A. K. Pillai (University of Wisconsin, La Crosse)
John Polo (Edinboro University of Pennsylvania)
Michael Ram (University of Buffalo)
John Reading (Texas A&M University)
David Reid (Eastern Michigan University)
Charles Richardson (University of Arkansas)
William Riley (Ohio State University)
Larry Rowan (University of North Carolina)
D. Lee Rutledge (Oklahoma State University)
Hajime Sakai (University of Massachusetts, Amherst)
Thomas Sayetta (East Carolina University)
Neil Schiller (Ocean County College)
Ann Schmiedekamp (Pennsylvania State University, Ogontz)
Juergen Schroeer (Illinois State University)
Mark Semon (Bates College)
James P. Sheerin (Eastern Michigan University)
Eric Sheldon (University of Massachusetts, Lowell)
K. Y. Shen (California State University, Long Beach)
Marc Sher (College of William and Mary)

Joseph Shinar (Iowa State University)
Thomas W. Sills (Wilbur Wright College)
Anthony A. Siluidi (Kent State University)
Michael A. Simon (Housatonic Community College)
Upindranath Singh (Embry-Riddle)
Michael I. Sobel (Brooklyn College)
Donald Sparks (Los Angeles Pierce College)
Thor F. Stromberg (New Mexico State University)
James F. Sullivan (University of Cincinnati)
Kenneth Swinney (Bevill State Community College)
Harold E. Taylor (Stockton State University)
John E. Teggins (Auburn University at Montgomery)
Colin Terry (Ventura College)
Michael Thoennessen (Michigan State University)
Kwok Yeung Tsang (Georgia Institute of Technology)
Jagdish K. Tuli (Brookhaven National Laboratory)
Paul Urone (CSU, Sacramento)
Linn D. Van Woerkom (Ohio State University)
S. L. Varghese (University of South Alabama)
Jearl Walker (Cleveland State University)
Robert A. Walking (University of Southern Maine)
Jai-Ching Wang (Alabama A&M University)
Thomas A. Weber (Iowa State University)
John C. Wells (Tennessee Technological)
Gareth Williams (San Jose State University)
Wendall S. Williams (Case Western Reserve University)
Jerry Wilson (Metropolitan State College at Denver)
Lowell Wood (University of Houston)
David Wright (Tidewater Community College)
Peter Zimmerman (Louisiana State University)

I owe special thanks to Profs. Bob Davis and J. Erik Hendrickson for much valuable input, and especially for working out all the Problems and producing the Solutions Manual with solutions to all Problems and Questions, as well as for providing the answers to odd-numbered Problems at the end of this book. Thanks as well to the team they managed (Profs. David Curott, Bryan Long, and Richard Louie) who also worked out all the Problems and Questions, each checking the others.

I am grateful to Profs. Robert Coakley, Lisa Morris, Kathryn Dimiduk, Robert Pelcovits, Raymond Turner, Cornelius Bennhold, Gerald Feldman, Alan Pepper, Michael Strauss, and Zaven Altounian, who inspired many of the Examples, Questions, Problems, and significant clarifications.

Chapter 33 on Cosmology and Astrophysics absorbed more time by far than any other Chapter because of the very recent, and ongoing, "revolutionary" results that I wanted to present. I was fortunate to receive generous input from some of the top experts in the field, to whom I owe a debt of gratitude: Paul Richards and Alex Filippenko (U.C. Berkeley), Lyman Page (Princeton and WMAP), Edward Wright (U.C.L.A. and WMAP), Mitchell Begelman (U. Colorado), Bruce Partridge (Haverford College), Arthur Kosowsky (Rutgers), and Michael Strauss (Princeton and SDSS).

I especially wish to thank Profs. Howard Shugart, Chris McKee, and many others at the University of California, Berkeley, Physics Department for helpful discussions, and for hospitality. Thanks also to Prof. Tito Arecchi and others at the Istituto Nazionale di Ottica, Florence, Italy.

Finally, I am most grateful to the many people at Prentice Hall with whom I worked on this project, especially Paul Corey, Erik Fahlgren, Andrew Sobel, Chirag Thakkar, John Challice, and above all to the highly professional and wonderfully dedicated Karen Karlin and Susan Fisher. The final responsibility for all errors lies with me. I welcome comments, corrections, and suggestions† as soon as possible to benefit students for the next reprint.

D.C.G.

† Please send to:
email: physics_service@prenhall.com
or by postal service: Physics Editor
Prentice Hall Inc.
One Lake Street
Upper Saddle River, NJ 07458

NOTES TO STUDENTS (AND INSTRUCTORS) ON THE FORMAT

1. Sections marked with a star (*) are considered optional. They can be omitted without interrupting the main flow of topics. No later material depends on them except possibly later starred Sections. They may be fun to read.
2. The customary conventions are used: symbols for quantities (such as m for mass) are italicized, whereas units (such as m for meter) are not italicized. Symbols for vectors are shown in boldface with a small arrow above: $\vec{\mathbf{F}}$.
3. Few equations are valid in all situations. Where practical, the limitations of important equations are stated in square brackets next to the equation. The equations that represent the great laws of physics are displayed with a tan background, as are a few other indispensable equations.
4. The number of significant figures (Section 1–4) should not be assumed to be greater or less than given: if a number is stated as (say) 6, with its units, it is meant to be 6 and not 6.0 or 6.00.
5. At the end of each Chapter is a set of Questions that students should attempt to answer (to themselves at least). These are followed by Problems which are ranked as level I, II, or III, according to estimated difficulty, with level I Problems being the easiest. Level II are normal Problems, and level III are for "extra credit." These ranked Problems are arranged by Section, but Problems for a given Section may depend on earlier material as well. There follows a group of General Problems, which are not arranged by Section nor ranked as to difficulty. Questions and Problems that relate to optional Sections are starred (*). Answers to odd-numbered Problems are given at the end of the book.
6. Being able to solve problems is a crucial part of learning physics, and provides a powerful means for understanding the concepts and principles. This book contains many aids to problem solving: (a) worked-out Examples and their solutions in the text (set off with a vertical blue line in the margin) which should be studied as an integral part of the text; (b) some of the worked-out Examples are Estimation Examples, which show how rough or approximate results can be obtained even if the given data are sparse (see Section 1–7); (c) special "Problem Solving Boxes" placed throughout the text to suggest a step-by-step approach to problem solving for a particular topic—but don't get the idea that every topic has its own "techniques," because the basics remain the same; some of these "Boxes" are followed by an Example that is solved by explicitly following the suggested steps; (d) special problem-solving Sections; (e) "Problem Solving" marginal notes (see point 9 below) which refer to hints for solving problems within the text; (f) Exercises within the text that you should work out immediately, and then check your response against the answer given at the bottom of the last page of that Chapter; (g) the Problems themselves at the end of each Chapter (point 5 above).
7. Conceptual Examples are conceptual rather than numerical. Each poses a question or two, which hopefully starts you to think and come up with a response. Give yourself a little time to come up with your own response before reading the Response given.
8. "Additional Examples" subheadings contain Examples that you could skip on a first reading, in case you are feeling overwhelmed. But a day or two later, when you read the Chapter a second time, try to work through these Examples too because they can give you more power in doing a wide range of Problems.
9. Margin notes: brief notes in the margin of almost every page are printed in blue and are of five types: (a) ordinary notes (the majority) that serve as a sort of outline of the text and can help you later locate important concepts and equations; (b) notes that refer to the great laws and principles of physics, and these are in capital letters and in a box for emphasis; (c) notes that refer to a problem-solving hint or technique treated in the text, and these say "Problem Solving"; (d) notes that refer to an application of physics in the text or an Example, and these say "Physics Applied"; (e) "Caution" notes that point out a possible misconception spelled out in the adjacent text.
10. This book is printed in full color—but not simply to make it more attractive. The color is used above all in the Figures, to give them greater clarity for our analysis. The Table on the next page is a summary of which colors are used for the different kinds of vectors, for field lines, and for other symbols and objects. These colors are used consistently throughout the book.
11. Math review, plus some additional topics, are found in Appendices. Useful data, conversion factors, and math formulas are found inside the front and back covers.

USE OF COLOR

Vectors

A general vector	
resultant vector (sum) is slightly thicker	
components of any vector are dashed	
Displacement ($\vec{\mathbf{D}}, \vec{\mathbf{r}}$)	
Velocity ($\vec{\mathbf{v}}$)	
Acceleration ($\vec{\mathbf{a}}$)	
Force ($\vec{\mathbf{F}}$)	
Force on second or third object in same figure	
Momentum ($\vec{\mathbf{p}}$ or $m\vec{\mathbf{v}}$)	
Angular momentum ($\vec{\mathbf{L}}$)	
Angular velocity ($\vec{\boldsymbol{\omega}}$)	
Torque ($\vec{\boldsymbol{\tau}}$)	
Electric field ($\vec{\mathbf{E}}$)	
Magnetic field ($\vec{\mathbf{B}}$)	

Electricity and magnetism

- Electric field lines
- Equipotential lines
- Magnetic field lines
- Electric charge (+) — + or ● +
- Electric charge (–) — – or ● –

Electric circuit symbols

- Wire
- Resistor
- Capacitor
- Inductor
- Battery

Optics

- Light rays
- Object
- Real image (dashed)
- Virtual image (dashed and paler)

Other

- Energy level (atom, etc.)
- Measurement lines — ⊢1.0 m⊣
- Path of a moving object
- Direction of motion or current

This comb has acquired a static electric charge, either from passing through hair, or being rubbed by a cloth or paper towel. The electrical charge on the comb induces a polarization (separation of charge) in scraps of paper, and thus attracts them.

Our introduction to electricity in this Chapter covers conductors and insulators, and Coulomb's law which relates the force between two point charges as a function of their distance apart. We also introduce the powerful concept of electric field.

CHAPTER 16

Electric Charge and Electric Field

The word "electricity" may evoke an image of complex modern technology: lights, motors, electronics, and computers. But the electric force is thought to play an even deeper role in our lives. According to atomic theory, electric forces between atoms and molecules hold them together to form liquids and solids, and electric forces are also involved in the metabolic processes that occur within our bodies. Many of the forces we have dealt with so far, such as elastic forces, the normal force, and friction and other contact forces (pushes and pulls), are now considered to result from electric forces acting at the atomic level. Gravity, on the other hand, is a separate force.[†]

The earliest studies on electricity date back to the ancients, but it has been only in the past two centuries that electricity was studied in detail. We will discuss the development of ideas about electricity, including practical devices, as well as the relation to magnetism, in the next seven Chapters.

[†]As we discussed in Section 5–10, physicists in the twentieth century came to recognize four different fundamental forces in nature: (1) gravitational force, (2) electromagnetic force (we will see later that electric and magnetic forces are intimately related), (3) strong nuclear force, and (4) weak nuclear force. The last two forces operate at the level of the nucleus of an atom. Recent theory has combined the electromagnetic and weak nuclear forces so they are now considered to have a common origin known as the electroweak force. We will discuss these forces in later Chapters.

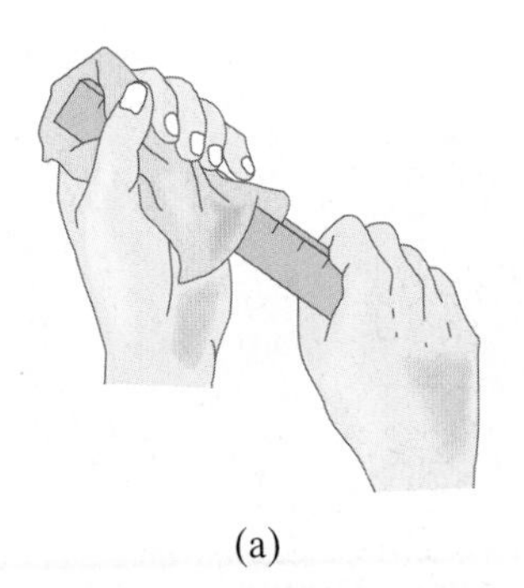

(a)

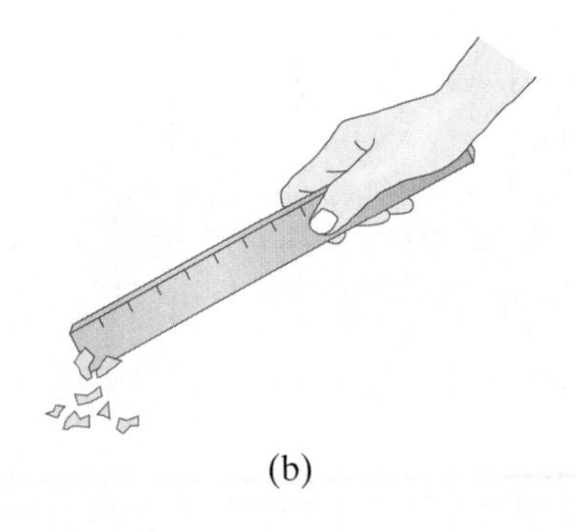

(b)

FIGURE 16–1 (a) Rub a plastic ruler and (b) bring it close to some tiny pieces of paper.

16–1 Static Electricity; Electric Charge and Its Conservation

FIGURE 16–2 Like charges repel one another; unlike charges attract.

(a) Two charged plastic rulers repel

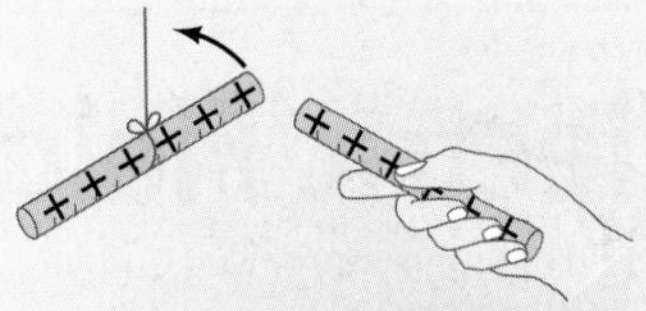

(b) Two charged glass rods repel

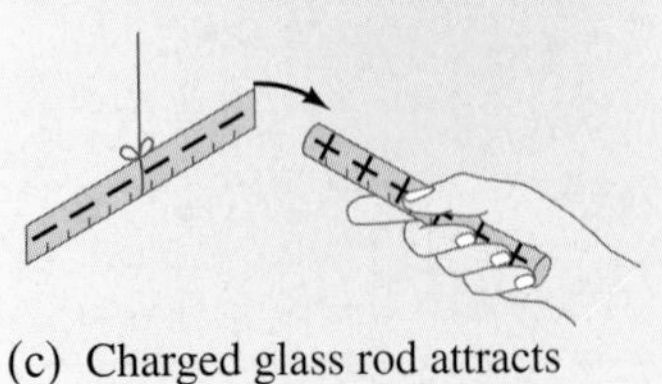

(c) Charged glass rod attracts charged plastic ruler

Like charges repel; unlike charges attract

The word *electricity* comes from the Greek word *elektron*, which means "amber." Amber is petrified tree resin, and the ancients knew that if you rub a piece of amber with a cloth, the amber attracts small pieces of leaves or dust. A piece of hard rubber, a glass rod, or a plastic ruler rubbed with a cloth will also display this "amber effect," or **static electricity** as we call it today. You can readily pick up small pieces of paper with a plastic comb or ruler that you've just vigorously rubbed with even a paper towel. See the photo on the previous page and Fig. 16–1. You have probably experienced static electricity when combing your hair or when taking a synthetic blouse or shirt from a clothes dryer. And you may have felt a shock when you touched a metal doorknob after sliding across a car seat or walking across a nylon carpet. In each case, an object becomes "charged" as a result of rubbing, and is said to possess a net **electric charge**.

Is all electric charge the same, or is there more than one type? In fact, there are *two* types of electric charge, as the following simple experiments show. A plastic ruler suspended by a thread is vigorously rubbed with a cloth to charge it. When a second plastic ruler, which has also been charged in the same way, is brought close to the first, it is found that the one ruler *repels* the other. This is shown in Fig. 16–2a. Similarly, if a rubbed glass rod is brought close to a second charged glass rod, again a repulsive force is seen to act, Fig. 16–2b. However, if the charged glass rod is brought close to the charged plastic ruler, it is found that they *attract* each other, Fig. 16–2c. The charge on the glass must therefore be different from that on the plastic. Indeed, it is found experimentally that all charged objects fall into one of two categories. Either they are attracted to the plastic and repelled by the glass; or they are repelled by the plastic and attracted to the glass. Thus there seem to be two, and only two, types of electric charge. Each type of charge repels the same type but attracts the opposite type. That is: **unlike charges attract; like charges repel**.

The two types of electric charge were referred to as *positive* and *negative* by the American statesman, philosopher, and scientist Benjamin Franklin (1706–1790). The choice of which name went with which type of charge was arbitrary. Franklin's choice set the charge on the rubbed glass rod to be positive charge, so the charge on a rubbed plastic ruler (or amber) is called negative charge. We still follow this convention today.

Franklin argued that whenever a certain amount of charge is produced on one object, an equal amount of the opposite type of charge is produced on another object. The positive and negative are to be treated *algebraically*, so during any process, the net change in the amount of charge produced is zero. For example, when a plastic ruler is rubbed with a paper towel, the plastic acquires a negative charge and the towel acquires an equal amount of positive charge. The charges are separated, but the sum of the two is zero.

This is an example of a law that is now well established: the **law of conservation of electric charge**, which states that

the net amount of electric charge produced in any process is zero,

or said another way,

no net electric charge can be created or destroyed.

LAW OF CONSERVATION OF ELECTRIC CHARGE

If one object (or a region of space) acquires a positive charge, then an equal amount of negative charge will be found in neighboring areas or objects. No violations have ever been found, and this conservation law is as firmly established as those for energy and momentum.

16–2 Electric Charge in the Atom

Only within the past century has it become clear that an understanding of electricity originates inside the atom itself. In later Chapters we will discuss atomic structure and the ideas that led to our present view of the atom in more detail. But it will help our understanding of electricity if we discuss it briefly now.

Electrons, protons, neutrons

A simplified model of an atom shows it as having a tiny but heavy, positively charged nucleus surrounded by one or more negatively charged electrons (Fig. 16–3). The nucleus contains protons, which are positively charged, and neutrons, which have no net electric charge. All protons and all electrons have exactly the same magnitude of electric charge; but their signs are opposite. Hence neutral atoms, having no net charge, contain equal numbers of protons and electrons. Sometimes, an atom may lose one or more of its electrons, or may gain extra electrons, in which case it will have a net positive or negative charge and is called an **ion**.

Ion

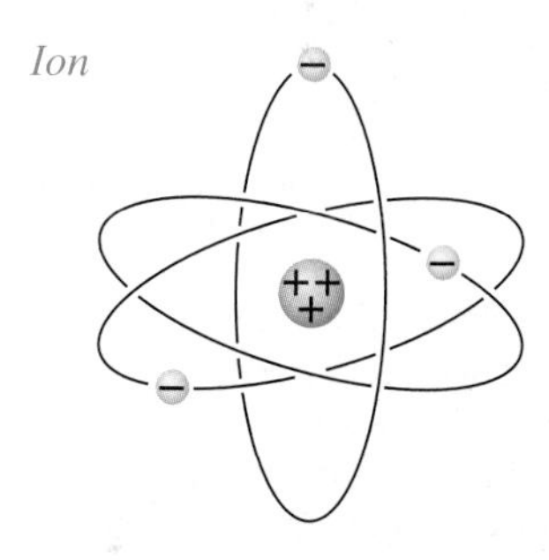

FIGURE 16–3 Simple model of the atom.

In solid materials the nuclei tend to remain close to fixed positions, whereas some of the electrons may move quite freely. When an object is *neutral*, it contains equal amounts of positive and negative charge. The charging of a solid object by rubbing can be explained by the transfer of electrons from one object to the other. When a plastic ruler becomes negatively charged by rubbing with a paper towel, the transfer of electrons from the towel to the plastic leaves the towel with a positive charge equal in magnitude to the negative charge acquired by the plastic. In liquids and gases, nuclei or ions can move as well as electrons.

Normally when objects are charged by rubbing, they hold their charge only for a limited time and eventually return to the neutral state. Where does the charge go? Usually the charge "leaks off" onto water molecules in the air. This is because water molecules are **polar**—that is, even though they are neutral, their charge is not distributed uniformly, Fig. 16–4. Thus the extra electrons on, say, a charged plastic ruler can "leak off" into the air because they are attracted to the positive end of water molecules. A positively charged object, on the other hand, can be neutralized by transfer of loosely held electrons from water molecules in the air. On dry days, static electricity is much more noticeable since the air contains fewer water molecules to allow leakage. On humid or rainy days, it is difficult to make any object hold a net charge for long.

Polar molecule

FIGURE 16–4 Diagram of a water molecule. Because it has opposite charges on different ends, it is called a "polar" molecule.

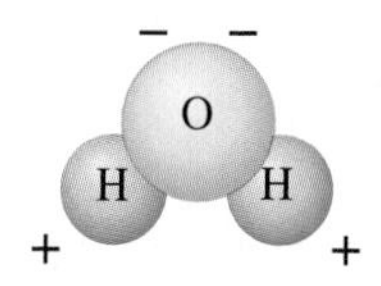

16–3 Insulators and Conductors

Suppose we have two metal spheres, one highly charged and the other electrically neutral (Fig. 16–5a). If we now place a metal object, such as a nail, so that it touches both spheres (Fig. 16–5b), the previously uncharged sphere quickly becomes charged. If, instead, we had connected the two spheres by a wooden rod or a piece of rubber (Fig. 16–5c), the uncharged ball would not become noticeably charged. Materials like the iron nail are said to be **conductors** of electricity, whereas wood and rubber are **nonconductors** or **insulators**.

Conductors and insulators

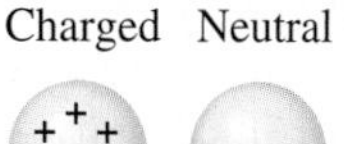

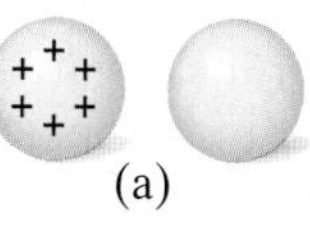

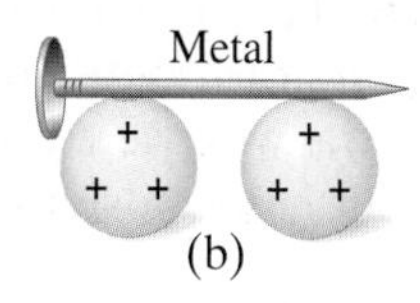

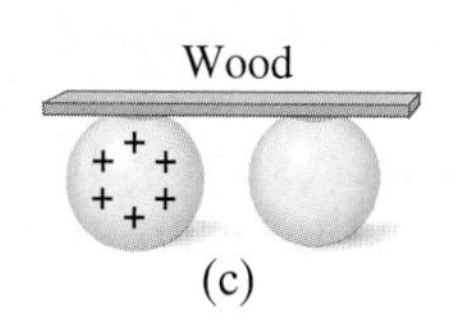

FIGURE 16–5 (a) A charged metal sphere and a neutral metal sphere. (b) The two spheres connected by a conductor (a metal nail), which conducts charge from one sphere to the other. (c) The two spheres connected by an insulator (wood); almost no charge is conducted.

Metals are good conductors

Metals are generally good conductors, whereas most other materials are insulators (although even insulators conduct electricity very slightly). Nearly all natural materials fall into one or the other of these two quite distinct categories. However, a few materials (notably silicon and germanium) fall into an intermediate category known as **semiconductors**.

From the atomic point of view, the electrons in an insulating material are bound very tightly to the nuclei. In a good conductor, on the other hand, some of the electrons are bound very loosely and can move about freely within the material (although they cannot *leave* the object easily) and are often referred to as *free electrons* or *conduction electrons*. When a positively charged object is brought close to or touches a conductor, the free electrons in the conductor are attracted by this positively charged object and move quickly toward it. On the other hand, the free electrons move swiftly away from a negatively charged object that is brought close. In a semiconductor, there are many fewer free electrons, and in an insulator, almost none.

16–4 Induced Charge; the Electroscope

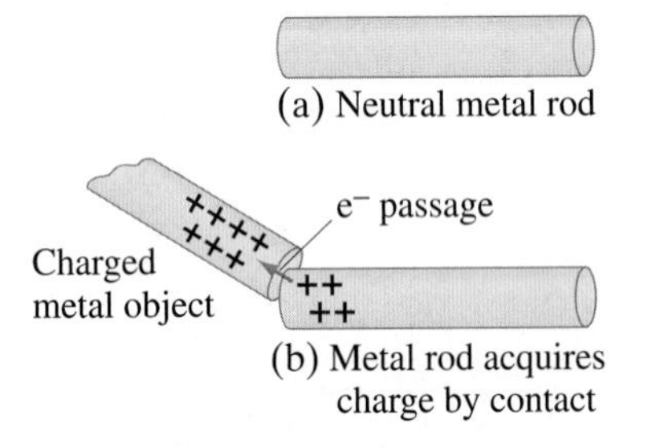

FIGURE 16–6 A neutral metal rod in (a) will acquire a positive charge if placed in contact (b) with a positively charged metal object. (Electrons move as shown by the orange arrow.) This is called charging by conduction.

Suppose a positively charged metal object is brought close to an uncharged metal object. If the two touch, the free electrons in the neutral one are attracted to the positively charged object and some will pass over to it, Fig. 16–6. Since the second object, originally neutral, is now missing some of its negative electrons, it will have a net positive charge. This process is called "charging by conduction," or "by contact," and the two objects end up with the same sign of charge.

Now suppose a positively charged object is brought close to a neutral metal rod, but does not touch it. Although the free electrons of the metal rod do not leave the rod, they still move within the metal toward the external positive charge, leaving a positive charge at the opposite end of the rod (Fig. 16–7). A charge is said to have been *induced* at the two ends of the metal rod. No net charge has been created in the rod: charges have merely been *separated*. The net charge on the metal rod is still zero. However, if the metal is broken into two pieces, we would have two charged objects: one charged positively and one charged negatively.

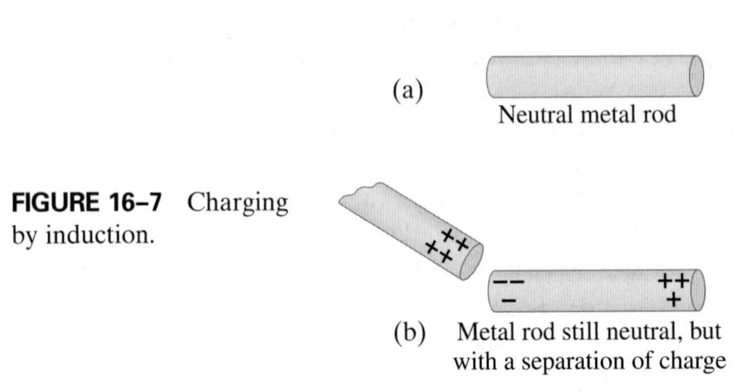

FIGURE 16–7 Charging by induction.

FIGURE 16–8 Inducing a charge on an object connected to ground.

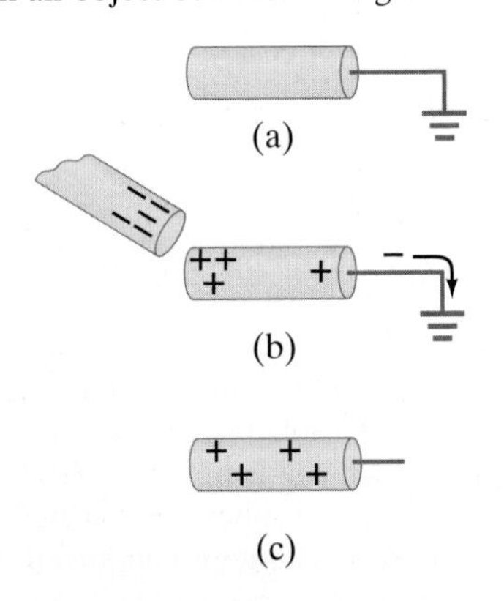

Another way to induce a net charge on a metal object is to first connect it with a conducting wire to the ground (or a conducting pipe leading into the ground) as shown in Fig. 16–8a (the symbol ⏚ means connected to "ground"). The object is then said to be "grounded" or "earthed." The Earth, because it is so large and can conduct, easily accepts or gives up electrons; hence it acts like a reservoir for charge. If a charged object—say negative this time—is brought up close to the metal object, free electrons in the metal are repelled and many of them move down the wire into the Earth, Fig. 16–8b. This leaves the metal positively charged. If the wire is now cut, the metal object will have a positive induced charge on it (Fig. 16–8c). If the wire were cut after the negative object is moved away, the electrons would all have moved back into the metal object and it would be neutral.

Charge separation can also be done in nonconductors. If you bring a positively charged object close to a neutral nonconductor as shown in Fig. 16–9, almost no electrons can move about freely within the nonconductor. But they can move slightly within their own atoms and molecules. Each oval in Fig. 16–9 represents a molecule (not to scale); the negatively charged electrons, attracted to the external positive charge, tend to move in its direction within their molecules. Because the negative charges in the nonconductor are nearer to the external positive charge, the nonconductor as a whole is attracted to the external positive charge (see the Chapter-opening photo, p. 439).

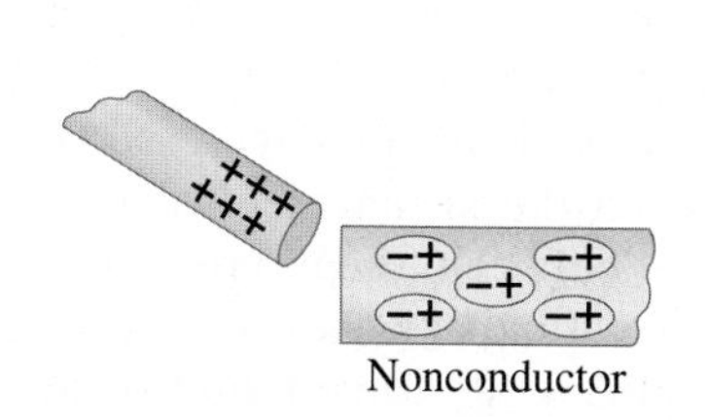

FIGURE 16–9 A charged object brought near an insulator causes a charge separation within the insulator's molecules.

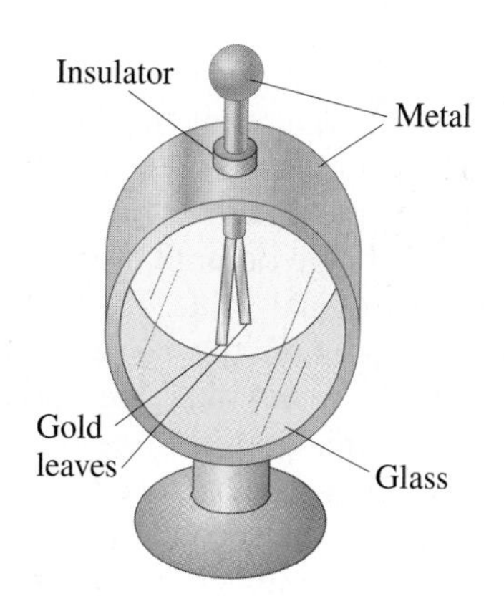

FIGURE 16–10 Electroscope.

An **electroscope** is a device that can be used for detecting charge. As shown in Fig. 16–10, inside of a case are two movable metal leaves, often made of gold. (Sometimes only one leaf is movable.) The leaves are connected by a conductor to a metal knob on the outside of the case, but are insulated from the case itself. If a positively charged object is brought close to the knob, a separation of charge is induced: electrons are attracted up into the knob, leaving the leaves positively charged, Fig. 16–11a. The two leaves repel each other as shown, because they are both positively charged. If, instead, the knob is charged by conduction, the whole apparatus acquires a net charge as shown in Fig. 16–11b. In either case, the greater the amount of charge, the greater the separation of the leaves.

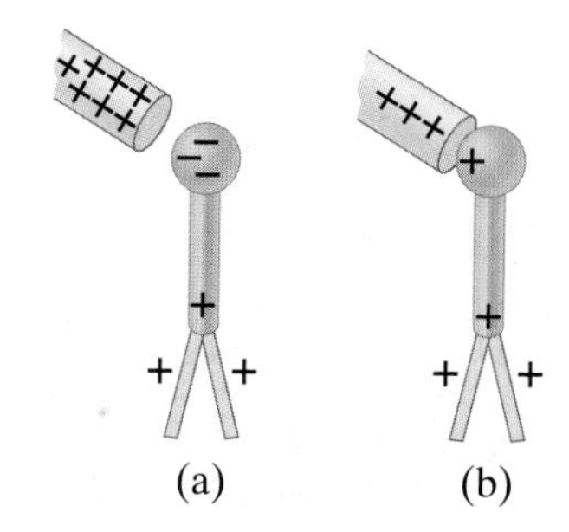

FIGURE 16–11 Electroscope charged (a) by induction, (b) by conduction.

Note that you cannot tell the sign of the charge in this way, since negative charge will cause the leaves to separate just as much as an equal amount of positive charge; in either case, the two leaves repel each other. An electroscope can, however, be used to determine the sign of the charge if it is first charged by conduction, say, negatively, as in Fig. 16–12a. Now if a negative object is brought close, as in Fig. 16–12b, more electrons are induced to move down into the leaves and they separate further. If a positive charge is brought close instead, the electrons are induced to flow upward, leaving the leaves less negative and their separation is reduced, Fig. 16–12c.

The electroscope was much used in the early studies of electricity. The same principle, aided by some electronics, is used in much more sensitive modern **electrometers**.

Electrometer

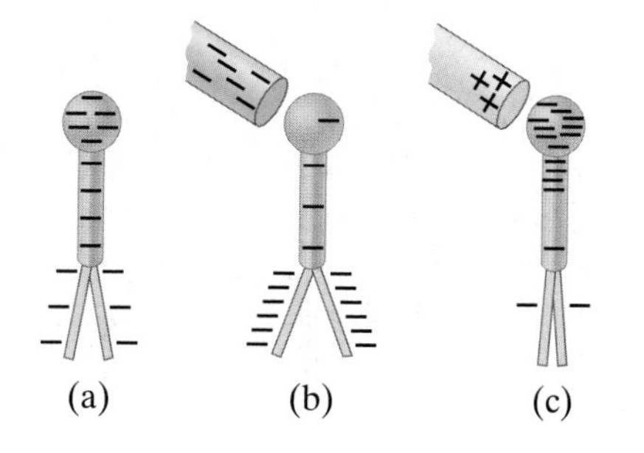

FIGURE 16–12 A previously charged electroscope can be used to determine the sign of a charged object.

16–5 Coulomb's Law

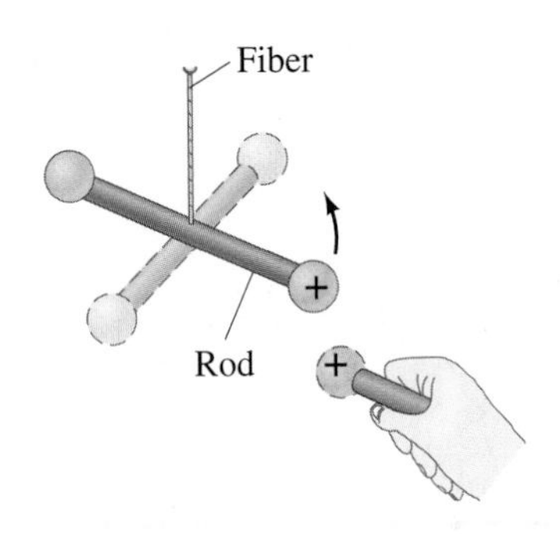

FIGURE 16–13 Principle of Coulomb's apparatus. It is similar to Cavendish's, which was used for the gravitational force. When an external charged sphere is placed close to the charged one on the suspended bar, the bar rotates slightly. The suspending fiber resists the twisting motion, and the angle of twist is proportional to the force applied. With this apparatus, Coulomb investigated how the electric force varies as a function of the magnitude of the charges and of the distance between them.

We have seen that an electric charge exerts a force of attraction or repulsion on other electric charges. What factors affect the magnitude of this force? To find an answer, the French physicist Charles Coulomb (1736–1806) investigated electric forces in the 1780s using a torsion balance (Fig. 16–13) much like that used by Cavendish for his studies of the gravitational force (Chapter 5).

Precise instruments for the measurement of electric charge were not available in Coulomb's time. Nonetheless, Coulomb was able to prepare small spheres with different magnitudes of charge in which the *ratio* of the charges was known.† Although he had some difficulty with induced charges, Coulomb was able to argue that the force one tiny charged object exerted on a second tiny charged object is directly proportional to the charge on each of them. That is, if the charge on either one of the objects was doubled, the force was doubled; and if the charge on both of the objects was doubled, the force increased to four times the original value. This was the case when the distance between the two charges remained the same. If the distance between them was allowed to increase, he found that the force decreased with the *square of the distance* between them. That is, if the distance was doubled, the force fell to one-fourth of its original value. Thus, Coulomb concluded, the force one small charged object exerts on a second one is proportional to the product of the magnitude of the charge on one, Q_1, times the magnitude of the charge on the other, Q_2, and inversely proportional to the square of the distance r between them (Fig. 16–14). As an equation, we can write **Coulomb's law** as

COULOMB'S LAW

$$F = k\frac{Q_1 Q_2}{r^2}, \qquad \text{[magnitudes]} \quad \textbf{(16–1)}$$

where k is a proportionality constant.‡

Force direction

Coulomb's law, Eq. 16–1, gives the *magnitude* of the electric force that either object exerts on the other. The *direction* of the electric force *is always along the line joining the two objects*. If the two charges have the same sign, the force on either object is directed away from the other (they repel each other). If the two charges have opposite signs, the force on one is directed toward the other (they attract). See Fig. 16–15. Notice that the force one charge exerts on the second is equal but opposite to that exerted by the second on the first, in accord with Newton's third law.

FIGURE 16–14 Coulomb's law, Eq. 16–1, gives the force between two point charges, Q_1 and Q_2, a distance r apart.

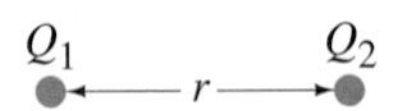

†Coulomb reasoned that if a charged conducting sphere is placed in contact with an identical uncharged sphere, the charge on the first would be shared equally by the two of them because of symmetry. He thus had a way to produce charges equal to $\frac{1}{2}, \frac{1}{4}$, and so on, of the original charge.

‡The validity of Coulomb's law today rests on precision measurements that are much more sophisticated than Coulomb's original experiment. The exponent, 2, in Coulomb's law has been shown to be accurate to 1 part in 10^{16} [that is, $2 \pm (1 \times 10^{-16})$].

FIGURE 16–15 Direction of the force depends on whether the charges have the same sign as in (a) and (b), or opposite signs (c).

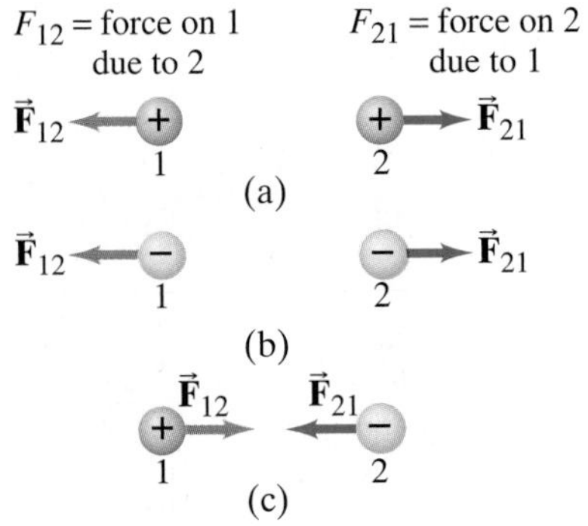

F_{12} = force on 1 due to 2
F_{21} = force on 2 due to 1

The SI unit of charge is the **coulomb** (C).[†] The precise definition of the coulomb today is in terms of electric current and magnetic field, and will be discussed later (Section 20–6). In SI units, k has the value

Unit for charge: the coulomb

$$k = 8.988 \times 10^9 \, \text{N}\cdot\text{m}^2/\text{C}^2$$

or, when we only need two significant figures,

$$k \approx 9.0 \times 10^9 \, \text{N}\cdot\text{m}^2/\text{C}^2.$$

Thus, 1 C is that amount of charge which, if placed on each of two point objects that are 1.0 m apart, will result in each object exerting a force of $(9.0 \times 10^9 \, \text{N}\cdot\text{m}^2/\text{C}^2)(1.0\,\text{C})(1.0\,\text{C})/(1.0\,\text{m})^2 = 9.0 \times 10^9\,\text{N}$ on the other. This would be an enormous force, equal to the weight of almost a million tons. We don't normally encounter charges as large as a coulomb.

Charges produced by rubbing ordinary objects (such as a comb or plastic ruler) are typically around a microcoulomb $(1\,\mu\text{C} = 10^{-6}\,\text{C})$ or less. Objects that carry a positive charge have a deficit of electrons, whereas negatively charged objects have an excess of electrons. The charge on one electron has been determined to have a magnitude of about 1.602×10^{-19} C, and is negative. This is the smallest charge found in nature,[‡] and because it is fundamental, it is given the symbol e and is often referred to as the *elementary charge*:

$$e = 1.602 \times 10^{-19}\,\text{C}.$$

Charge on electron (the elementary charge)

Note that e is defined as a positive number, so the charge on the electron is $-e$. (The charge on a proton, on the other hand, is $+e$.) Since an object cannot gain or lose a fraction of an electron, the net charge on any object must be an integral multiple of this charge. Electric charge is thus said to be **quantized** (existing only in discrete amounts: $1e$, $2e$, $3e$, etc.). Because e is so small, however, we normally don't notice this discreteness in macroscopic charges (1 μC requires about 10^{13} electrons), which thus seem continuous.

Electric charge is quantized

Coulomb's law looks a lot like the *law of universal gravitation*, $F = G m_1 m_2/r^2$, which expresses the gravitational force a mass m_1 exerts on a mass m_2 (Eq. 5–4). Both are inverse square laws $(F \propto 1/r^2)$. Both also have a proportionality to a property of each object—mass for gravity, electric charge for electricity. And both act over a distance (that is, there is no need for contact). A major difference between the two laws is that gravity is always an attractive force, whereas the electric force can be either attractive or repulsive. Electric charge comes in two types, positive and negative; gravitational mass is only positive.

Coulomb's law and the law of universal gravitation

The constant k in Eq. 16–1 is often written in terms of another constant, ϵ_0, called the **permittivity of free space**. It is related to k by $k = 1/4\pi\epsilon_0$. Coulomb's law can then be written

$$F = \frac{1}{4\pi\epsilon_0}\frac{Q_1 Q_2}{r^2}, \qquad \textbf{(16–2)}$$

COULOMB'S LAW (in terms of ϵ_0)

where

$$\epsilon_0 = \frac{1}{4\pi k} = 8.85 \times 10^{-12}\,\text{C}^2/\text{N}\cdot\text{m}^2.$$

Equation 16–2 looks more complicated than Eq. 16–1, but other fundamental equations we haven't seen yet are simpler in terms of ϵ_0 rather than k. It doesn't matter which form we use since Eqs. 16–1 and 16–2 are equivalent. (The latest precise values of e and ϵ_0 are given inside the front cover.)

[Our convention for units, such as $\text{C}^2/\text{N}\cdot\text{m}^2$ for ϵ_0, means m^2 is in the denominator. That is, $\text{C}^2/\text{N}\cdot\text{m}^2$ does *not* mean $\text{C}^2\cdot\text{m}^2/\text{N}$.]

Writing units

[†] In the once common cgs system of units, k is set equal to 1, and the unit of electric charge is called the *electrostatic unit* (esu) or the statcoulomb. One esu is defined as that charge, on each of two point objects 1 cm apart, that gives rise to a force of 1 dyne.

[‡] According to the standard model of elementary particle physics, subnuclear particles called quarks (Chapter 32) have a smaller charge than that on the electron, equal to $\frac{1}{3}e$ or $\frac{2}{3}e$. Quarks have not been detected directly as isolated objects, and theory indicates that free quarks may not be detectable.

16–7 The Electric Field

Many common forces might be referred to as "contact forces," such as your hands pushing or pulling a cart, or a tennis racket hitting a tennis ball.

In contrast, both the gravitational force and the electrical force act over a distance: there is a force between two objects even when the objects are not touching. The idea of a force *acting at a distance* was a difficult one for early thinkers. Newton himself felt uneasy with this idea when he published his law of universal gravitation. A helpful way to look at the situation uses the idea of the

Field

field, developed by the British scientist Michael Faraday (1791–1867). In the electrical case, according to Faraday, an *electric field* extends outward from every charge and permeates all of space (Fig. 16–22). If a second charge (call it Q_2) is placed near the first charge, it feels a force exerted by the electric field that is there (say, at point P in Fig. 16–22). The electric field at point P is considered to interact directly with charge Q_2 to produce the force on Q_2.

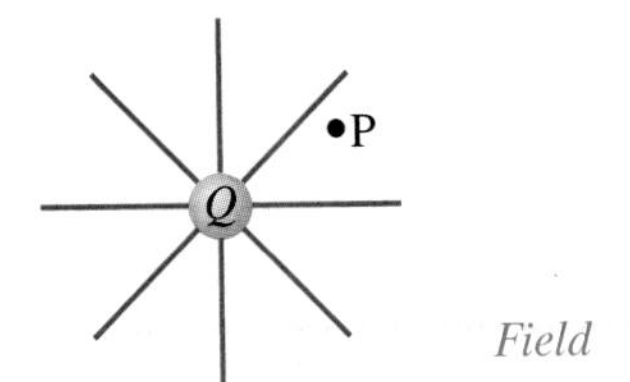

FIGURE 16–22 An electric field surrounds every charge. P is an arbitrary point.

Test charge

We can in principle investigate the electric field surrounding a charge or group of charges by measuring the force on a small positive **test charge**. By a test charge we mean a charge so small that the force it exerts does not significantly alter the distribution of those other charges that create the field. If a tiny positive test charge q is placed at various locations in the vicinity of a single positive charge Q as shown in Fig. 16–23 (points a, b, c), the force exerted on q is as shown. The force at b is less than at a because b's distance from Q is greater (Coulomb's law); and the force at c is smaller still. In each case, the force on q is directed radially away from Q. The electric field is defined in terms of the force on such a positive test charge. In particular, the **electric field**, $\vec{\mathbf{E}}$, at any point in space is defined as the force $\vec{\mathbf{F}}$ exerted on a tiny positive test charge placed at that point divided by the magnitude of the test charge q:

FIGURE 16–23 Force exerted by charge $+Q$ on a small test charge, q, placed at points a, b, and c.

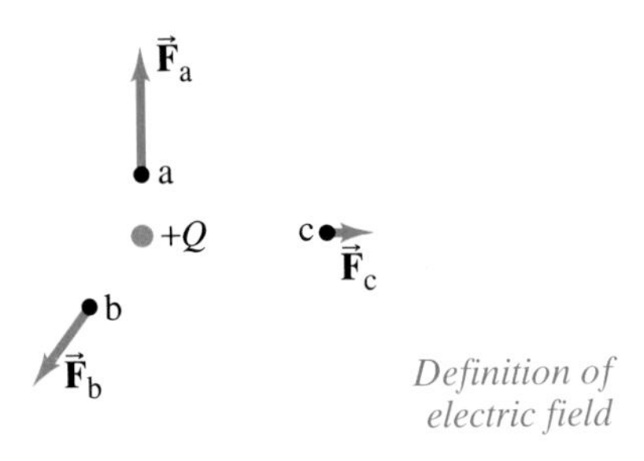

Definition of electric field

$$\vec{\mathbf{E}} = \frac{\vec{\mathbf{F}}}{q}. \tag{16–3}$$

More precisely, $\vec{\mathbf{E}}$ is defined as the limit of $\vec{\mathbf{F}}/q$ as q is taken smaller and smaller, approaching zero. That is, q is so tiny that it exerts essentially no force on the other charges which created the field. From this definition (Eq. 16–3), we see that the electric field at any point in space is a vector whose direction is the direction of the force on a tiny positive test charge at that point, and whose magnitude is the *force per unit charge*. Thus $\vec{\mathbf{E}}$ has SI units of newtons per coulomb (N/C).

$\vec{\mathbf{E}}$ is a vector

The reason for defining $\vec{\mathbf{E}}$ as $\vec{\mathbf{F}}/q$ (with $q \to 0$) is so that $\vec{\mathbf{E}}$ does not depend on the magnitude of the test charge q. This means that $\vec{\mathbf{E}}$ describes only the effect of the charges creating the electric field at that point.

The electric field at any point in space can be measured, based on the definition, Eq. 16–3. For simple situations involving one or several point charges, we can calculate $\vec{\mathbf{E}}$. For example, the electric field at a distance r from a single point charge Q would have magnitude

$$E = \frac{F}{q} = \frac{kqQ/r^2}{q}$$

Electric field due to one point charge

$$E = k\frac{Q}{r^2}; \qquad \text{[single point charge]} \tag{16–4a}$$

or, in terms of ϵ_0 as in Eq. 16–2 ($k = 1/4\pi\epsilon_0$):

$$E = \frac{1}{4\pi\epsilon_0}\frac{Q}{r^2}. \qquad \text{[single point charge]} \tag{16–4b}$$

Notice that E is independent of the test charge q—that is, E depends only on the charge Q which produces the field, and not on the value of the test charge q. Equations 16–4 are referred to as the electric field form of Coulomb's law.

If we are given the electric field $\vec{\mathbf{E}}$ at a given point in space, then we can calculate the force $\vec{\mathbf{F}}$ on any charge q placed at that point by writing (see Eq. 16–3):

$$\vec{\mathbf{F}} = q\vec{\mathbf{E}}. \tag{16–5}$$

This is valid even if q is not small as long as q does not cause the charges creating $\vec{\mathbf{E}}$ to move. If q is positive, $\vec{\mathbf{F}}$ and $\vec{\mathbf{E}}$ point in the same direction. If q is negative, $\vec{\mathbf{F}}$ and $\vec{\mathbf{E}}$ point in opposite directions. See Fig. 16–24.

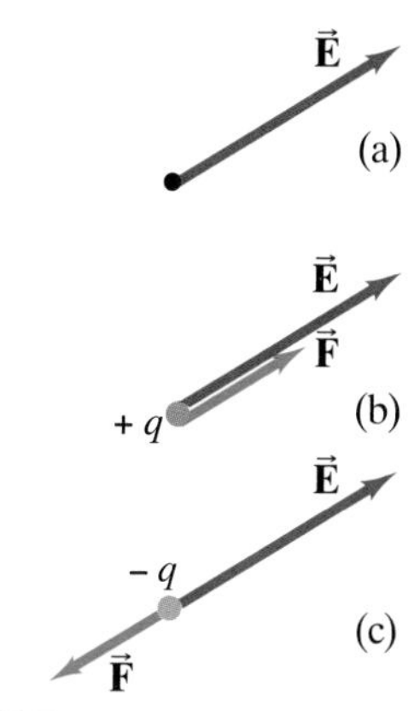

FIGURE 16–24 (a) Electric field at a given point in space. (b) Force on a positive charge at that point. (c) Force on a negative charge at that point.

EXAMPLE 16–6 **Photocopy machine.** A photocopy machine works by arranging positive charges (in the pattern to be copied) on the surface of a drum, then gently sprinkling negatively charged dry toner (ink) particles onto the drum. The toner particles temporarily stick to the pattern on the drum (Fig. 16–25) and are later transferred to paper and "melted" to produce the copy. Suppose each toner particle has a mass of 9.0×10^{-16} kg and carries an average of 20 extra electrons to provide an electric charge. Assuming that the electric force on a toner particle must exceed twice its weight in order to ensure sufficient attraction, compute the required electric field strength near the surface of the drum.

PHYSICS APPLIED

Photocopier

APPROACH The electric force on a toner particle of charge $q = 20e$ is $F = qE$, where E is the needed electric field. This force needs to be at least as great as twice the weight (mg) of the particle.

SOLUTION The minimum value of electric field satisfies the relation

$$qE = 2mg$$

where $q = 20e$. Hence

$$E = \frac{2mg}{q} = \frac{2(9.0 \times 10^{-16}\,\text{kg})(9.8\,\text{m/s}^2)}{20(1.6 \times 10^{-19}\,\text{C})} = 5.5 \times 10^3\,\text{N/C}.$$

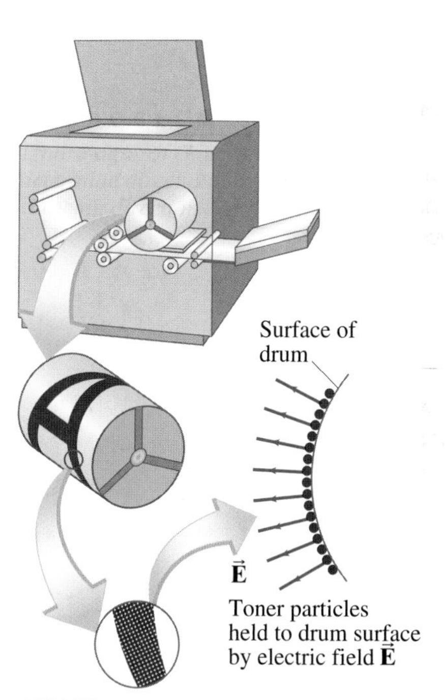

FIGURE 16–25 Example 16–6.

EXAMPLE 16–7 **Electric field of a single point charge.** Calculate the magnitude and direction of the electric field at a point P which is 30 cm to the right of a point charge $Q = -3.0 \times 10^{-6}\,\text{C}$.

APPROACH The magnitude of the electric field due to a single point charge is given by Eq. 16–4. The direction is found using the sign of the charge Q.

SOLUTION The magnitude of the electric field is:

$$E = k\frac{Q}{r^2} = \frac{(9.0 \times 10^9\,\text{N}\cdot\text{m}^2/\text{C}^2)(3.0 \times 10^{-6}\,\text{C})}{(0.30\,\text{m})^2} = 3.0 \times 10^5\,\text{N/C}.$$

The direction of the electric field is *toward* the charge Q, to the left as shown in Fig. 16–26a, since we defined the direction as that of the force on a positive test charge which here would be attractive. If Q had been positive, the electric field would have pointed away, as in Fig. 16–26b.

NOTE There is no electric charge at point P. But there is an electric field there. The only real charge is Q.

FIGURE 16–26 Example 16–7. Electric field at point P (a) due to a negative charge Q, and (b) due to a positive charge Q, each 30 cm from P.

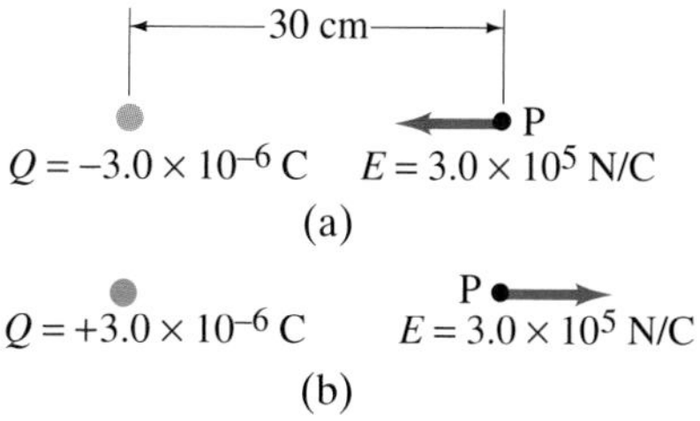

30 cm

P

$Q = -3.0 \times 10^{-6}$ C $E = 3.0 \times 10^5$ N/C

(a)

P

$Q = +3.0 \times 10^{-6}$ C $E = 3.0 \times 10^5$ N/C

(b)

This Example illustrates a general result: The electric field $\vec{\mathbf{E}}$ due to a positive charge points away from the charge, whereas $\vec{\mathbf{E}}$ due to a negative charge points toward that charge.

EXERCISE F What is the magnitude and the direction of the electric field due to a $+2.5\,\mu$C charge at a point 50 cm below it?

If the electric field at a given point in space is due to more than one charge, the individual fields (call them $\vec{\mathbf{E}}_1$, $\vec{\mathbf{E}}_2$, etc.) due to each charge are added vectorially to get the total field at that point:

Superposition principle for electric fields

$$\vec{\mathbf{E}} = \vec{\mathbf{E}}_1 + \vec{\mathbf{E}}_2 + \cdots.$$

The validity of this **superposition principle** for electric fields is fully confirmed by experiment.

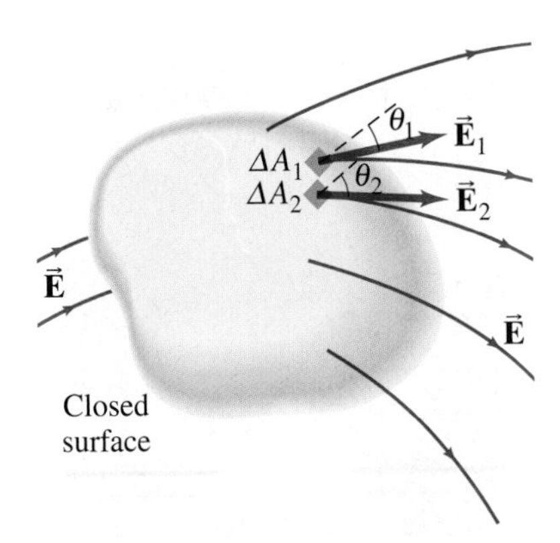

FIGURE 16–38 Electric field lines passing through a closed surface. The surface is divided up into many tiny areas, $\Delta A_1, \Delta A_2, \cdots$, and so on, of which only two are shown.

Gauss's law involves the *total* flux through a closed surface—a surface of any shape that encloses a volume of space. For any such surface, such as that shown in Fig. 16–38, we divide the surface up into many tiny areas, $\Delta A_1, \Delta A_2, \Delta A_3, \cdots$, and so on. We make the division so that each ΔA is small enough that it can be considered flat and so that the electric field can be considered constant within each ΔA. Then the *total* flux through the entire surface is the sum over all the individual fluxes through each of the tiny areas:

$$\begin{aligned}\Phi_E &= E_1\,\Delta A_1 \cos\theta_1 + E_2\,\Delta A_2 \cos\theta_2 + \cdots \\ &= \sum E\,\Delta A \cos\theta = \sum E_\perp\,\Delta A,\end{aligned}$$

where the symbol Σ means "sum of." We saw in Section 16–8 that the number of field lines starting on a positive charge or ending on a negative charge is proportional to the magnitude of the charge. Hence, the *net* number of lines N pointing out of any closed surface (number of lines pointing out minus the number pointing in) must be proportional to the net charge enclosed by the surface, Q_{encl}. But from Eq. 16–8, we have that the net number of lines N is proportional to the total flux Φ_E. Therefore,

$$\Phi_E = \sum_{\substack{\text{closed}\\\text{surface}}} E_\perp\,\Delta A \propto Q_{\text{encl}}.$$

The constant of proportionality is $1/\epsilon_0$, consistent with Coulomb's law, so we have

GAUSS'S LAW

$$\sum_{\substack{\text{closed}\\\text{surface}}} E_\perp\,\Delta A = \frac{Q_{\text{encl}}}{\epsilon_0}, \qquad \textbf{(16–9)}$$

where the sum (Σ) is over any closed surface, and Q_{encl} is the net charge enclosed within that surface. This is **Gauss's law**.

Coulomb's law and Gauss's law can be used to determine the electric field due to a given (static) charge distribution. Gauss's law is useful when the charge distribution is simple and symmetrical. However, we must choose the closed "gaussian" surface very carefully so we can determine $\vec{\mathbf{E}}$. We normally choose a surface that has just the symmetry needed so that E will be constant on all or on parts of its surface.

EXAMPLE 16–11 Charged spherical shell. A thin spherical shell of radius r_0 possesses a total net charge Q that is uniformly distributed on it, Fig. 16–39. Determine the electric field at points (*a*) outside the shell, and (*b*) inside the shell.

APPROACH Because the charge is distributed symmetrically, the electric field must be symmetric. Thus the field outside the shell must be directed radially outward (inward if $Q < 0$) and must depend only on r.

SOLUTION (*a*) The electric field will have the same magnitude at all points on an imaginary gaussian surface, if we choose it as a sphere of radius r $(r > r_0)$ concentric with the shell, and shown in Fig. 16–39 as the dashed circle A_1. Because $\vec{\mathbf{E}}$ is perpendicular to this surface, Gauss's law gives (with $Q_{\text{encl}} = Q$ in Eq. 16–9)

$$\sum E_\perp\,\Delta A = E \sum \Delta A = E\left(4\pi r^2\right) = \frac{Q}{\epsilon_0},$$

where $4\pi r^2$ is the surface area of our sphere (Gaussian surface) of radius r. Thus

$$E = \frac{1}{4\pi\epsilon_0}\frac{Q}{r^2}. \qquad [r > r_0]$$

Thus the field outside a uniformly charged spherical shell is the same as if all the charge were concentrated at the center as a point charge.

FIGURE 16–39 Cross-sectional drawing of a thin spherical shell of radius r_0, carrying a net charge Q uniformly distributed. A_1 and A_2 represent two gaussian surfaces we use to determine $\vec{\mathbf{E}}$. Example 16–11.

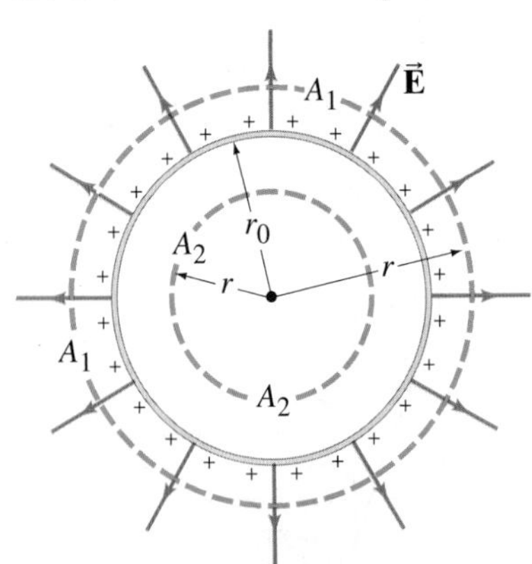

(*b*) Inside the shell, the field must also be symmetric. So E must again have the same value at all points on a spherical gaussian surface (A_2 in Fig. 16–39) concentric with the shell. Thus, E can be factored out of the sum and, with $Q_{\text{encl}} = 0$ since the charge inside the surface is zero, we have

$$\sum E_{\perp}\,\Delta A = E\sum \Delta A = E(4\pi r^2) = \frac{Q_{\text{encl}}}{\epsilon_0} = 0.$$

Hence

$$E = 0 \qquad [r < r_0]$$

inside a uniform spherical shell of charge.

The useful results of Example 16–11 also apply to a uniform *solid* spherical conductor that is charged, since all the charge would lie in a thin layer at the surface (Section 16–9).

EXERCISE H A very long, straight wire possesses a uniform charge per unit length, Q/L. Show that the electric field at points near (but outside) the wire, far from the ends, is given by

$$E = \frac{1}{2\pi\epsilon_0 r}\frac{Q}{L}$$

using the cylindrical gaussian surface shown (dashed) in Fig. 16–40. [*Hint:* there is no electric flux through the flat ends of the cylinder.]

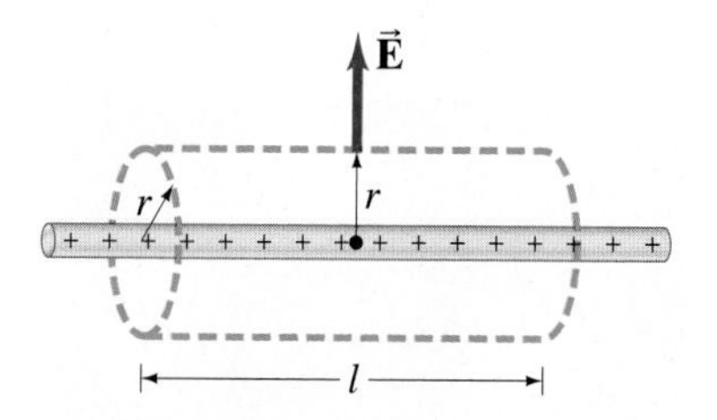

FIGURE 16–40 Calculation of $\vec{\mathbf{E}}$ due to a very long line of charge, Exercise H, where the cylinder shown (dashed) is the gaussian surface.

EXAMPLE 16–12 ***E* at surface of conductor.** Show that the electric field just outside the surface of any good conductor of arbitrary shape is given by

$$E = \frac{\sigma}{\epsilon_0},$$

where σ is the surface charge density (Q/A) on the conductor at that point.

APPROACH We choose as our gaussian surface a small cylindrical box, very small in height so that one of its circular ends is just above the conductor (Fig. 16–41). The other end is just below the conductor's surface, and the sides are perpendicular to it.

SOLUTION The electric field is zero inside a conductor and is perpendicular to the surface just outside it (Section 16–9), so electric flux passes only through the outside end of our cylindrical box; no flux passes through the short sides or inside end. We choose the area A (of the flat cylinder end above the conductor surface) small enough so that E is essentially uniform over it. Then Gauss's law gives

$$\sum E_{\perp}\,\Delta A = EA = \frac{Q_{\text{encl}}}{\epsilon_0} = \frac{\sigma A}{\epsilon_0},$$

so that

$$E = \frac{\sigma}{\epsilon_0}. \qquad \text{[at surface of conductor]}$$

This useful result applies for any shape conductor, including a large, uniformly charged flat sheet: the electric field will be constant and equal to σ/ϵ_0.

FIGURE 16–41 Electric field near the surface of a conductor. Two small cylindrical boxes are shown dashed. Either one can serve as our gaussian surface. Example 16–12.

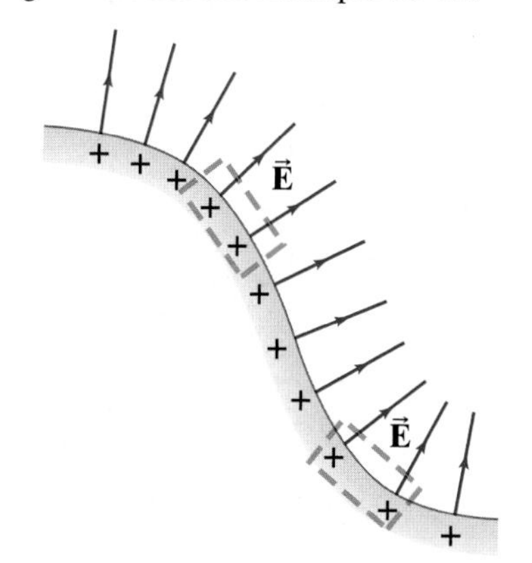

Electric field at surface of charged conductor

This last Example also gives us the field between the two parallel plates we discussed in Fig. 16–31d. If the plates are large compared to their separation, then the field lines are perpendicular to the plates and, except near the edges, they are parallel to each other. Therefore the electric field (see Fig. 16–42, which shows the same gaussian surface as Fig. 16–41) is also

$$E = \frac{\sigma}{\epsilon_0} = \frac{Q/A}{\epsilon_0}, \qquad \left[\begin{array}{l}\text{between two closely spaced}\\ \text{oppositely charged parallel plates}\end{array}\right] \qquad \textbf{(16–10)}$$

where $Q = \sigma A$ is the charge on one of the plates.

Oppositely charged parallel plates

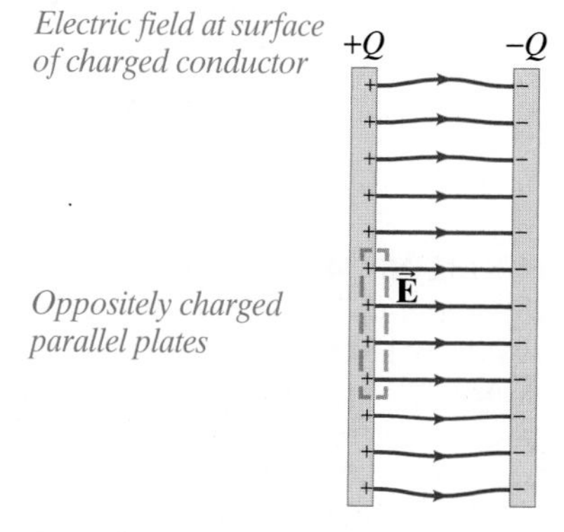

FIGURE 16–42 The electric field between two parallel plates is uniform and equal to $E = \sigma/\epsilon_0$.

* 16–11 Electric Forces in Molecular Biology: DNA Structure and Replication

PHYSICS APPLIED

Inside a cell: Kinetic theory plus electrostatic force

The study of the structure and functioning of a living cell at the molecular level is known as molecular biology. It is an important area for application of physics. Since the interior of a cell is mainly water, we can imagine it as a vast sea of molecules continually in motion (as in kinetic theory, Chapter 13), colliding with one another with various amounts of kinetic energy. These molecules interact with one another in various ways—chemical reactions (making and breaking of bonds between atoms) and more brief interactions or unions that occur because of *electrostatic attraction* between molecules.

The many processes that occur within the cell are now considered to be the result of *random ("thermal") molecular motion plus the ordering effect of the electrostatic force*. We use these ideas now to analyze some basic cellular processes involving macromolecules (large molecules). The picture we present here has not been seen "in action." Rather, it is a model of what happens based on presently accepted physical theories and experimental results.

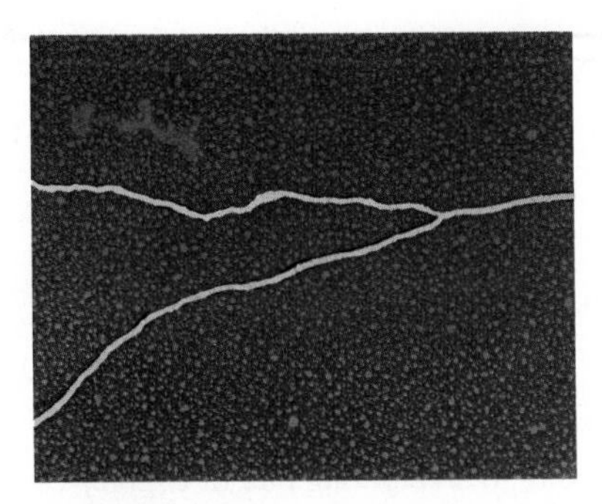

FIGURE 16–43 DNA replicating in a human HeLa cancer cell. This is a false-color image made by a transmission electron microscope (TEM; discussed in Chapter 27).

The genetic information that is passed on from generation to generation in all living cells is contained in the chromosomes, which are made up of genes. Each gene contains the information needed to produce a particular type of protein molecule. The genetic information contained in a gene is built into the principal molecule of a chromosome, DNA (deoxyribonucleic acid), Fig. 16–43. DNA molecules are made up of many small molecules known as nucleotide bases. There are four types of nucleotide bases in DNA: adenine (A), cytosine (C), guanine (G), and thymine (T).

PHYSICS APPLIED

DNA structure

The DNA of a chromosome generally consists of two long DNA strands wrapped about one another in the shape of a "double helix." The genetic information is contained in the specific order of the four bases (A, C, G, T) along the strand. As shown in Fig. 16–44, the two strands are attracted by electrostatic forces—that is, by the attraction of positive charges to negative charges. We see in Fig. 16–44a that an A (adenine) on one strand is always opposite a T on the

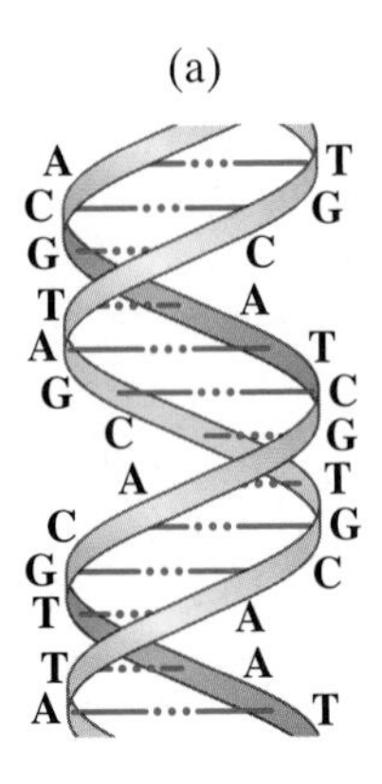

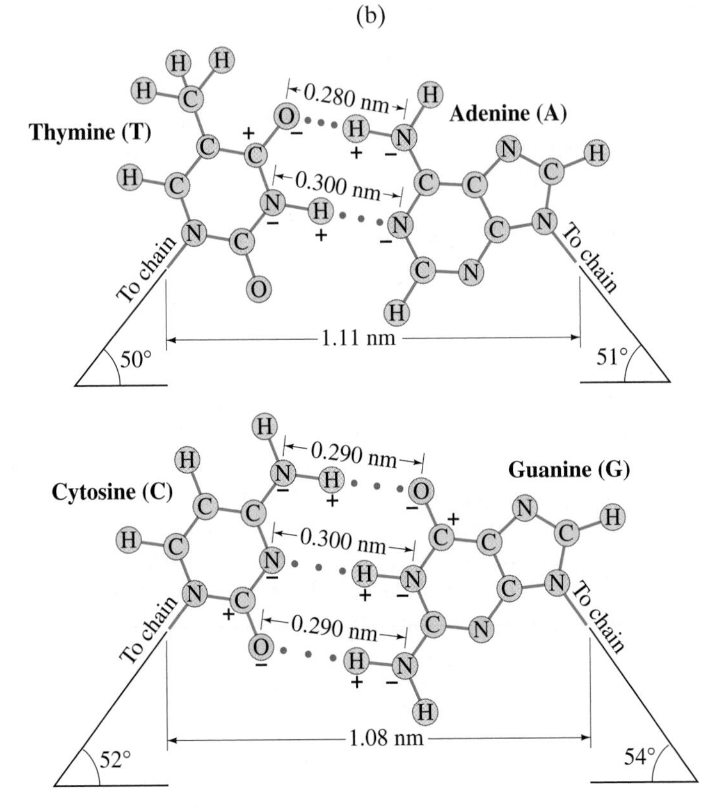

FIGURE 16–44 (a) Section of a DNA double helix. (b) "Close-up" view of the helix, showing how A and T attract each other and how G and C attract each other through electrostatic forces. The + and − signs indicated on certain atoms represent net charges, usually a fraction of e, due to uneven sharing of electrons. The red dots indicate the electrostatic attraction (often called a "weak bond" or "hydrogen bond"). Note that there are two weak bonds between A and T, and three between C and G.

other strand; similarly, a G is always opposite a C. This important ordering effect occurs because the shapes of A, T, C, and G are such that a T fits closely only into an A, and a G into a C; and only in the case of this close proximity of the charged portions is the electrostatic force great enough to hold them together even for a short time (Fig. 16–44b), forming what are referred to as "weak bonds." The electrostatic force between A and T, and between C and G, exists because these molecules have charged parts. These charges are due to some electrons in each of these molecules spending more time orbiting one atom than another. For example, the electron normally on the H atom of adenine (upper part of Fig. 16–44b) spends some of its time orbiting the adjacent N atom (more on this in Chapter 29), so the N has a net negative charge and the H a positive charge. This H^+ atom of adenine[†] is then attracted to the O^- atom of thymine. These net + and − charges usually have magnitudes of a fraction of e (charge on the electron) such as $0.2e$ or $0.4e$.

How does the arrangement shown in Fig. 16–44 come about? It occurs when the DNA replicates (duplicates) itself just before cell division. Indeed, the arrangement of A opposite T and G opposite C is crucial for ensuring that the genetic information is passed on accurately to the next generation. The process of replication is shown in a simplified form in Fig. 16–45. The two strands of DNA separate (with the help of enzymes, which also operate via the electrostatic force), leaving the charged parts of the bases exposed. Once replication starts, let us see how the correct order of bases occurs by focusing our attention on the G molecule indicated by the arrow on the lowest strand in Fig. 16–45. There are many unattached nucleotide bases of all four kinds bouncing around in the cellular fluid. The only one of the four bases that will experience attraction to our G, if it bounces close to it, will be a C. The charges on the other three bases are not arranged so that they can get close to those on the G, and thus there will be no significant attractive force exerted on them—remember that the force decreases rapidly with distance $(\propto 1/r^2)$. Because the G does not attract an A, T, or G appreciably, an A, T, or G will be knocked away by collisions with other molecules before enzymes can attach it to the growing chain (number 3). But the electrostatic force will often hold a C opposite our G long enough so that an enzyme can attach the C to the growing end of the new chain.

Thus we see that electrostatic forces are responsible for selecting the bases in the proper order during replication, so the genetic information is passed on accurately to the next generation. Note in Fig. 16–45 that the new number 4 strand has the same order of bases as the old number 1 strand; and the new number 3 strand is the same as the old number 2. So the two new double helixes, 1–3 and 2–4, are identical to the original 1–2 helix.

[†]When H^+ is involved, the weak bond it can make with a nearby negative charge, such as O^-, is relatively strong among weak bonds (partly because H^+ is so small) and is referred to as a "hydrogen bond" (Section 29–3).

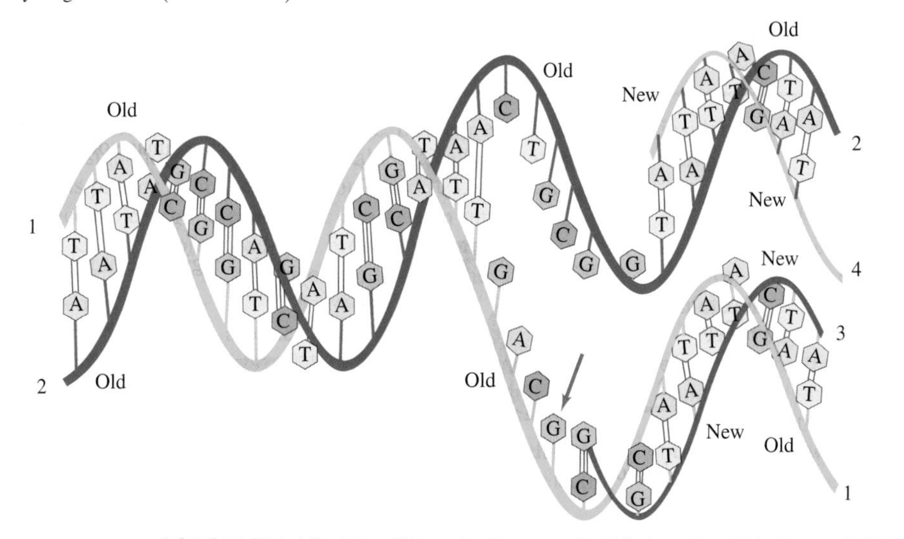

FIGURE 16–45 Replication of DNA.

A huge amount of electric potential energy can be stored in clouds. In lightning, the voltage (= potential difference = ΔPE per charge) between the clouds and the Earth can become so high that stray electrons in the air are accelerated to a KE large enough to knock electrons out of atoms of the air. The air becomes a conductor as the ionized atoms and freed electrons flow rapidly, colliding with more atoms, and causing more ionization. The massive flow of charge reduces the potential difference, and the "discharge" quickly ceases. The energy released when the ions and electrons recombine to form atoms appears as light.

CHAPTER 17

Electric Potential

We saw in Chapter 6 that the concept of energy was extremely valuable in dealing with the subject of mechanics. For one thing, energy is a conserved quantity and is thus an important tool for understanding nature. Furthermore, we saw that many Problems could be solved using the energy concept even though a detailed knowledge of the forces involved was not possible, or when a calculation involving Newton's laws would have been too difficult.

The energy point of view can be used in electricity, and it is especially useful. It not only extends the law of conservation of energy, but it gives us another way to view electrical phenomena. Energy is also a tool in solving Problems more easily in many cases than by using forces and electric fields.

17–1 Electric Potential Energy and Potential Difference

Electric Potential Energy

To apply conservation of energy, we need to define electric potential energy as for other types of potential energy. As we saw in Chapter 6, potential energy can be defined only for a conservative force. The work done by a conservative force in moving an object between any two positions is independent of the path taken. The electrostatic force between any two charges (Eq. 16–1, $F = kQ_1Q_2/r^2$) is conservative since the dependence is on position just like the gravitational force, which is conservative. Hence we can define potential energy PE for the electrostatic force.

We saw in Chapter 6 that the change in potential energy between two points a and b equals the negative of the work done by the conservative force to move an object from a to b: $\Delta \text{PE} = -W$.

Thus we define the change in electric potential energy, $\text{PE}_b - \text{PE}_a$, when a point charge q moves from some point a to another point b, as the negative of the work done by the electric force to move the charge from a to b. For example, consider the electric field between two equally but oppositely charged parallel plates; we assume their separation is small compared to their width and height, so the field $\vec{\mathbf{E}}$ will be uniform over most of the region, Fig. 17–1. Now consider a tiny positive point charge q placed at point a very near the positive plate as shown. This charge q is so small it doesn't affect $\vec{\mathbf{E}}$. If this charge q at point a is released, the electric force will do work on the charge and accelerate it toward the negative plate. The work W done by the electric field E to move the charge a distance d is

$$W = Fd = qEd$$

where we used Eq. 16–5, $F = qE$. The change in electric potential energy equals the negative of the work done by the electric force:

$$\text{PE}_b - \text{PE}_a = -qEd \qquad [\text{uniform } \vec{\mathbf{E}}] \qquad \textbf{(17–1)}$$

for this case of uniform electric field $\vec{\mathbf{E}}$. In the case illustrated, the potential energy decreases (ΔPE is negative); and as the charged particle accelerates from point a to point b in Fig. 17–1, the particle's kinetic energy KE increases—by an equal amount. In accord with the conservation of energy, electric potential energy is transformed into kinetic energy, and the total energy is conserved. Note that the positive charge q has its greatest potential energy at point a, near the positive plate.† The reverse is true for a negative charge: its potential energy is greatest near the negative plate.

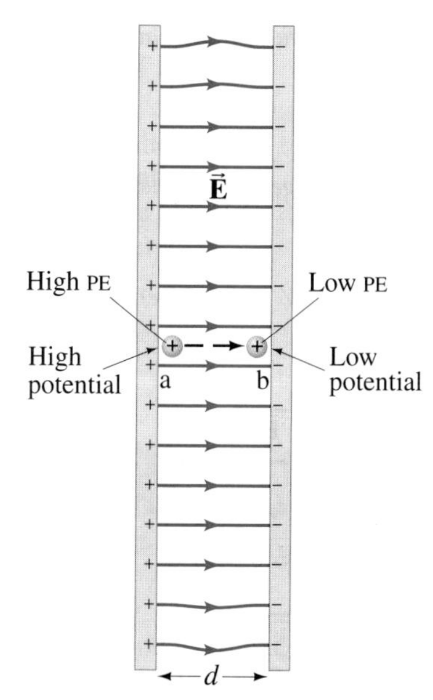

FIGURE 17–1 Work is done by the electric field in moving the positive charge from position a to position b.

Electric Potential and Potential Difference

In Chapter 16, we found it useful to define the electric field as the force per unit charge. Similarly, it is useful to define the **electric potential** (or simply the **potential** when "electric" is understood) as the *electric potential energy per unit charge*. Electric potential is given the symbol V. If a positive test charge q has electric potential energy PE_a at some point a (relative to some zero potential energy), the electric potential V_a at this point is

Potential is potential energy per unit charge

$$V_a = \frac{\text{PE}_a}{q}. \qquad \textbf{(17–2a)}$$

As we discussed in Chapter 6, only differences in potential energy are physically meaningful. Hence only the **difference in potential**, or the **potential difference**, between two points a and b (such as between a and b in Fig. 17–1) is measurable. When the electric force does positive work on a charge, the kinetic energy increases and the potential energy decreases. The difference in potential energy, $\text{PE}_b - \text{PE}_a$, is equal to the negative of the work, W_{ba}, done by the electric field to move the charge from a to b; so the potential difference V_{ba} is

Potential difference

$$V_{ba} = V_b - V_a = \frac{\text{PE}_b - \text{PE}_a}{q} = -\frac{W_{ba}}{q}. \qquad \textbf{(17–2b)}$$

Note that electric potential, like electric field, does not depend on our test charge q. V depends on the other charges that create the field, not on q; q acquires potential energy by being in the potential V due to the other charges.

We can see from our definition that the positive plate in Fig. 17–1 is at a higher potential than the negative plate. Thus a positively charged object moves naturally from a high potential to a low potential. A negative charge does the reverse.

† At this point the charge has its greatest ability to do work (on some other object or system).

17–4 The Electron Volt, a Unit of Energy

The joule is a very large unit for dealing with energies of electrons, atoms, or molecules. For this purpose, the unit **electron volt** (eV) is used. One electron volt is defined as the energy acquired by a particle carrying a charge whose magnitude equals that on the electron ($q = e$) as a result of moving through a potential difference of 1 V. Since the charge on an electron has magnitude 1.6×10^{-19} C, and since the change in potential energy equals qV, 1 eV is equal to $(1.6 \times 10^{-19}\,\text{C})(1.0\,\text{V}) = 1.6 \times 10^{-19}$ J:

Electron volt (unit)

$$1\,\text{eV} = 1.6 \times 10^{-19}\,\text{J}.$$

An electron that accelerates through a potential difference of 1000 V will lose 1000 eV of potential energy and will thus gain 1000 eV, or 1 keV (kiloelectron volt) of kinetic energy. On the other hand, if a particle with a charge equal to twice the magnitude of the charge on the electron $(= 2e = 3.2 \times 10^{-19}\,\text{C})$ moves through a potential difference of 1000 V, its energy will change by 2000 eV.

Although the electron volt is handy for *stating* the energies of molecules and elementary particles, it is *not* a proper SI unit. For calculations, electron volts should be converted to joules using the conversion factor just given. In Example 17–2, for example, the electron acquired a kinetic energy of 8.0×10^{-16} J. We normally would quote this energy as 5000 eV $(= 8.0 \times 10^{-16}\,\text{J}/1.6 \times 10^{-19}\,\text{J/eV})$. But when determining the speed of a particle in SI units, we must use the KE in joules (J).

17–5 Electric Potential Due to Point Charges

The electric potential at a distance r from a single point charge Q can be derived from the expression for its electric field (Eq. 16–4) using calculus. The potential in this case is usually taken to be zero at infinity (∞); this is also where the electric field $(E = kQ/r^2)$ is zero. The result is

Electric potential of a point charge ($V = 0$ at $r = \infty$)

$$V = k\frac{Q}{r} = \frac{1}{4\pi\epsilon_0}\frac{Q}{r}, \qquad \text{[single point charge]} \quad \textbf{(17–5)}$$

where $k = 8.99 \times 10^9\,\text{N}\cdot\text{m}^2/\text{C}^2$. We can think of V here as representing the *absolute potential* at a distance r from the charge Q, where $V = 0$ at $r = \infty$, or we can think of V as the potential difference between r and infinity. Notice that the potential V decreases with the first power of the distance, whereas the electric field (Eq. 16–4) decreases as the *square* of the distance. The potential near a positive charge is large and positive, and it decreases toward zero at very large distances. The potential near a negative charge is negative and increases toward zero at large distances (Fig. 17–9).

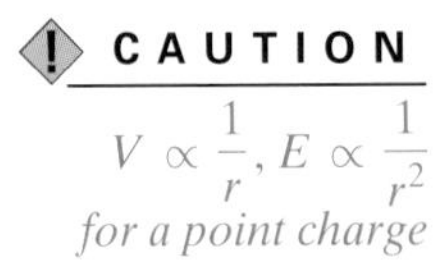

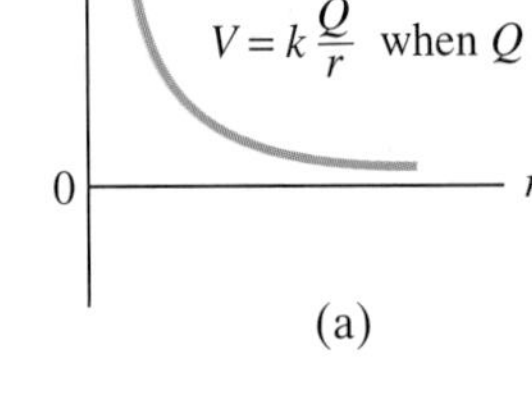

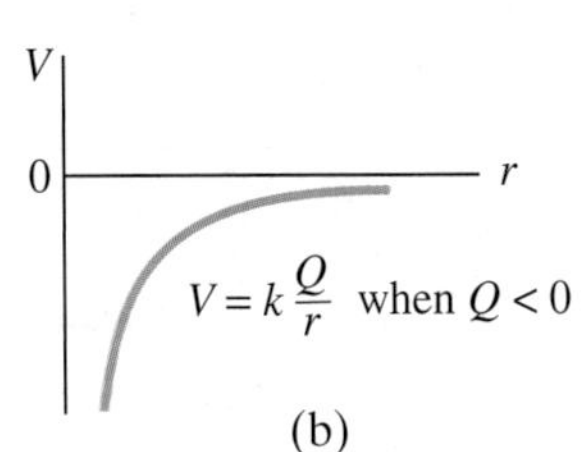

FIGURE 17–9 Potential V as a function of distance r from a single point charge Q when the charge is (*a*) positive, (*b*) negative.

EXAMPLE 17–4 Potential due to a positive or a negative charge. Determine the potential at a point 0.50 m (*a*) from a $+20\,\mu\text{C}$ point charge, (*b*) from a $-20\,\mu\text{C}$ point charge.

APPROACH The potential due to a point charge is given by Eq. 17–5, $V = kQ/r$.

SOLUTION (*a*) At a distance of 0.50 m from a positive $20\,\mu\text{C}$ charge, the potential is

$$V = k\frac{Q}{r}$$

$$= \left(9.0 \times 10^9\,\text{N}\cdot\text{m}^2/\text{C}^2\right)\left(\frac{20 \times 10^{-6}\,\text{C}}{0.50\,\text{m}}\right) = 3.6 \times 10^5\,\text{V}.$$

(*b*) For the negative charge,

$$V = \left(9.0 \times 10^9\,\text{N}\cdot\text{m}^2/\text{C}^2\right)\left(\frac{-20 \times 10^{-6}\,\text{C}}{0.50\,\text{m}}\right) = -3.6 \times 10^5\,\text{V}.$$

NOTE Potential can be positive or negative. In contrast to calculations of electric field magnitudes, for which we usually ignore the sign of the charges, it is important to include a charge's sign when we find electric potential.

➡ PROBLEM SOLVING

Keep track of charge signs for electric potential

EXAMPLE 17–5 Work done to bring two positive charges close together. What minimum work must be done by an external force to bring a charge $q = 3.00\,\mu\text{C}$ from a great distance away (take $r = \infty$) to a point 0.500 m from a charge $Q = 20.0\,\mu\text{C}$?

APPROACH To find the work we cannot simply multiply the force times distance because the force is not constant. Instead we can set the change in potential energy equal to the (positive of the) work required of an *external* force (Chapter 6), and Eq. 17–3: $W = \Delta\text{PE} = q(V_b - V_a)$. We get the potentials V_b and V_a using Eq. 17–5.

⚠ CAUTION

We cannot use $W = Fd$ if F is not constant

SOLUTION The work required is equal to the change in potential energy:

$$W = q(V_b - V_a) = q\left(\frac{kQ}{r_b} - \frac{kQ}{r_a}\right),$$

where $r_b = 0.500\,\text{m}$ and $r_a = \infty$. The right-hand term within the parentheses is zero $(1/\infty = 0)$ so

$$W = \left(3.00 \times 10^{-6}\,\text{C}\right)\frac{\left(8.99 \times 10^9\,\text{N}\cdot\text{m}^2/\text{C}^2\right)\left(2.00 \times 10^{-5}\,\text{C}\right)}{(0.500\,\text{m})} = 1.08\,\text{J}.$$

NOTE We could not use Eqs. 17–4 here because they apply *only* to uniform fields. But we did use Eq. 17–3 because it is always valid.

EXERCISE B What work is required to bring a charge $q = 3.00\,\mu\text{C}$ originally a distance of 1.50 m from a charge $Q = 20.0\,\mu\text{C}$ until it is 0.50 m away?

To determine the electric field at points near a collection of two or more point charges requires adding up the electric fields due to each charge. Since the electric field is a vector, this can be time consuming or complicated. To find the electric potential at a point due to a collection of point charges is far easier, since the electric potential is a scalar, and hence you only need to add numbers together without concern for direction. This is a major advantage in using electric potential for solving Problems. We do have to include the signs of charges, however.

Potentials add as scalars (fields add as vectors)

27. (III) In the Bohr model of the hydrogen atom, an electron orbits a proton (the nucleus) in a circular orbit of radius 0.53×10^{-10} m. (*a*) What is the electric potential at the electron's orbit due to the proton? (*b*) What is the kinetic energy of the electron? (*c*) What is the total energy of the electron in its orbit? (*d*) What is the *ionization energy*—that is, the energy required to remove the electron from the atom and take it to $r = \infty$, at rest? Express the results of parts b, c and d in joules and eV.

* 17–6 Electric Dipoles

* 28. (I) An electron and a proton are 0.53×10^{-10} m apart. What is their dipole moment if they are at rest?

* 29. (II) Calculate the electric potential due to a dipole whose dipole moment is $4.8 \times 10^{-30}\,\mathrm{C \cdot m}$ at a point 1.1×10^{-9} m away if this point is (*a*) along the axis of the dipole nearer the positive charge; (*b*) 45° above the axis but nearer the positive charge; (*c*) 45° above the axis but nearer the negative charge.

* 30. (III) The dipole moment, considered as a vector, points from the negative to the positive charge. The water molecule, Fig. 17–28, has a dipole moment $\vec{\mathbf{p}}$ which can be considered as the vector sum of the two dipole moments, $\vec{\mathbf{p}}_1$ and $\vec{\mathbf{p}}_2$, as shown. The distance between each H and the O is about 0.96×10^{-10} m. The lines joining the center of the O atom with each H atom make an angle of 104°, as shown, and the net dipole moment has been measured to be $p = 6.1 \times 10^{-30}\,\mathrm{C \cdot m}$. Determine the charge q on each H atom.

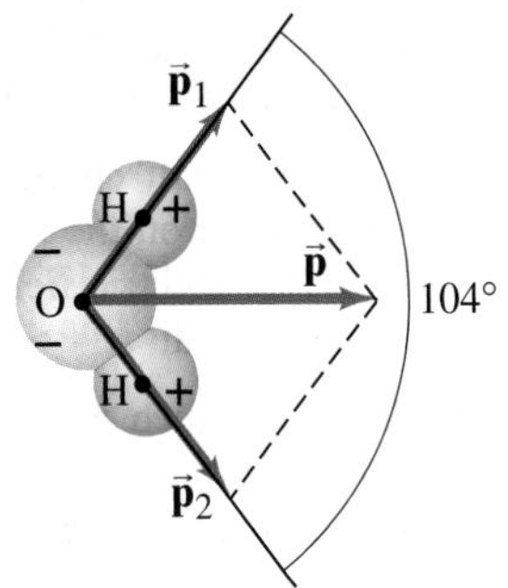

FIGURE 17–28 Problem 30.

17–7 Capacitance

31. (I) The two plates of a capacitor hold $+2500\,\mu$C and $-2500\,\mu$C of charge, respectively, when the potential difference is 850 V. What is the capacitance?

32. (I) A 9500-pF capacitor holds plus and minus charges of 16.5×10^{-8} C. What is the voltage across the capacitor?

33. (I) The potential difference between two short sections of parallel wire in air is 120 V. They carry equal and opposite charge of magnitude 95 pC. What is the capacitance of the two wires?

34. (I) How much charge flows from each terminal of a 12.0-V battery when it is connected to a 7.00-μF capacitor?

35. (I) A 0.20-F capacitor is desired. What area must the plates have if they are to be separated by a 2.2-mm air gap?

36. (II) The charge on a capacitor increases by 18 μC when the voltage across it increases from 97 V to 121 V. What is the capacitance of the capacitor?

37. (II) An electric field of 8.50×10^5 V/m is desired between two parallel plates, each of area 35.0 cm^2 and separated by 2.45 mm of air. What charge must be on each plate?

38. (II) If a capacitor has opposite 5.2 μC charges on the plates, and an electric field of 2.0 kV/mm is desired between the plates, what must each plate's area be?

39. (II) How strong is the electric field between the plates of a 0.80-μF air-gap capacitor if they are 2.0 mm apart and each has a charge of 72 μC?

40. (III) A 7.7-μF capacitor is charged by a 125-V battery (Fig. 17–29a) and then is disconnected from the battery. When this capacitor (C_1) is then connected (Fig. 17–29b) to a second (initially uncharged) capacitor, C_2, the final voltage on each capacitor is 15 V. What is the value of C_2? [*Hint*: charge is conserved.]

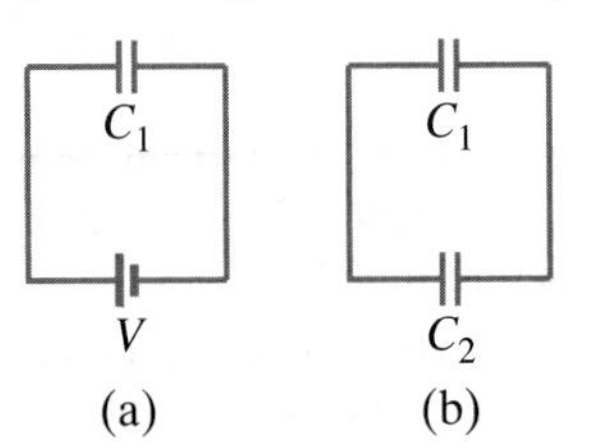

FIGURE 17–29 Problems 40 and 52.

41. (III) A 2.50-μF capacitor is charged to 857 V and a 6.80-μF capacitor is charged to 652 V. These capacitors are then disconnected from their batteries. Next the positive plates are connected to each other and the negative plates are connected to each other. What will be the potential difference across each and the charge on each? [*Hint*: charge is conserved.]

17–8 Dielectrics

42. (I) What is the capacitance of two square parallel plates 5.5 cm on a side that are separated by 1.8 mm of paraffin?

43. (I) What is the capacitance of a pair of circular plates with a radius of 5.0 cm separated by 3.2 mm of mica?

44. (II) A 3500-pF air-gap capacitor is connected to a 22-V battery. If a piece of mica is placed between the plates, how much charge will flow from the battery?

* 45. (II) The electric field between the plates of a paper-separated ($K = 3.75$) capacitor is 8.24×10^4 V/m. The plates are 1.95 mm apart, and the charge on each plate is 0.775 μC. Determine the capacitance of this capacitor and the area of each plate.

17–9 Electric Energy Storage

46. (I) 650 V is applied to a 2200-pF capacitor. How much energy is stored?

47. (I) A cardiac defibrillator is used to shock a heart that is beating erratically. A capacitor in this device is charged to 5.0 kV and stores 1200 J of energy. What is its capacitance?

48. (II) How much energy is stored by the electric field between two square plates, 8.0 cm on a side, separated by a 1.5-mm air gap? The charges on the plates are equal and opposite and of magnitude 420 μC.

49. (II) A homemade capacitor is assembled by placing two 9-in. pie pans 5 cm apart and connecting them to the opposite terminals of a 9-V battery. Estimate (*a*) the capacitance, (*b*) the charge on each plate, (*c*) the electric field halfway between the plates, and (*d*) the work done by the battery to charge the plates. (*e*) Which of the above values change if a dielectric is inserted?

50. (II) A parallel-plate capacitor has fixed charges $+Q$ and $-Q$. The separation of the plates is then doubled. (*a*) By what factor does the energy stored in the electric field change? (*b*) How much work must be done in doubling the plate separation from d to $2d$? The area of each plate is A.

51. (II) How does the energy stored in a capacitor change if (*a*) the potential difference is doubled, and (*b*) the charge on each plate is doubled, as the capacitor remains connected to a battery?

52. (III) A 2.70-μF capacitor is charged by a 12.0-V battery. It is disconnected from the battery and then connected to an uncharged 4.00-μF capacitor (Fig. 17–29). Determine the total stored energy (*a*) before the two capacitors are connected, and (*b*) after they are connected. (*c*) What is the change in energy?

* 17–10 Cathode Ray Tube

* 53. (III) In a given CRT, electrons are accelerated horizontally by 7.0 kV. They then pass through a uniform electric field E for a distance of 2.8 cm, which deflects them upward so they reach the screen top 22 cm away, 11 cm above the center. Estimate the value of E.

* 54. (III) Electrons are accelerated by 6.0 kV in a CRT. The screen is 30 cm wide and is 34 cm from the 2.6-cm-long deflection plates. Over what range must the horizontally deflecting electric field vary to sweep the beam fully across the screen?

General Problems

55. An electron starting from rest acquires 6.3 keV of KE in moving from point A to point B. (*a*) How much KE would a proton acquire, starting from rest at B and moving to point A? (*b*) Determine the ratio of their speeds at the end of their respective trajectories.

56. A lightning flash transfers 4.0 C of charge and 4.2 MJ of energy to the Earth. (*a*) Across what potential difference did it travel? (*b*) How much water could this boil and vaporize, starting from room temperature?

57. There is an electric field near the Earth's surface whose magnitude is about 150 V/m. How much energy is stored per cubic meter in this field?

58. In a television picture tube, electrons are accelerated by thousands of volts through a vacuum. If a television set were laid on its back, would electrons be able to move upward against the force of gravity? What potential difference, acting over a distance of 3.0 cm, would be needed to balance the downward force of gravity so that an electron would remain stationary? Assume that the electric field is uniform.

59. A huge 4.0-F capacitor has enough stored energy to heat 2.5 kg of water from 21° C to 95° C. What is the potential difference across the plates?

60. An uncharged capacitor is connected to a 24.0-V battery until it is fully charged, after which it is disconnected from the battery. A slab of paraffin is then inserted between the plates. What will now be the voltage between the plates?

61. Dry air will break down if the electric field exceeds 3.0×10^6 V/m. What amount of charge can be placed on a parallel-plate capacitor if the area of each plate is 56 cm^2?

62. Three charges are at the corners of an equilateral triangle (side L) as shown in Fig. 17–30. Determine the potential at the midpoint of each of the sides.

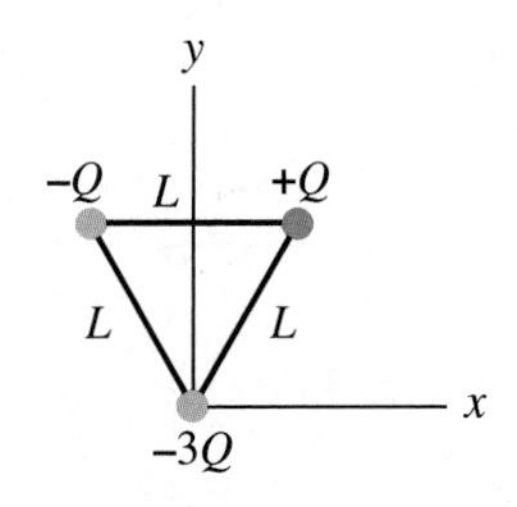

FIGURE 17–30 Problem 62.

63. A 3.4 μC and a $-2.6\,\mu$C charge are placed 1.6 cm apart. At what points along the line joining them is (*a*) the electric field zero, and (*b*) the electric potential zero?

64. A 2600-pF air-gap capacitor is connected to a 9.0-V battery. If a piece of Pyrex glass is placed between the plates, how much charge will then flow from the battery?

65. An electron is accelerated horizontally from rest in a television picture tube by a potential difference of 5500 V. It then passes between two horizontal plates 6.5 cm long and 1.3 cm apart that have a potential difference of 250 V (Fig. 17–31). At what angle θ will the electron be traveling after it passes between the plates?

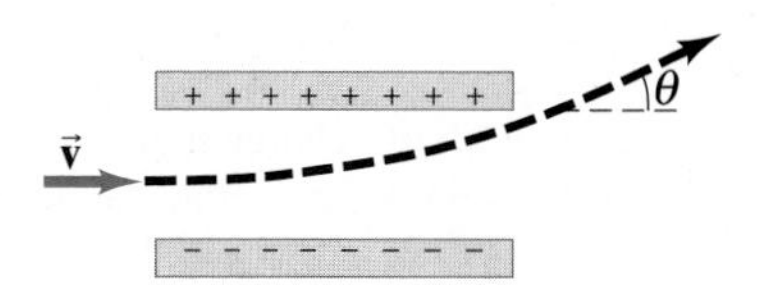

FIGURE 17–31 Problem 65.

66. A capacitor of capacitance C_1 carries a charge Q_0. It is then connected directly to a second, uncharged, capacitor of capacitance C_2, as shown in Fig. 17–32. What charge will each carry now? What will be the potential difference across each?

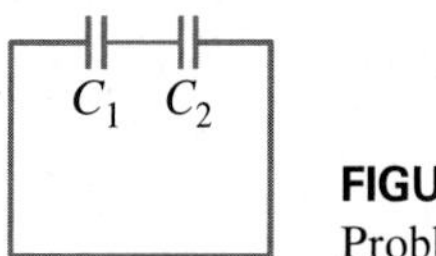

FIGURE 17–32 Problem 66.

67. To get an idea how big a farad is, suppose you want to make a 1-F air-filled parallel-plate capacitor for a circuit you are building. To make it a reasonable size, suppose you limit the plate area to 1.0 cm^2. What would the gap have to be between the plates? Is this practically achievable?

68. Near the surface of the Earth there is an electric field of about 150 V/m which points downward. Two identical balls with mass $m = 0.540$ kg are dropped from a height of 2.00 m, but one of the balls is positively charged with $q_1 = 650\,\mu$C, and the second is negatively charged with $q_2 = -650\,\mu$C. Use conservation of energy to determine the difference in the speed of the two balls when they hit the ground. (Neglect air resistance.)

69. The power supply for a pulsed nitrogen laser has a 0.050-μF capacitor with a maximum voltage rating of 30 kV. (*a*) Estimate how much energy could be stored in this capacitor. (*b*) If 12% of this stored electrical energy is converted to light energy in a pulse that is 8.0 microseconds long, what is the power of the laser pulse?

70. In lightning storms, the potential difference between the Earth and the bottom of the thunderclouds can be as high as 35,000,000 V. The bottoms of the thunderclouds are typically 1500 m above the Earth, and can have an area of 110 km^2. Modeling the Earth–cloud system as a huge capacitor, calculate (*a*) the capacitance of the Earth–cloud system, (*b*) the charge stored in the "capacitor," and (*c*) the energy stored in the "capacitor."

71. In a photocell, ultraviolet (UV) light provides enough energy to some electrons in barium metal to eject them from a surface at high speed. See Fig. 17–33. To measure the maximum energy of the electrons, another plate above the barium surface is kept at a negative enough potential that the emitted electrons are slowed down and stopped, and return to the barium surface. If the plate voltage is -3.02 V (compared to the barium) when the fastest electrons are stopped, what was the speed of these electrons when they were emitted?

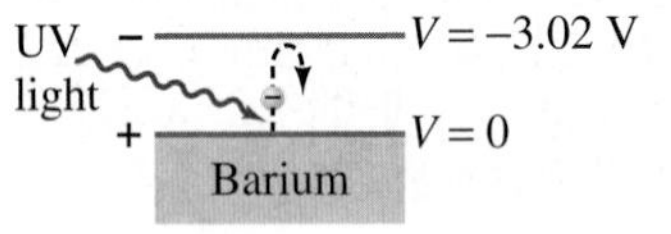

UV light; $V = -3.02$ V; $V = 0$; Barium

FIGURE 17–33 Problem 71.

72. A $+33\ \mu$C point charge is placed 36 cm from an identical $+33\ \mu$C charge. A $-1.5\ \mu$C charge is moved from point a to point b in Fig. 17–34. What is the change in potential energy?

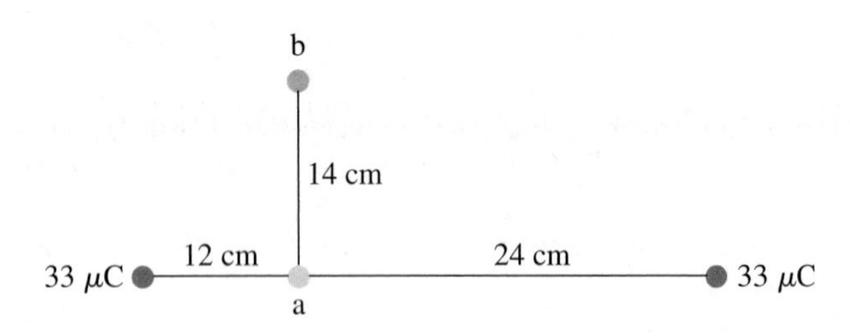

FIGURE 17–34 Problem 72.

73. A capacitor is made from two 1.1-cm-diameter coins separated by a 0.15-mm-thick piece of paper ($K = 3.7$). A 12-V battery is connected to the capacitor. How much charge is on each coin?

74. A $+4.5\ \mu$C charge is 23 cm to the right of a $-8.2\ \mu$C charge. At the midpoint between the two charges, (*a*) what are the potential and (*b*) the electric field?

75. A parallel-plate capacitor with plate area 2.0 cm^2 and air-gap separation 0.50 mm is connected to a 12-V battery, and fully charged. The battery is then disconnected. (*a*) What is the charge on the capacitor? (*b*) The plates are now pulled to a separation of 0.75 mm. What is the charge on the capacitor now? (*c*) What is the potential difference across the plates now? (*d*) How much work was required to pull the plates to their new separation?

76. A 2.5-μF capacitor is fully charged by a 6.0-V battery. The battery is then disconnected. The capacitor is not ideal and the charge slowly leaks out from the plates. The next day, the capacitor has lost half its stored energy. Calculate the amount of charge lost.

77. Two point charges are fixed 4.0 cm apart from each other. Their charges are $Q_1 = Q_2 = 5.0\ \mu$C, and their masses are $m_1 = 1.5$ mg and $m_2 = 2.5$ mg. (*a*) If Q_1 is released from rest, what will be its speed after a very long time? (*b*) If both charges are released from rest at the same time, what will be the speed of Q_1 after a very long time?

78. Two charges are placed as shown in Fig. 17–35 with $q_1 = 1.5\ \mu$C and $q_2 = -3.3\ \mu$C. Find the potential difference between points A and B.

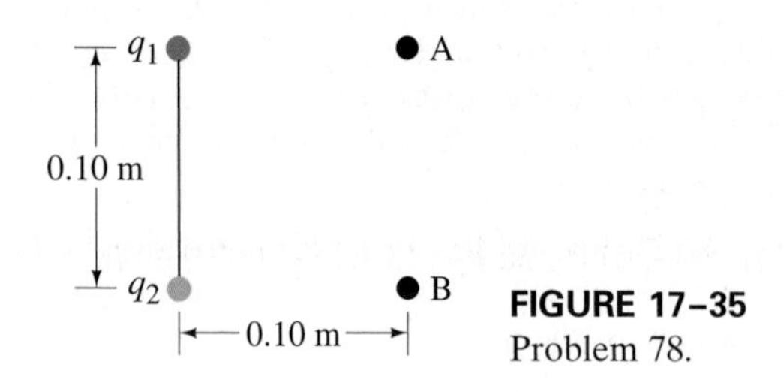

FIGURE 17–35 Problem 78.

Answers to Exercises

A: (*a*) -8.0×10^{-16} J; (*b*) 9.8×10^5 m/s.
B: 0.72 J.
C: 8.3×10^{-9} C.
D: (*a*) 3 times greater; (*b*) 3 times greater.
E: 12 mF.

The glow of the thin wire filament of a lightbulb is caused by the electric current passing through it. Electric energy is transformed to thermal energy (via collisions between moving electrons and atoms of the wire), which causes the wire's temperature to become so high that it glows. Electric current and electric power in electric circuits are of basic importance in everyday life. We examine both dc and ac in this Chapter, and include the microscopic analysis of electric current.

CHAPTER 18

Electric Currents

In the previous two Chapters we have been studying static electricity: electric charges at rest. In this Chapter we begin our study of charges in motion, and we call a flow of charge an electric current.

In everyday life we are familiar with electric currents in wires and other conductors. Indeed, most practical electrical devices depend on electric current: current through a lightbulb, current in the heating element of a stove or electric heater, and of course currents in electronic devices. Electric currents can exist in conductors such as wires, and also in other devices such as the CRT of a television or computer monitor whose charged electrons flow through space (Section 17–10).

In electrostatic situations, we saw in Section 16–9 that the electric field must be zero inside a conductor (if it weren't, the charges would move). But when charges are *moving* in a conductor, there usually *is* an electric field in the conductor. Indeed, an electric field is needed to set charges into motion, and to keep them in motion in any normal conductor. We can control the flow of charge using electric fields and electric potential (voltage), concepts we have just been discussing. In order to have a current in a wire, a potential difference is needed, which can be provided by a battery.

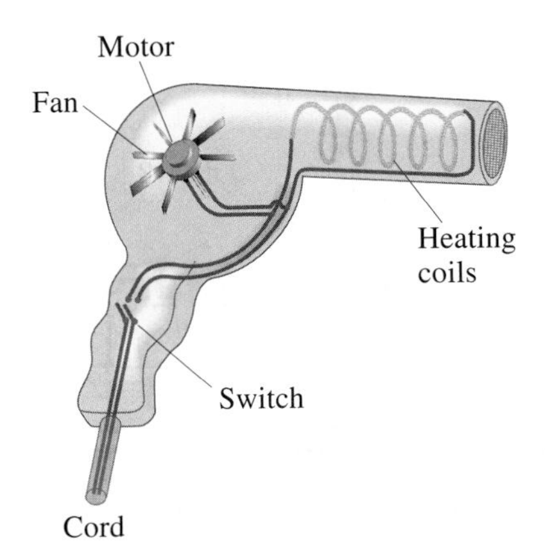

FIGURE 18–23 A hair dryer. Most of the current goes through the heating coils, a pure resistance; a small part goes to the motor to turn the fan. Example 18–12.

EXAMPLE 18–12 **Hair dryer.** (*a*) Calculate the resistance and the peak current in a 1000-W hair dryer (Fig. 18–23) connected to a 120-V line. (*b*) What happens if it is connected to a 240-V line in Britain?

APPROACH We are given $\overline{P}$ and V_{rms}, so $I_{\mathrm{rms}} = \overline{P}/V_{\mathrm{rms}}$ (Eq. 18–9a or 18–5), and $I_0 = \sqrt{2}\, I_{\mathrm{rms}}$. Then we find R from $V = IR$.

SOLUTION (*a*) We solve Eq. 18–9a for the rms current:

$$I_{\mathrm{rms}} = \frac{\overline{P}}{V_{\mathrm{rms}}} = \frac{1000\ \mathrm{W}}{120\ \mathrm{V}} = 8.33\ \mathrm{A}.$$

Then

$$I_0 = \sqrt{2}\, I_{\mathrm{rms}} = 11.8\ \mathrm{A}.$$

The resistance is

$$R = \frac{V_{\mathrm{rms}}}{I_{\mathrm{rms}}} = \frac{120\ \mathrm{V}}{8.33\ \mathrm{A}} = 14.4\ \Omega.$$

The resistance could equally well be calculated using peak values:

$$R = \frac{V_0}{I_0} = \frac{170\ \mathrm{V}}{11.8\ \mathrm{A}} = 14.4\ \Omega.$$

(*b*) When connected to a 240-V line, more current would flow and the resistance would change with the increased temperature (Section 18–4). But let us make an estimate of the power transformed based on the same 14.4-Ω resistance. The average power would be

$$\overline{P} = \frac{V_{\mathrm{rms}}^2}{R} = \frac{(240\ \mathrm{V})^2}{(14.4\ \Omega)} = 4000\ \mathrm{W}.$$

This is four times the dryer's power rating and would undoubtedly melt the heating element or the wire coils of the motor.

EXAMPLE 18–13 **Stereo power.** Each channel of a stereo receiver is capable of an average power output of 100 W into an 8-Ω loudspeaker (see Fig. 18–14). What are the rms voltage and the rms current fed to the speaker (*a*) at the maximum power of 100 W, and (*b*) at 1.0 W when the volume is turned down?

APPROACH We assume that the loudspeaker can be treated as a simple resistance (not quite true—see Chapter 21) with $R = 8.0\ \Omega$. We are given the power P, so we can determine V_{rms} and I_{rms} using the power equations, Eqs. 18–9.

SOLUTION (*a*) We solve Eq. 18–9c for V_{rms} and set $\overline{P} = 100\ \mathrm{W}$ (at the maximum):

$$V_{\mathrm{rms}} = \sqrt{\overline{P}R} = \sqrt{(100\ \mathrm{W})(8.0\ \Omega)} = 28\ \mathrm{V}.$$

Next we solve Eq. 18–9b for I_{rms} and obtain

$$I_{\mathrm{rms}} = \sqrt{\frac{\overline{P}}{R}} = \sqrt{\frac{100\ \mathrm{W}}{8.0\ \Omega}} = 3.5\ \mathrm{A}.$$

Or we could use Ohm's law ($V = IR$):

$$I_{\mathrm{rms}} = \frac{V_{\mathrm{rms}}}{R} = \frac{28\ \mathrm{V}}{8.0\ \Omega} = 3.5\ \mathrm{A}.$$

(*b*) At $\overline{P} = 1.0\ \mathrm{W}$,

$$V_{\mathrm{rms}} = \sqrt{(1.0\ \mathrm{W})(8.0\ \Omega)} = 2.8\ \mathrm{V}$$

$$I_{\mathrm{rms}} = \frac{2.8\ \mathrm{V}}{8.0\ \Omega} = 0.35\ \mathrm{A}.$$

EXERCISE F What would be the rms voltage and rms current of the stereo in Example 18–13 if the 100 W was connected to a loudspeaker rated at 4 Ω?

This Section has given a brief introduction to the simpler aspects of alternating currents. We will discuss ac circuits in more detail in Chapter 21. In Chapter 19 we will deal with the details of dc circuits only.

* 18–8 Microscopic View of Electric Current

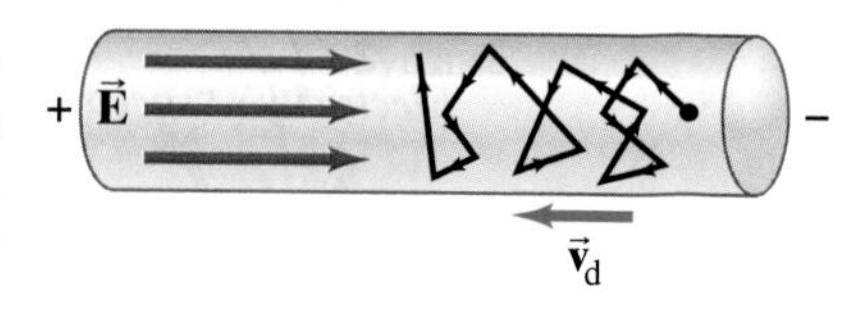

FIGURE 18–24 Electric field $\vec{\mathbf{E}}$ in a wire gives electrons in random motion a drift speed v_d.

It can be useful to analyze a simple model of electric current at the microscopic level of atoms and electrons. In a conducting wire, for example, we can imagine the free electrons as moving about randomly at high speeds, bouncing off the atoms of the wire (somewhat like the molecules of a gas—Sections 13–9 to 13–11). When an electric field exists in the wire (Fig. 18–24) due to a potential difference applied between its ends, the electrons feel a force and initially begin to accelerate. But they soon reach a more or less steady average speed (due to collisions with atoms in the wire), known as their **drift speed**, v_d. The drift speed is normally very much smaller than the electrons' average random speed.

Drift speed

We can relate v_d to the macroscopic current I in the wire. In a time Δt, the electrons will travel a distance $l = v_d\,\Delta t$ on average. Suppose the wire has cross-sectional area A. Then in time Δt, all electrons in a volume $V = Al = Av_d\,\Delta t$ will pass through the cross section A of wire, as shown in Fig. 18–25. If there are n free electrons (each of charge e) per unit volume, then the total number of electrons is $N = nV$ (V is volume, not voltage) and the total charge ΔQ that passes through the area A in a time Δt is

FIGURE 18–25 Electrons in the volume Al will all pass through the cross section indicated in a time Δt, where $l = v_d\,\Delta t$.

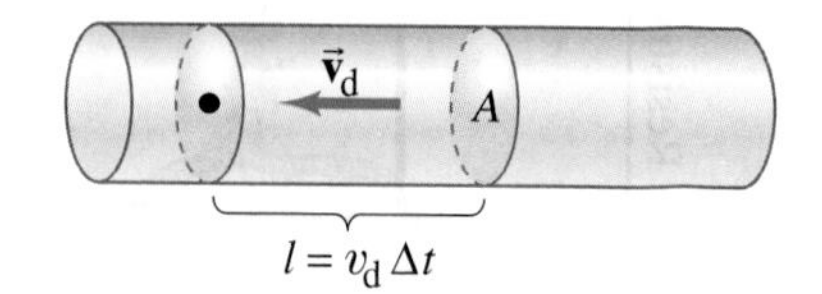

$$\begin{aligned}\Delta Q &= (\text{number of charges, } N) \times (\text{charge per particle})\\ &= (nV)(e) = (nAv_d\,\Delta t)(e).\end{aligned}$$

The current I in the wire is thus

$$I = \frac{\Delta Q}{\Delta t} = neAv_d. \qquad \textbf{(18–10)}$$

Current (microscopic variables)

EXAMPLE 18–14 Electron speeds in a wire. A copper wire, 3.2 mm in diameter, carries a 5.0-A current. Determine the drift speed of the free electrons. Assume that one electron per Cu atom is free to move (the others remain bound to the atom).

APPROACH We can apply Eq. 18–10 to find the drift speed if we can determine the number n of free electrons per unit volume. Since we assume there is one free electron per atom, the density of free electrons, n, is the same as the density of Cu atoms. The atomic mass of Cu is 63.5 u (see Periodic Table inside the back cover), so 63.5 g of Cu contains one mole or 6.02×10^{23} free electrons. We then use the mass density of copper (Table 10–1), $\rho_D = 8.9 \times 10^3\,\text{kg/m}^3$, to find the volume of this amount of copper, and then $n = N/V$. (We use ρ_D to distinguish it here from ρ for resistivity.)

SOLUTION The mass density $\rho_D = m/V$ is related to the number of free electrons per unit volume, $n = N/V$, by

$$\begin{aligned}n = \frac{N}{V} = \frac{N}{m/\rho_D} &= \frac{N\,(1\text{ mole})}{m\,(1\text{ mole})}\,\rho_D\\ &= \left(\frac{6.02 \times 10^{23}\text{ electrons}}{63.5 \times 10^{-3}\text{ kg}}\right)\left(8.9 \times 10^3\,\text{kg/m}^3\right)\\ &= 8.4 \times 10^{28}\,\text{m}^{-3}.\end{aligned}$$

The cross-sectional area of the wire is

$$A = \pi r^2 = (3.14)\left(1.6 \times 10^{-3}\,\text{m}\right)^2 = 8.0 \times 10^{-6}\,\text{m}^2.$$

Then, by Eq. 18–10, the drift speed is

$$\begin{aligned}v_d = \frac{I}{neA} &= \frac{5.0\text{ A}}{\left(8.4 \times 10^{28}\,\text{m}^{-3}\right)\left(1.6 \times 10^{-19}\,\text{C}\right)\left(8.0 \times 10^{-6}\,\text{m}^2\right)}\\ &= 4.7 \times 10^{-5}\,\text{m/s},\end{aligned}$$

which is only about 0.05 mm/s.

NOTE We can compare this drift speed to the actual speed of free electrons bouncing around inside the metal like molecules in a gas, calculated to be about $1.6 \times 10^6\,\text{m/s}$ at 20°C.

We can write a similar formula for the charge $Q\,(= CV_C)$ on the capacitor:

$$Q = Q_0\left(1 - e^{-t/RC}\right),$$

where Q_0 represents the maximum charge.

The product of the resistance R times the capacitance C, which appears in the exponent, is called the **time constant** τ of the circuit:

Time constant $\tau = RC$

$$\tau = RC. \qquad \textbf{(19–7)}$$

The time constant is a measure of how quickly the capacitor becomes charged. [The units of RC are $\Omega\cdot\text{F} = (\text{V/A})(\text{C/V}) = \text{C/(C/s)} = \text{s}$.] Specifically, it can be shown that the product RC gives the time required for the capacitor's voltage (and charge) to reach 63% of the maximum. This can be checked† using any calculator with an e^x key: $e^{-1} = 0.37$, so for $t = RC$, then $\left(1 - e^{-t/RC}\right) = \left(1 - e^{-1}\right) = (1 - 0.37) = 0.63$. In a circuit, for example, where $R = 200\,\text{k}\Omega$ and $C = 3.0\,\mu\text{F}$, the time constant is $\left(2.0 \times 10^5\,\Omega\right)\left(3.0 \times 10^{-6}\,\text{F}\right) = 0.60\,\text{s}$. If the resistance is much smaller, the time constant is much smaller and the capacitor becomes charged almost instantly. This makes sense, since a lower resistance will retard the flow of charge less. All circuits contain some resistance (if only in the connecting wires), so a capacitor can never be charged instantaneously when connected to a battery.

Capacitor discharges

The circuit just discussed involved the *charging* of a capacitor by a battery through a resistance. Now let us look at another situation: a capacitor is already charged (say, to a voltage V_0 and charge Q_0), and it is then allowed to *discharge* through a resistance R as shown in Fig. 19–21a. (In this case there is no battery.) When the switch S is closed, charge begins to flow through resistor R from one side of the capacitor toward the other side, until it is fully discharged. The voltage across the capacitor decreases, as shown in Fig. 19–21b. This "exponential decay" curve is given by

$$V_C = V_0 e^{-t/RC},$$

where V_0 is the initial voltage across the capacitor. The voltage falls 63% of the way to zero (to $0.37V_0$) in a time $\tau = RC$. Because the charge Q on the capacitor is $Q = CV$, we can write

$$Q = Q_0 e^{-t/RC}$$

for a discharging capacitor, where Q_0 is the initial charge.

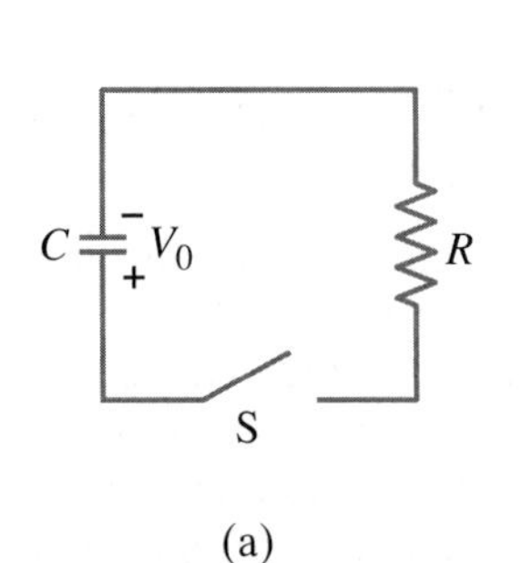

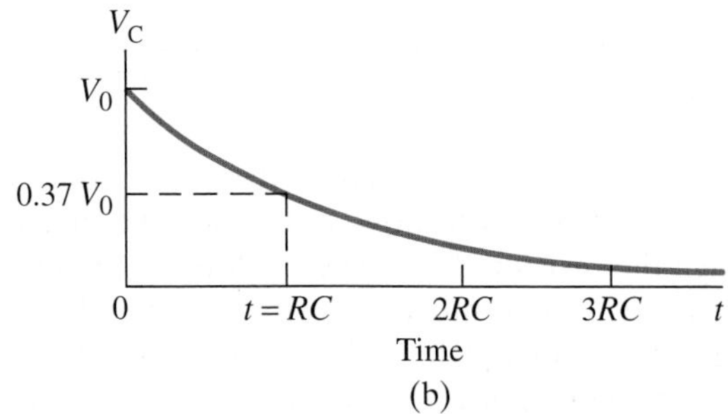

FIGURE 19–21 For the RC circuit shown in (a), the voltage V_C on the capacitor decreases with time, as shown in (b), after the switch S is closed. The charge on the capacitor follows the same curve since $Q \propto V$.

EXAMPLE 19–12 A discharging *RC* circuit. If a charged capacitor, $C = 35\,\mu\text{F}$, is connected to a resistance $R = 120\,\Omega$ as in Fig. 19–21a, how much time will elapse until the voltage falls to 10% of its original (maximum) value?

APPROACH The voltage across the capacitor decreases according to $V_C = V_0 e^{-t/RC}$. We set $V_C = 0.10V_0$ (10% of V_0), but first we need to calculate $\tau = RC$.

†More simply, since $e = 2.718\cdots$, then $e^{-1} = 1/e = 1/2.718 = 0.37$. Note that e is the inverse operation to the natural logarithm ln: $\ln(e) = 1$, and $\ln\left(e^x\right) = x$.

SOLUTION The time constant for this circuit is given by

$$\tau = RC = (120\,\Omega)(35 \times 10^{-6}\,\text{F}) = 4.2 \times 10^{-3}\,\text{s}.$$

After a time t the voltage across the capacitor will be

$$V_C = V_0\left(e^{-t/RC}\right).$$

We want to know the time t for which $V_C = 0.10V_0$. We substitute into the above equation

$$0.10V_0 = V_0 e^{-t/RC}$$

so

$$e^{-t/RC} = 0.10.$$

The inverse operation to the exponential e is the natural log, ln. Thus

$$\ln\left(e^{-t/RC}\right) = -\frac{t}{RC} = \ln 0.10 = -2.3.$$

Solving for t, we find the elapsed time is

$$t = 2.3(RC) = (2.3)\left(4.2 \times 10^{-3}\,\text{s}\right) = 9.7 \times 10^{-3}\,\text{s}$$

or 9.7 ms.

NOTE We can find the time for any specified voltage across a capacitor by using $t = RC\ln(V_0/V_C)$.

EXERCISE E For the same 35-μF capacitor as in Example 19–12, what value of resistance R would produce a voltage reduction to 10% of V_0 in exactly 1.0 s?

CONCEPTUAL EXAMPLE 19–13 **Bulb in *RC* circuit.** In the circuit of Fig. 19–22, the capacitor is originally uncharged. Describe the behavior of the lightbulb from the instant switch S is closed until a long time later.

RESPONSE When the switch is first closed, the current in the circuit is high and the lightbulb burns brightly. As the capacitor charges, the voltage across the capacitor increases and the current is reduced, causing the lightbulb to dim. As the potential difference across the capacitor approaches the same voltage as the battery, the current decreases toward zero and the lightbulb goes out.

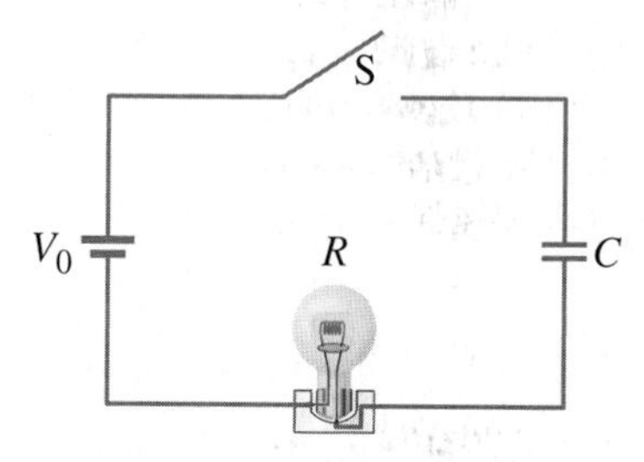

FIGURE 19–22 Example 19–13.

* Medical and Other Applications of *RC* Circuits

The charging and discharging in an RC circuit can be used to produce voltage pulses at a regular frequency. The charge on the capacitor increases to a particular voltage, and then discharges. A simple way of initiating the discharge of the capacitor is by the use of a gas-filled tube which has an electrical breakdown when the voltage across it reaches a certain value V_0. After the discharge is finished, the tube no longer conducts current and the recharging process repeats itself, starting at a lower voltage V_0'. Figure 19–23 shows a possible circuit, and the "sawtooth" voltage it produces.

Sawtooth voltage
blinking flashers

A simple blinking light can be an application of a sawtooth oscillator circuit. Here the emf is supplied by a battery; the neon bulb flashes on at a rate of perhaps 1 cycle per second. The main component of a "flasher unit" is a moderately large capacitor.

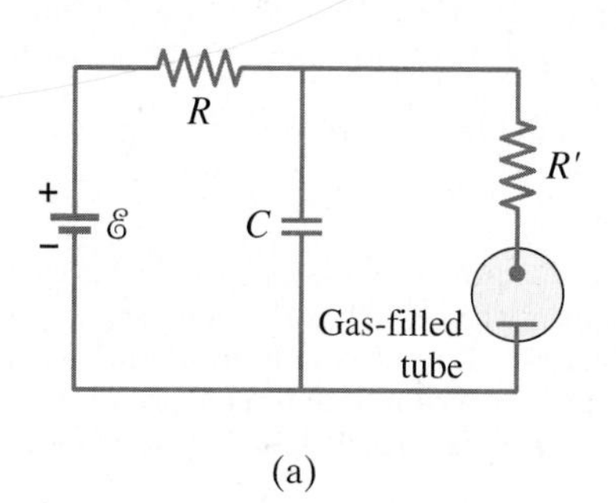

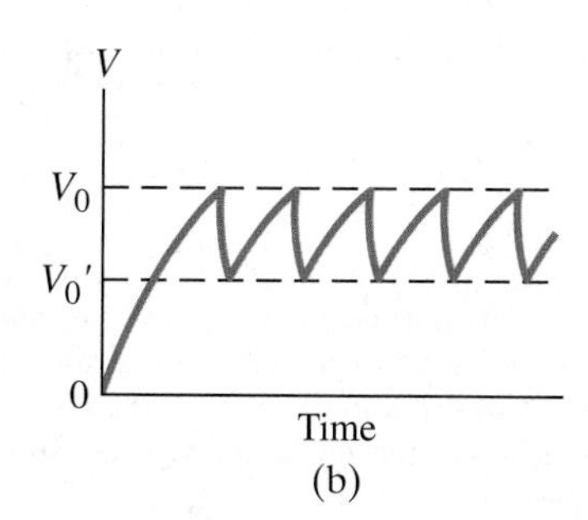

FIGURE 19–23 (a) An RC circuit, coupled with a gas-filled tube as a switch, can produce a repeating "sawtooth" voltage, as shown in (b).

PHYSICS APPLIED
Windshield wipers

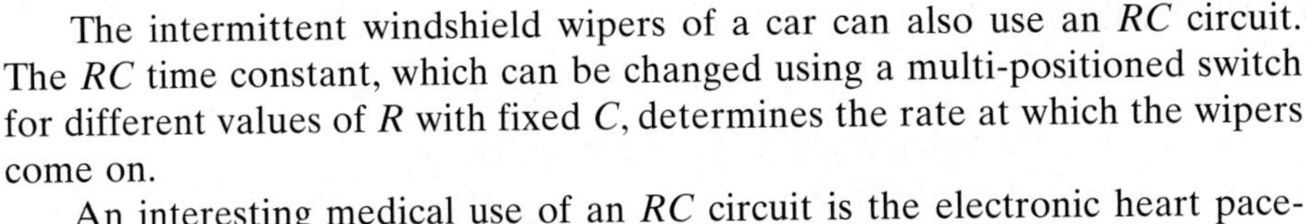

The intermittent windshield wipers of a car can also use an RC circuit. The RC time constant, which can be changed using a multi-positioned switch for different values of R with fixed C, determines the rate at which the wipers come on.

PHYSICS APPLIED
Heart pacemaker

An interesting medical use of an RC circuit is the electronic heart pacemaker, which can make a stopped heart start beating again by applying an electric stimulus through electrodes attached to the chest. The stimulus can be repeated at the normal heartbeat rate if necessary. The heart itself contains *pacemaker* cells, which send out tiny electric pulses at a rate of 60 to 80 per minute. These signals induce the start of each heartbeat. In some forms of heart disease, the natural pacemaker fails to function properly, and the heart loses its beat. Such patients use *electronic pacemakers* which produce a regular voltage pulse that starts and controls the frequency of the heartbeat. The electrodes are implanted in or near the heart (Fig. 19–24), and the circuit contains a capacitor and a resistor. The charge on the capacitor increases to a certain point and then discharges. Then it starts charging again. The pulsing rate depends on the values of R and C.

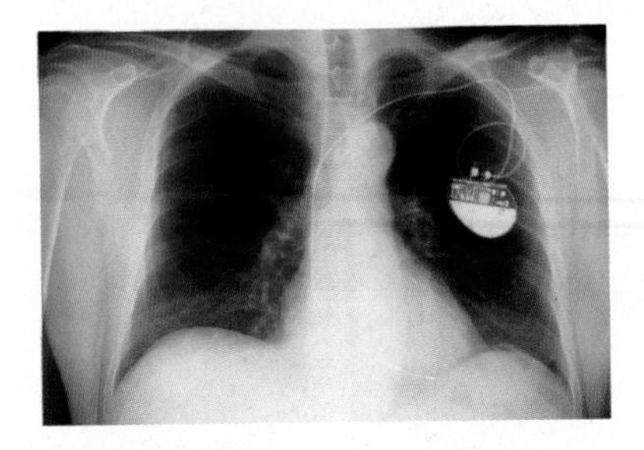

FIGURE 19–24 Electronic battery-powered pacemaker can be seen on the rib cage in this X-ray.

19–7 Electric Hazards

PHYSICS APPLIED
Dangers of electricity

Excess electric current can heat wires in buildings and cause fires, as discussed in Section 18–6. Electric current can also damage the human body or even be fatal. Electric current through the human body can cause damage in two ways: (1) Electric current heats tissue and can cause burns; (2) electric current stimulates nerves and muscles (whose operation, as we saw in Sections 17–11 and 18–10, is electrical), and we feel a "shock." The severity of a shock depends on the magnitude of the current, how long it acts, and through what part of the body it passes. A current passing through vital organs such as the heart or brain is especially serious for it can interfere with their operation.

Most people can "feel" a current of about 1 mA. Currents of a few mA cause pain but rarely cause much damage in a healthy person. Currents above 10 mA cause severe contraction of the muscles, and a person may not be able to release the source of the current (say, a faulty appliance or wire). Death from paralysis of the respiratory system can occur. Artificial respiration, however, can sometimes revive a victim. If a current above about 80 to 100 mA passes across the torso, so that a portion passes through the heart for more than a second or two, the heart muscles will begin to contract irregularly and blood will not be properly pumped. This condition is called *ventricular fibrillation*. If it lasts for long, death results. Strangely enough, if the current is much larger, on the order of 1 A, death by heart failure may be less likely,† but such currents can cause serious burns, especially if concentrated through a small area of the body.

The seriousness of a shock depends on the applied voltage and on the effective resistance of the body. Living tissue has low resistance since the fluid of cells contains ions that can conduct quite well. However, the outer layer of skin, when dry, offers high resistance and is thus protective. The effective resistance between two points on opposite sides of the body when the skin is dry is in the range of 10^4 to $10^6\ \Omega$. But when the skin is wet, the resistance may be $10^3\ \Omega$ or less.

†Larger currents apparently bring the entire heart to a standstill. Upon release of the current, the heart returns to its normal rhythm. This may not happen when fibrillation occurs because, once started, it can be hard to stop. Fibrillation may also occur as a result of a heart attack or during heart surgery. A device known as a *defibrillator* (described in Section 17–9) can apply a brief high current to the heart, causing complete heart stoppage which is often followed by resumption of normal beating.

A person who is barefoot or wearing thin-soled shoes will be in good contact with the ground, and touching a 120-V line with a wet hand can result in a current

$$I = \frac{120\ \text{V}}{1000\ \Omega} = 120\ \text{mA}.$$

As we saw, this could be lethal.

A person who has received a shock has become part of a complete circuit. Figure 19–25 shows two ways the circuit might be completed when a person accidentally touches a "hot" electric wire—"hot" meaning a high potential such as 120 V (normal household voltage) relative to ground. The other side of building wiring is connected to ground—either by a wire connected to a buried conductor, or via a water pipe into the ground. In Fig. 19–25a, the current passes from the high-voltage wire, through the person, to the ground through his bare feet, and back along the ground (a fair conductor) to the ground terminal of the source. If the person stands on a good insulator—thick rubber-soled shoes or a dry wood floor—there will be much more resistance in the circuit and consequently much less current through the person. If the person stands with bare feet on the ground, or is in a bathtub, there is lethal danger because the resistance is much less and the current greater. In a bathtub (or swimming pool), not only are you wet, but the water is in contact with the drain pipe that leads to the ground. It is strongly recommended that you not touch anything electrical when wet or in bare feet.

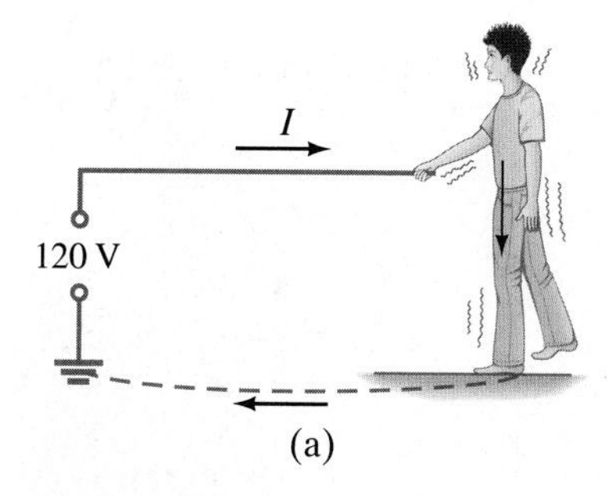

(a)

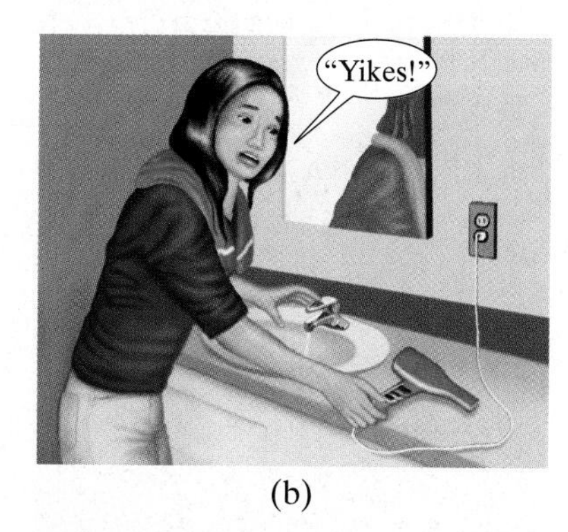

(b)

FIGURE 19–25 A person receives an electric shock when the circuit is completed.

In Fig. 19–25b, a person touches a faulty "hot" wire with one hand, and the other hand touches a sink faucet (connected to ground via the pipe). The current is particularly dangerous because it passes across the chest, through the heart and lungs. A rule of thumb: if one hand is touching something electrical, keep your other hand in your pocket (don't use it!), and wear thick rubber-soled shoes. It is also a good idea to remove metal jewelry, especially rings (your finger is usually moist under a ring).

CAUTION

Keep one hand in your pocket when other touches electricity

You can come into contact with a hot wire by touching a bare wire whose insulation has worn off, or from a bare wire inside an appliance when you're tinkering with it. (Always unplug an electrical device before investigating[†] its insides!) Another possibility is that a wire inside a device may break or lose its insulation and come in contact with the case. If the case is metal, it will conduct electricity. A person could then suffer a severe shock merely by touching the case, as shown in Fig. 19–26b. To prevent an accident, metal cases are supposed to be connected directly to ground by a separate ground wire. Then if a "hot" wire touches the grounded case, a short circuit to ground immediately occurs internally, as shown in Fig. 19–26c, and most of the current passes through the low-resistance ground wire rather than through the person. Furthermore, the high current should open the fuse or circuit breaker.

PHYSICS APPLIED

Grounding and shocks

[†]Even then you can get a bad shock from a capacitor that hasn't been discharged until you touch it.

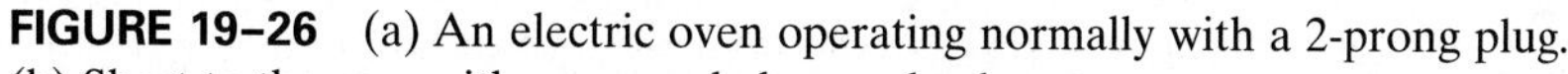

FIGURE 19–26 (a) An electric oven operating normally with a 2-prong plug. (b) Short to the case with ungrounded case: shock. (c) Short to the case with the case grounded by a 3-prong plug.

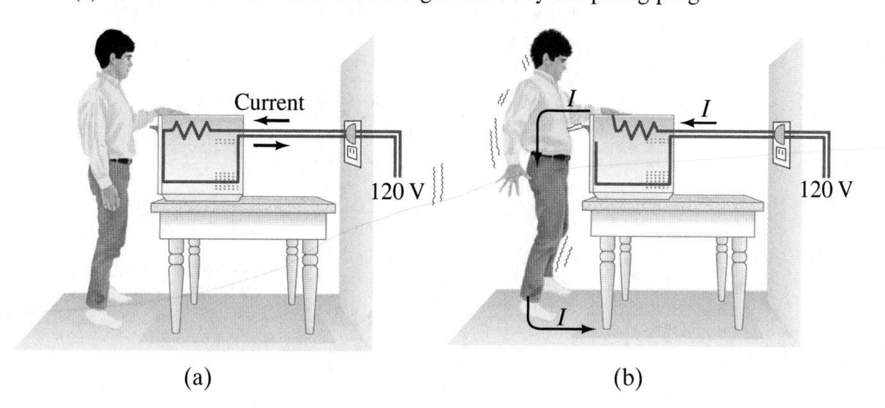

(a) (b)

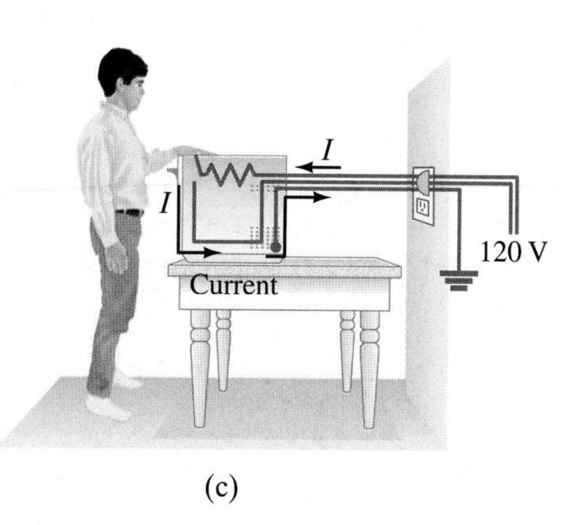

(c)

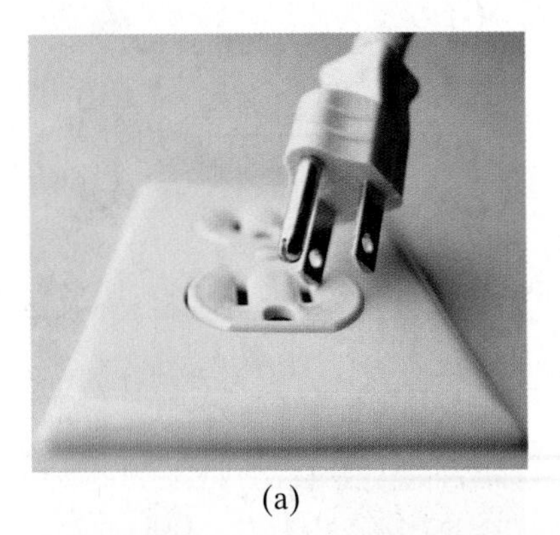

(a)

(b)

(c)

FIGURE 19–27 (a) A 3-prong plug, and (b) an adapter (gray) for old fashioned 2-prong outlets—be sure to screw down the ground tab. (c) A polarized 2-prong plug.

CAUTION

Black wire may be either ground or hot. Beware!

Grounding a metal case is done by a separate ground wire connected to the third (round) prong of a 3-prong plug (Fig. 19–27a). Never cut off the third prong of a plug—it could be deadly.

Why is a third wire needed? The 120 V is carried by the other two wires—one **hot** (120 V ac), the other **neutral**, which is itself grounded.† The third "dedicated" ground wire with the round prong may seem redundant. But it is protection for two reasons: (1) it protects against internal wiring that may have been done incorrectly; (2) the neutral wire carries normal current ("return" current from the 120 V) and it does have resistance; so there can be a voltage drop along it—normally small, but if connections are poor or corroded, or the plug is loose, the resistance could be large enough that you might feel that voltage if you touched the neutral wire some distance from its grounding point.

Some electrical devices come with only two wires, and the plug's two prongs are of different widths; the plug can be inserted only one way into the outlet so that the intended neutral (wider prong) in the device is connected to neutral in the wiring. For example, the screw threads of a lightbulb are meant to be connected to neutral (and the base contact to hot), to avoid shocks when changing a bulb in a possibly protruding socket. Devices with 2-prong plugs do *not* have their cases grounded; they are supposed to have double electric insulation. Take extra care anyway.

The insulation on a wire may be color coded. Hand-held meters may have red (hot) and black (ground) lead wires. But in a house, black is usually hot (or it may be red), whereas white is neutral and green is the dedicated ground. But beware: these color codes cannot always be trusted.

Normal circuit breakers (Sections 18–6 and 20–7) protect equipment and buildings from overload and fires. They protect humans only in some circumstances, such as the very high currents that result from a short, if they respond quickly enough. Ground fault circuit interrupters, described in Section 21–8, are designed to protect people from the much lower currents (10 mA to 100 mA) that are lethal but would not throw a 15-A circuit breaker or blow a 20-A fuse.

†In the U.S., three wires normally enter a house: two *hot* wires at 120 V each (which add together to 240 V for appliances or devices that run on 240 V) plus the grounded *neutral* (carrying return current for the two hots). See Fig. 19–28 below. The "dedicated" *ground* wire (non-current carrying) is a fourth wire that does not come from the electric company but enters the house from a nearby heavy stake in the ground or a buried metal pipe. The two hot wires can feed separate 120-V circuits in the house, so each 120-V circuit inside the house has only 3 wires as discussed in the text.

FIGURE 19–28 Four wires entering a typical house. The color codes for wires are not always as shown here—be careful!

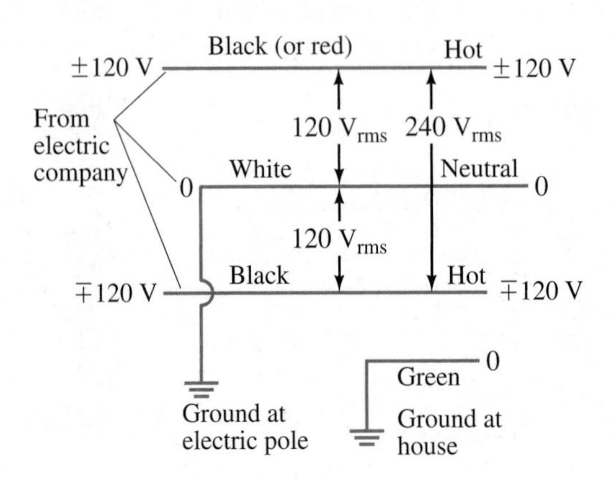

It is current that harms, but it is voltage that drives the current. 30 volts is sometimes said to be the threshhold for danger. But even a 12-V car battery (which can supply large currents) can cause nasty burns and shock.

Another danger is *leakage current*, by which we mean a current along an unintended path. Leakage currents are often "capacitively coupled." For example, a wire in a lamp forms a capacitor with the metal case; charges moving in one conductor attract or repel charge in the other, so there is a current. Typical electrical codes limit leakage currents to 1 mA for any device. A 1-mA leakage current is usually harmless. It can be very dangerous, however, to a hospital patient with implanted electrodes connected to ground through the apparatus. This is due to the absence of the protective skin layer and because the current can pass directly through the heart as compared to the usual situation where the current enters at the hands and spreads out through the body. Although 100 mA may be needed to cause heart fibrillation when entering through the hands (very little of it actually passes through the heart), as little as 0.02 mA has been known to cause fibrillation when passing directly to the heart. Thus, a "wired" patient is in considerable danger from leakage current even from as simple an act as touching a lamp.

Leakage current

Finally, don't touch a downed power line (lethal!) or even get near it. A hot power line is at thousands of volts. A huge current can flow along the ground or pavement, from where the high-voltage wire touches it over to the grounding point of the neutral line, enough that the voltage between your two legs could be large. Tip: stand on one foot or run (only one foot touches the ground at a time).

* 19–8 Ammeters and Voltmeters

DC meters

An **ammeter** is used to measure current, and a **voltmeter** measures potential difference or voltage. Measurements of current and voltage are made with meters that are of two types: (1) *analog* meters, which display numerical values by the position of a pointer that can move across a scale (Fig. 19–29a); and (2) *digital* meters, which display the numerical value in numbers (Fig. 19–29b). We now discuss the meters themselves and how they work, then how they are connected to circuits to make measurements. Finally we will discuss how using meters affects the circuit being measured, possibly causing erroneous results—and what to do about it.

(a)

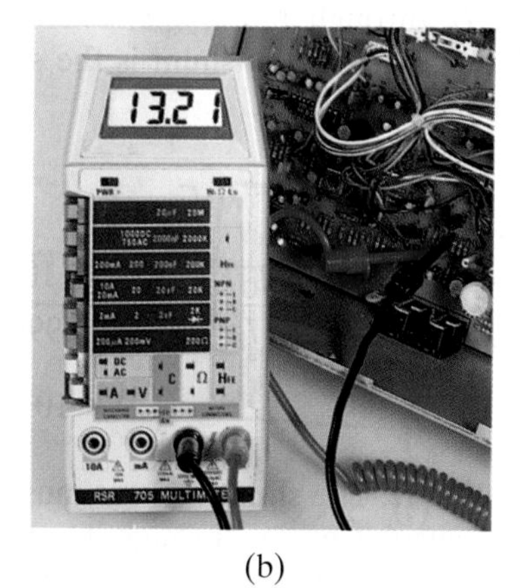
(b)

FIGURE 19–29 (a) An analog multimeter being used as a voltmeter. (b) An electronic digital meter.

* Analog Ammeters and Voltmeters

The crucial part of an analog ammeter or voltmeter, in which the reading is by a pointer on a scale (Fig. 19–29a), is a *galvanometer*. The galvanometer works on the principle of the force between a magnetic field and a current-carrying coil of wire, and will be discussed in Chapter 20. For now, we merely need to know that the deflection of the needle of a galvanometer is proportional to the current

* Ef

It

ta

PHYSICS APPLIED

Correcting for meter resistance

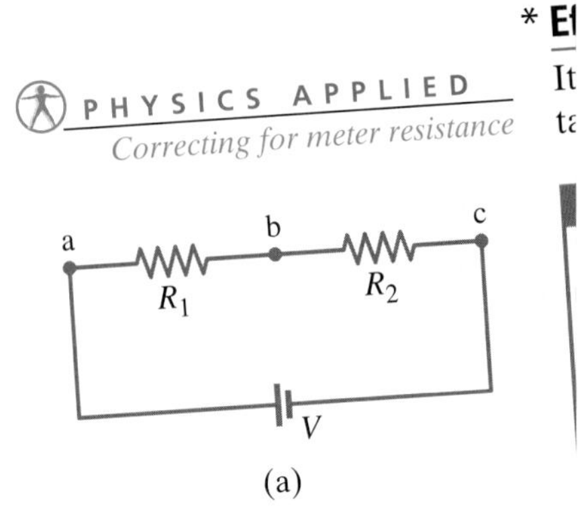

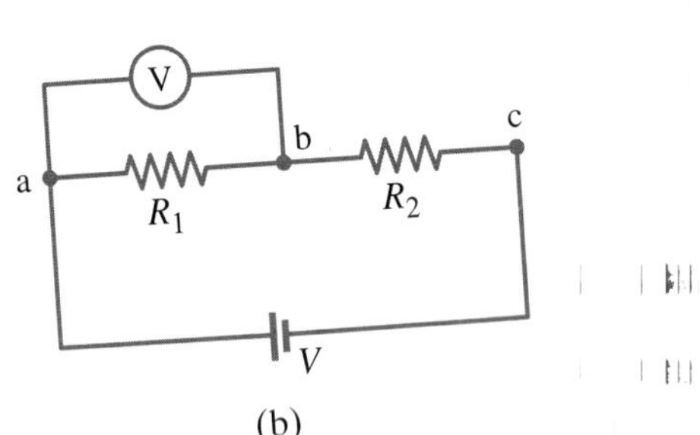

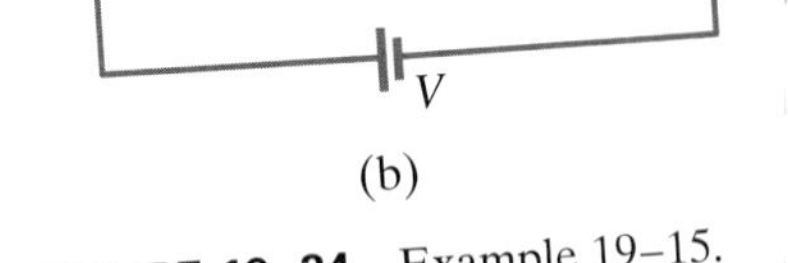

FIGURE 19–34 Example 19–15.

Magnetic field due to current in straight wire

The proportionality constant is written† as $\mu_0/2\pi$; thus,

$$B = \frac{\mu_0}{2\pi}\frac{I}{r}. \qquad \text{[near a long straight wire]} \qquad \textbf{(20–6)}$$

The value of the constant μ_0, which is called the **permeability of free space**, is $\mu_0 = 4\pi \times 10^{-7}\,\mathrm{T\cdot m/A}$.

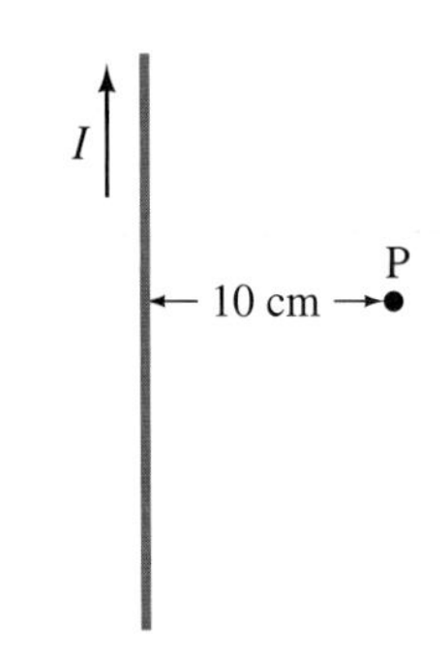

FIGURE 20–20 Example 20–7.

EXAMPLE 20–7 **Calculation of $\vec{\mathbf{B}}$ near a wire.** An electric wire in the wall of a building carries a dc current of 25 A vertically upward. What is the magnetic field due to this current at a point P 10 cm due north of the wire (Fig. 20–20)?

APPROACH We assume the wire is much longer than the 10-cm distance to the point P so we can apply Eq. 20–6.

SOLUTION According to Eq. 20–6:

$$B = \frac{\mu_0 I}{2\pi r} = \frac{(4\pi \times 10^{-7}\,\mathrm{T\cdot m/A})(25\,\mathrm{A})}{(2\pi)(0.10\,\mathrm{m})} = 5.0 \times 10^{-5}\,\mathrm{T},$$

or 0.50 G. By the right-hand rule (Fig. 20–8c), the field points to the west (into the page in Fig. 20–20) at this point.

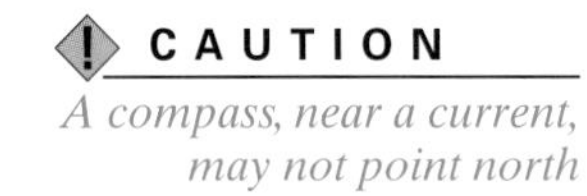

CAUTION

A compass, near a current, may not point north

NOTE The wire's field has about the same magnitude as Earth's, so a compass would not point north but in a northwesterly direction.

NOTE Most electrical wiring in buildings consists of cables with two wires in each cable. Since the two wires carry current in opposite directions, their magnetic fields will cancel to a large extent.

EXERCISE F At what distance from the wire in Example 20–7 is its magnetic field 5 times greater than the Earth's?

FIGURE 20–21 Example 20–8. Wire 1 carrying current I_1 out towards us, and wire 2 carrying current I_2 into the page, produce magnetic fields whose lines are circles around their respective wires.

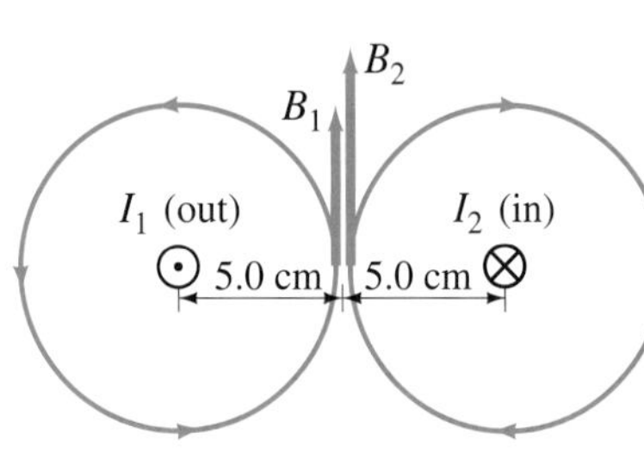

EXAMPLE 20–8 **Magnetic field midway between two currents.** Two parallel straight wires 10.0 cm apart carry currents in opposite directions (Fig. 20–21). Current $I_1 = 5.0\,\mathrm{A}$ is out of the page, and $I_2 = 7.0\,\mathrm{A}$ is into the page. Determine the magnitude and direction of the magnetic field halfway between the two wires.

APPROACH The magnitude of the field produced by each wire is calculated from Eq. 20–6. The direction of *each* wire's field is determined with the right-hand rule. The total field is the vector sum of the two fields at the midway point.

SOLUTION The magnetic field lines due to current I_1 form circles around the wire of I_1, and right-hand-rule-1 (Fig. 20–8c) tells us they point counterclockwise around the wire. The field lines due to I_2 form circles around the wire of I_2 and point clockwise, Fig. 20–21. At the midpoint, both fields point upward as shown, and so add together. The midpoint is 0.050 m from each wire, and from Eq. 20–6 the magnitudes of B_1 and B_2 are

$$B_1 = \frac{\mu_0 I_1}{2\pi r} = \frac{(4\pi \times 10^{-7}\,\mathrm{T\cdot m/A})(5.0\,\mathrm{A})}{2\pi(0.050\,\mathrm{m})} = 2.0 \times 10^{-5}\,\mathrm{T};$$

$$B_2 = \frac{\mu_0 I_2}{2\pi r} = \frac{(4\pi \times 10^{-7}\,\mathrm{T\cdot m/A})(7.0\,\mathrm{A})}{2\pi(0.050\,\mathrm{m})} = 2.8 \times 10^{-5}\,\mathrm{T}.$$

The total field is *up* with a magnitude of

$$B = B_1 + B_2 = 4.8 \times 10^{-5}\,\mathrm{T}.$$

†The constant is chosen in this complicated way so that Ampère's law (Section 20–8), which is considered more fundamental, will have a simple and elegant form.

CONCEPTUAL EXAMPLE 20–9 **Magnetic field due to four wires.** Figure 20–22 shows four long parallel wires which carry equal currents into or out of the page as shown. In which configuration, (*a*) or (*b*), is the magnetic field greater at the center of the square?

RESPONSE It is greater in (*a*). The arrows illustrate the directions of the field produced by each wire; check it out, using the right-hand rule to confirm these results. The net field at the center is the superposition of the four fields, which will point to the left in (*a*) and is zero in (*b*).

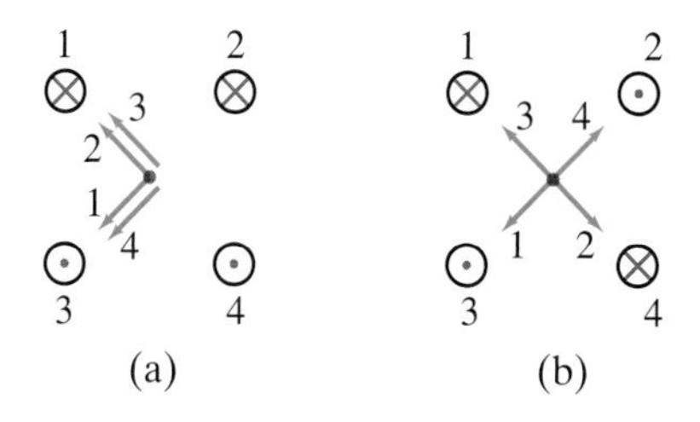

FIGURE 20–22 Example 20–9.

20–6 Force between Two Parallel Wires

We have seen that a wire carrying a current produces a magnetic field (magnitude given by Eq. 20–6 for a long straight wire). Also, a current-carrying wire feels a force when placed in a magnetic field (Section 20–3, Eq. 20–1). Thus, we expect that two current-carrying wires will exert a force on each other.

Consider two long parallel wires separated by a distance d, as in Fig. 20–23a. They carry currents I_1 and I_2, respectively. Each current produces a magnetic field that is "felt" by the other, so each must exert a force on the other. For example, the magnetic field B_1 produced by I_1 in Fig 20–23 is given by Eq. 20–6, which at the location of wire 2 is

$$B_1 = \frac{\mu_0}{2\pi}\frac{I_1}{d}.$$

See Fig. 20–23b, where the field due *only* to I_1 is shown. According to Eq. 20–2, the force F_2 exerted by B_1 on a length l_2 of wire 2, carrying current I_2, is

$$F_2 = I_2 B_1 l_2.$$

Note that the force on I_2 is due only to the field produced by I_1. Of course, I_2 also produces a field, but it does not exert a force on itself. We substitute B_1 into the formula for F_2 and find that the force on a length l_2 of wire 2 is

$$F_2 = \frac{\mu_0}{2\pi}\frac{I_1 I_2}{d} l_2. \qquad \textbf{(20–7)}$$

If we use right-hand-rule-1 of Fig. 20–8c, we see that the lines of B_1 are as shown in Fig. 20–23b. Then using right-hand-rule-2 of Fig. 20–11c, we see that the force exerted on I_2 will be to the left in Fig. 20–23b. That is, I_1 exerts an attractive force on I_2 (Fig. 20–24a). This is true as long as the currents are in the same direction. If I_2 is in the opposite direction, the right-hand rule indicates that the force is in the opposite direction. That is, I_1 exerts a repulsive force on I_2 (Fig. 20–24b).

Reasoning similar to that above shows that the magnetic field produced by I_2 exerts an equal but opposite force on I_1. We expect this to be true also from Newton's third law, of course. Thus, as shown in Fig. 20–24, parallel currents in the same directions attract each other, whereas parallel currents in opposite directions repel.

FIGURE 20–23 (a) Two parallel conductors carrying currents I_1 and I_2. (b) Magnetic field $\vec{\mathbf{B}}_1$ produced by I_1. (Field produced by I_2 is not shown.) $\vec{\mathbf{B}}_1$ points into page at position of I_2.

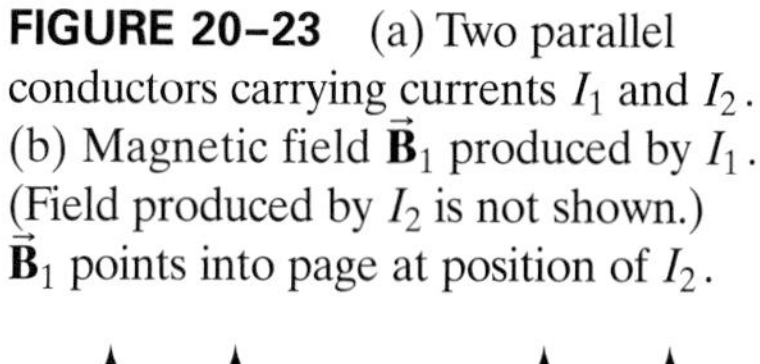
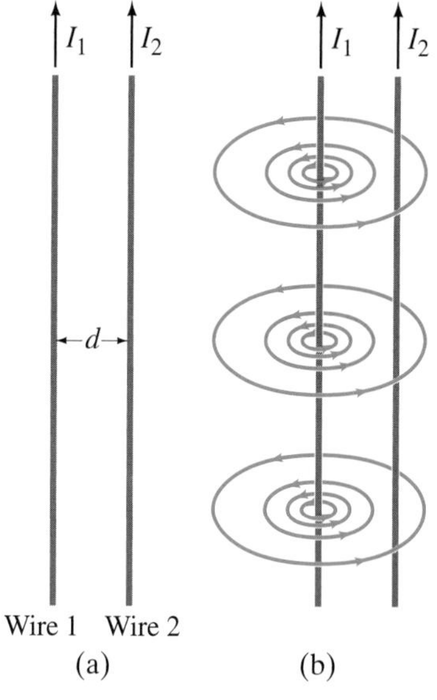

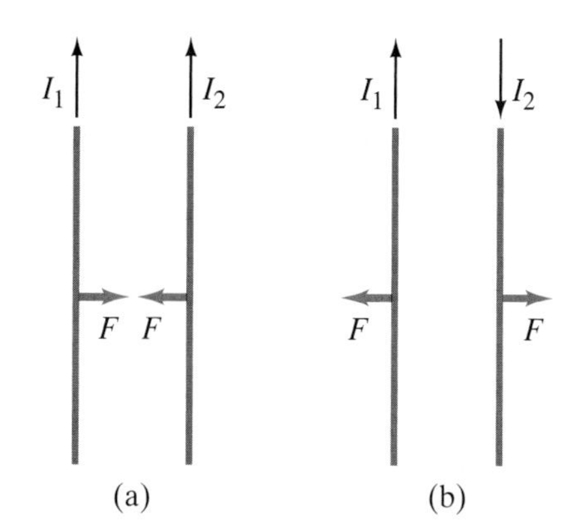

FIGURE 20–24 (a) Parallel currents in the same direction exert an attractive force on each other. (b) Antiparallel currents (in opposite directions) exert a repulsive force on each other.

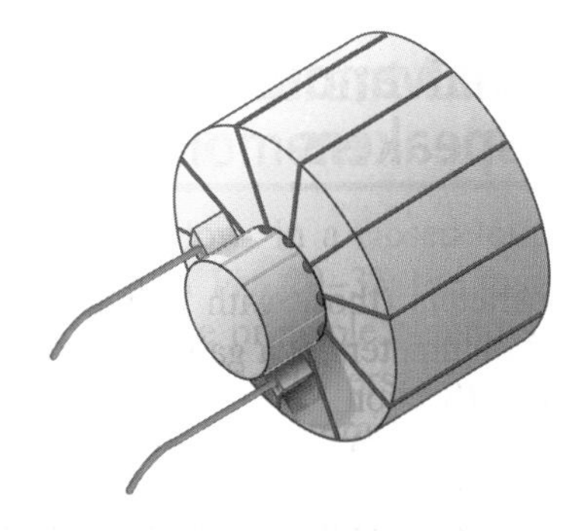

FIGURE 20–37 Motor with many windings.

Most motors contain several coils, called *windings*, each located in a different place on the armature, Fig. 20–37. Current flows through each coil only during a small part of a revolution, at the time when its orientation results in the maximum torque. In this way, a motor produces a much steadier torque than can be obtained from a single coil.

PHYSICS APPLIED
AC motor

An **ac motor**, with ac current as input, can work without commutators since the current itself alternates. Many motors use wire coils to produce the magnetic field (electromagnets) instead of a permanent magnet. Indeed the design of most motors is more complex than described here, but the general principles remain the same.

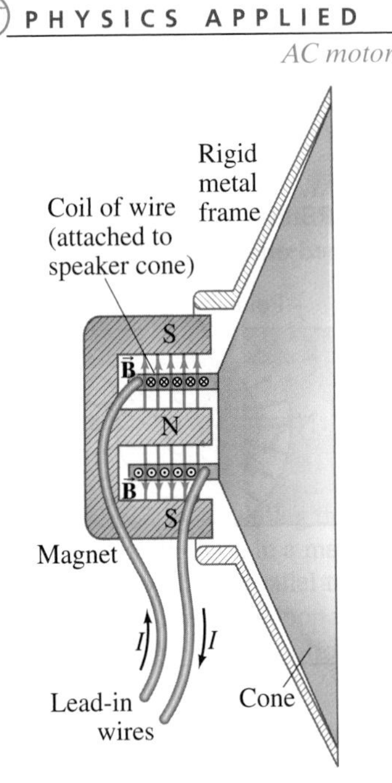

FIGURE 20–38 Loudspeaker.

* Loudspeakers

A **loudspeaker** also works on the principle that a magnet exerts a force on a current-carrying wire. The electrical output of a stereo or TV set is connected to the wire leads of the speaker. The speaker leads are connected internally to a coil of wire, which is itself attached to the speaker cone, Fig. 20–38. The speaker cone is usually made of stiffened cardboard and is mounted so that it can move back and forth freely. A permanent magnet is mounted directly in line with the coil of wire. When the alternating current of an audio signal flows through the wire coil, which is free to move within the magnet, the coil experiences a force due to the magnetic field of the magnet. As the current alternates at the frequency of the audio signal, the coil and attached speaker cone move back and forth at the same frequency, causing alternate compressions and rarefactions of the adjacent air, and sound waves are produced. A speaker thus changes electrical energy into sound energy, and the frequencies and intensities of the emitted sound waves can be an accurate reproduction of the electrical input.

* 20–11 Mass Spectrometer

PHYSICS APPLIED
The mass spectrometer

FIGURE 20–39 Bainbridge-type mass spectrometer. The magnetic fields B and B' point out of the paper (indicated by the dots).

A **mass spectrometer** is a device to measure masses of atoms. It is used today not only in physics but also in chemistry, geology, and medicine, often to identify atoms (and their concentration) in given samples. As shown in Fig. 20–39, ions are produced by heating, or by an electric current, in the source or sample S. They pass through slit s_1 and enter a region where there are crossed electric and magnetic fields. Ions follow a straight-line path in this region if the electric force qE (upward on a positive ion) is just balanced by the magnetic force qvB (downward on a positive ion): that is, if $qE = qvB$, or

$$v = \frac{E}{B}.$$

Only those ions whose speed is $v = E/B$ will pass through undeflected and emerge through slit s_2. (This arrangement is called a **velocity selector**.) In the semicircular region, after s_2, there is only a magnetic field, B', so the ions follow a circular path. The radius of the circular path is found from their mark on film (or detectors) if B' is fixed, or r is fixed by the position of a detector and B' is varied until detection occurs. Newton's second law, $\Sigma F = ma$, applied to an ion moving in a circle under the influence only of the magnetic field B' gives $qvB' = mv^2/r$. Since $v = E/B$, we have

$$m = \frac{qB'r}{v} = \frac{qBB'r}{E}.$$

All the quantities on the right side are known or can be measured, and thus m can be determined.

Historically, the masses of many atoms were measured this way. When a pure substance was used, it was sometimes found that two or more closely spaced marks would appear on the film. For example, neon produced two marks whose radii corresponded to atoms of mass 20 and 22 atomic mass units (u). Impurities were ruled out and it was concluded that there must be two types of neon with different masses. These different forms were called **isotopes**. It was soon found that most elements are mixtures of isotopes, and the difference in mass is due to different numbers of neutrons (discussed in Chapter 30).

Isotopes

EXAMPLE 20–13 **Mass spectrometry.** Carbon atoms of atomic mass 12.0 u are found to be mixed with another, unknown, element. In a mass spectrometer with fixed B', the carbon traverses a path of radius 22.4 cm and the unknown's path has a 26.2-cm radius. What is the unknown element? Assume they have the same charge.

APPROACH The carbon and unknown atoms pass through the same electric and magnetic fields. Hence their masses are proportional to the radius of their respective paths (see equation above).

SOLUTION We write a ratio for the masses, using the equation at the top of this page:

$$\frac{m_x}{m_C} = \frac{qBB'r_x/E}{qBB'r_C/E} = \frac{26.2\text{ cm}}{22.4\text{ cm}} = 1.17.$$

Thus $m_x = 1.17 \times 12.0\text{ u} = 14.0\text{ u}$. The other element is probably nitrogen (see the periodic table, inside the back cover).

NOTE The unknown could also be an isotope such as carbon-14 ($^{14}_{6}C$). See Appendix B. Further physical or chemical analysis would be needed.

20–12 Ferromagnetism: Domains and Hysteresis

We saw in Section 20–1 that iron (and a few other materials) can be made into strong magnets. These materials are said to be **ferromagnetic**.

Sources of Ferromagnetism

Microscopic examination reveals that a magnet is made up of tiny regions known as **domains**, at most about 1 mm in length or width. Each domain behaves like a tiny magnet with a north and a south pole. In an unmagnetized piece of iron, the domains are arranged randomly, Fig. 20–40a. The magnetic effects of the domains cancel each other out, so this piece of iron is not a magnet. In a magnet, the domains are preferentially aligned in one direction as shown in Fig. 20–40b (downward in this case). A magnet can be made from an unmagnetized piece of iron by placing it in a strong magnetic field. (You can make a needle magnetic, for example, by stroking it with one pole of a strong magnet.) The magnetization direction of domains may actually rotate slightly to be more nearly parallel to an external field, or the borders of domains move so domains with magnetic orientation parallel to the external field grow larger (compare Figs. 20–40a and b). This explains how a magnet can pick up unmagnetized pieces of iron like paper clips. The magnet's field causes a slight alignment of the domains in the unmagnetized object, which becomes a temporary magnet with its north pole facing the south pole of the permanent magnet, and vice versa; thus, attraction results. Similarly, elongated iron filings in a magnetic field

Domains

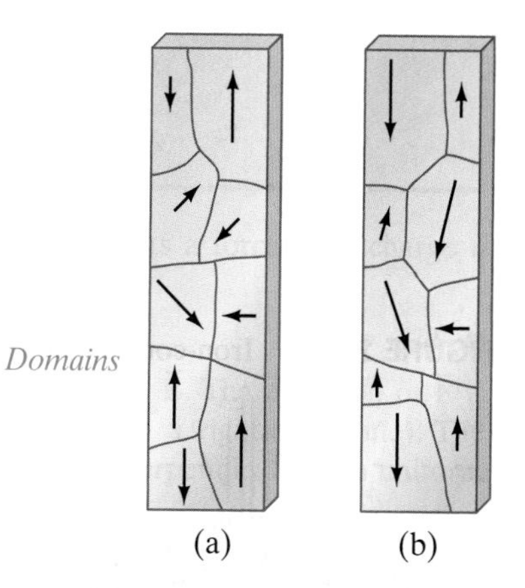

FIGURE 20–40 (a) An unmagnetized piece of iron is made up of domains that are randomly arranged. Each domain is like a tiny magnet; the arrows represent the magnetization direction, with the arrowhead being the N pole. (b) In a magnet, the domains are preferentially aligned in one direction (down in this case), and may be altered in size by the magnetization process.

[*Ampère's law states that around any chosen closed loop path, the sum of each path segment Δl times the component of $\vec{\mathbf{B}}$ parallel to the segment equals μ_0 times the current I enclosed by the closed path:

$$\Sigma B_{\parallel} \Delta l = \mu_0 I_{\text{encl}}. \tag{20–9}$$]

[*The torque τ on N loops of current I in a magnetic field $\vec{\mathbf{B}}$ is

$$\tau = NIAB \sin\theta. \tag{20–10}$$

The force exerted on a current-carrying wire by a magnetic field is the basis for operation of many devices, such as meters, motors, and loudspeakers.]

[*A **mass spectrometer** uses electric and magnetic fields to determine the masses of atoms.]

Iron and a few other materials that are **ferromagnetic** can be made into strong permanent magnets. Ferromagnetic materials are made up of tiny **domains**—each a tiny magnet—which are preferentially aligned in a permanent magnet.

[*When iron or another ferromagnetic material is placed in a magnetic field B_0 due to a current, the iron becomes magnetized. When the current is turned off, the material remains magnetized; when the current is increased in the opposite direction, a graph of the total field B versus B_0 is a **hysteresis loop**, and the fact that the curve does not retrace itself is called **hysteresis**.]

Questions

1. A compass needle is not always balanced parallel to the Earth's surface, but one end may dip downward. Explain.
2. Draw the magnetic field lines around a straight section of wire carrying a current horizontally to the left.
3. In what direction are the magnetic field lines surrounding a straight wire carrying a current that is moving directly away from you?
4. A horseshoe magnet is held vertically with the north pole on the left and south pole on the right. A wire passing between the poles, equidistant from them, carries a current directly away from you. In what direction is the force on the wire?
5. Will a magnet attract any metallic object, or only those made of iron? (Try it and see.) Why is this so?
6. Two iron bars attract each other no matter which ends are placed close together. Are both magnets? Explain.
7. The magnetic field due to current in wires in your home can affect a compass. Discuss the effect in terms of currents, including if they are ac or dc.
8. If a negatively charged particle enters a region of uniform magnetic field which is perpendicular to the particle's velocity, will the kinetic energy of the particle increase, decrease, or stay the same? Explain your answer. (Neglect gravity and assume there is no electric field.)
9. In Fig. 20–45, charged particles move in the vicinity of a current-carrying wire. For each charged particle, the arrow indicates the direction of motion of the particle, and the + or − indicates the sign of the charge. For each of the particles, indicate the direction of the magnetic force due to the magnetic field produced by the wire.

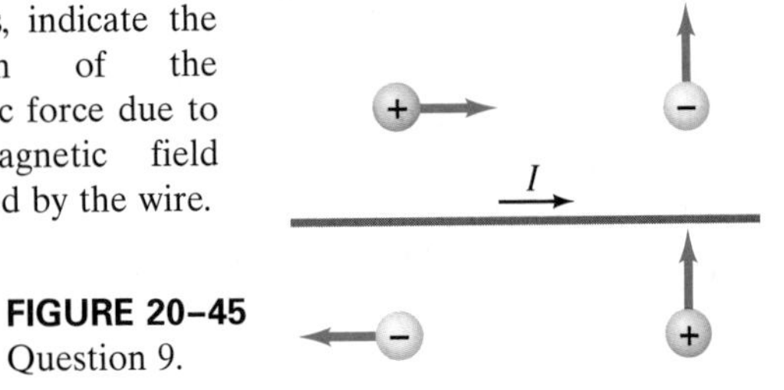

FIGURE 20–45 Question 9.

10. Three particles, a, b, and c, enter a magnetic field as shown in Fig. 20–46. What can you say about the charge on each particle?

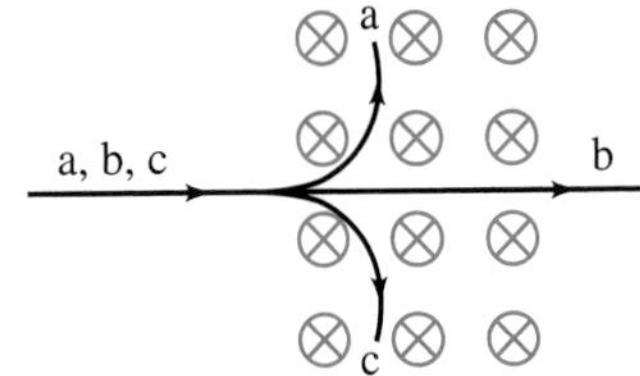

FIGURE 20–46 Question 10.

11. A positively charged particle in a nonuniform magnetic field follows the trajectory shown in Fig. 20–47. Indicate the direction of the magnetic field everywhere in space, assuming the path is always in the plane of the page, and indicate the relative magnitudes of the field in each region.

FIGURE 20–47 Question 11.

12. Can an iron rod attract a magnet? Can a magnet attract an iron rod? What must you consider to answer these questions?
13. Explain why a strong magnet held near a CRT television screen (Section 17–10) causes the picture to become distorted. Also, explain why the picture sometimes goes completely black where the field is the strongest. [But don't risk damage to your TV by trying this.]
14. Suppose you have three iron rods, two of which are magnetized but the third is not. How would you determine which two are the magnets without using any additional objects?
15. Can you set a resting electron into motion with a magnetic field? With an electric field? Explain.
16. A charged particle is moving in a circle under the influence of a uniform magnetic field. If an electric field that points in the same direction as the magnetic field is turned on, describe the path the charged particle will take.
17. The force on a particle in a magnetic field is the idea behind *electromagnetic pumping*. It is used to pump metallic fluids (such as sodium) and to pump blood in artificial heart machines. The basic design is shown in Fig. 20–48. An electric field is applied perpendicular to a blood vessel and to a magnetic field. Explain how ions are caused to move. Do positive and negative ions feel a force in the same direction?

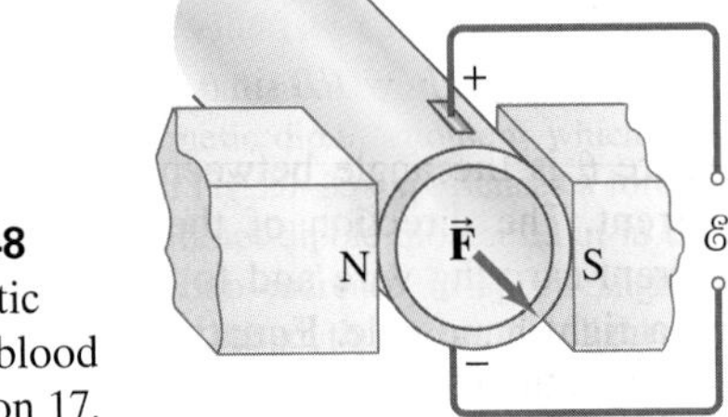

FIGURE 20–48 Electromagnetic pumping in a blood vessel. Question 17.

18. A beam of electrons is directed toward a horizontal wire carrying a current from left to right (Fig. 20–49). In what direction is the beam deflected?

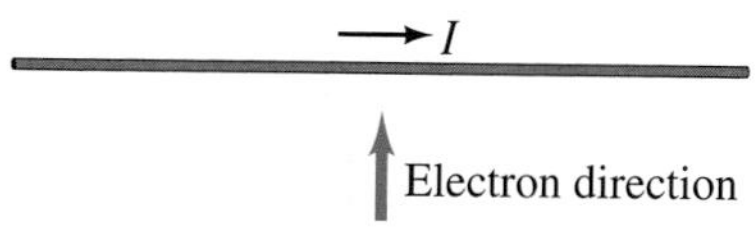

$\longrightarrow I$

Electron direction

FIGURE 20–49 Question 18.

19. Describe electric and/or magnetic fields that surround a moving electric charge.

20. A charged particle moves in a straight line through a particular region of space. Could there be a nonzero magnetic field in this region? If so, give two possible situations.

21. If a moving charged particle is deflected sideways in some region of space, can we conclude, for certain, that $\vec{\mathbf{B}} \neq 0$ in that region? Explain.

22. In a particular region of space there is a uniform magnetic field $\vec{\mathbf{B}}$. Outside this region, $B = 0$. Can you inject an electron from outside into the field perpendicularly so that it will move in a closed circular path in the field? What if the electron is injected near the center?

23. How could you tell whether moving electrons in a certain region of space are being deflected by an electric field or by a magnetic field (or by both)?

24. How can you make a compass without using iron or other ferromagnetic material?

25. Two long wires carrying equal currents I are at right angles to each other, but don't quite touch. Describe the magnetic force one exerts on the other.

26. A horizontal current-carrying wire, free to move in Earth's gravitational field, is suspended directly above a second, parallel, current-carrying wire. (*a*) In what direction is the current in the lower wire? (*b*) Can the upper wire be held in stable equilibrium due to the magnetic force of the lower wire? Explain.

27. Why will either pole of a magnet attract an unmagnetized piece of iron?

28. An unmagnetized nail will not attract an unmagnetized paper clip. However, if one end of the nail is in contact with a magnet, the other end *will* attract a paper clip. Explain.

* 29. Two ions have the same mass, but one is singly ionized and the other is doubly ionized. How will their positions on the film of a mass spectrometer (Fig. 20–39) differ?

30. What would be the effect on B inside a long solenoid if (*a*) the diameter of all the loops was doubled, (*b*) the spacing between loops was doubled, or (*c*) the solenoid's length was doubled along with a doubling in the total number of loops?

31. A type of magnetic switch similar to a solenoid is a **relay** (Fig. 20–50). A relay is an electromagnet (the iron rod inside the coil does not move) which, when activated, attracts a piece of iron on a pivot. Design a relay to close an electrical switch. A relay is used when you need to switch on a circuit carrying a very large current but you do not want that large current flowing through the main switch. For example, the starter switch of a car is connected to a relay so that the large current needed for the starter doesn't pass to the dashboard switch.

FIGURE 20–50 Question 31.

Problems

20–3 Force on Electric Current in Magnetic Field

1. (I) (*a*) What is the magnitude of the force per meter of length on a straight wire carrying an 8.40-A current when perpendicular to a 0.90-T uniform magnetic field? (*b*) What if the angle between the wire and field is 45.0°?

2. (I) Calculate the magnitude of the magnetic force on a 160-m length of straight wire stretched between two towers carrying a 150-A current. The Earth's magnetic field of 5.0×10^{-5} T makes an angle of 65° with the wire.

3. (I) How much current is flowing in a wire 4.80 m long if the maximum force on it is 0.750 N when placed in a uniform 0.0800-T field?

4. (II) A 1.5-m length of wire carrying 4.5 A of current is oriented horizontally. At that point on the Earth's surface, the dip angle of the Earth's magnetic field makes an angle of 38° to the wire. Estimate the magnitude of the magnetic force on the wire due to the Earth's magnetic field of 5.5×10^{-5} T at this point.

5. (II) The force on a wire carrying 8.75 A is a maximum of 1.28 N when placed between the pole faces of a magnet. If the pole faces are 55.5 cm in diameter, what is the approximate strength of the magnetic field?

6. (II) The magnetic force per meter on a wire is measured to be only 35% of its maximum possible value. Sketch the relationship of the wire and the field if the force had been a maximum, and sketch the relationship as it actually is, calculating the angle between the wire and the magnetic field.

7. (II) The force on a wire is a maximum of 6.50×10^{-2} N when placed between the pole faces of a magnet. The current flows horizontally to the right and the magnetic field is vertical. The wire is observed to "jump" toward the observer when the current is turned on. (*a*) What type of magnetic pole is the top pole face? (*b*) If the pole faces have a diameter of 10.0 cm, estimate the current in the wire if the field is 0.16 T. (*c*) If the wire is tipped so that it makes an angle of 10.0° with the horizontal, what force will it now feel?

8. (II) Suppose a straight 1.00-mm-diameter copper wire could just "float" horizontally in air because of the force due to the Earth's magnetic field $\vec{\mathbf{B}}$, which is horizontal, perpendicular to the wire, and of magnitude 5.0×10^{-5} T. What current would the wire carry? Does the answer seem feasible? Explain briefly.

20–4 Force on Charge Moving in Magnetic Field

9. (I) Alpha particles of charge $q = +2e$ and mass $m = 6.6 \times 10^{-27}$ kg are emitted from a radioactive source at a speed of 1.6×10^7 m/s. What magnetic field strength would be required to bend them into a circular path of radius $r = 0.25$ m?
10. (I) Determine the magnitude and direction of the force on an electron traveling 8.75×10^5 m/s horizontally to the east in a vertically upward magnetic field of strength 0.75 T.
11. (I) Find the direction of the force on a negative charge for each diagram shown in Fig. 20–51, where $\vec{\mathbf{v}}$ (green) is the velocity of the charge and $\vec{\mathbf{B}}$ (blue) is the direction of the magnetic field. (⊗ means the vector points inward. ⊙ means it points outward, toward you.)

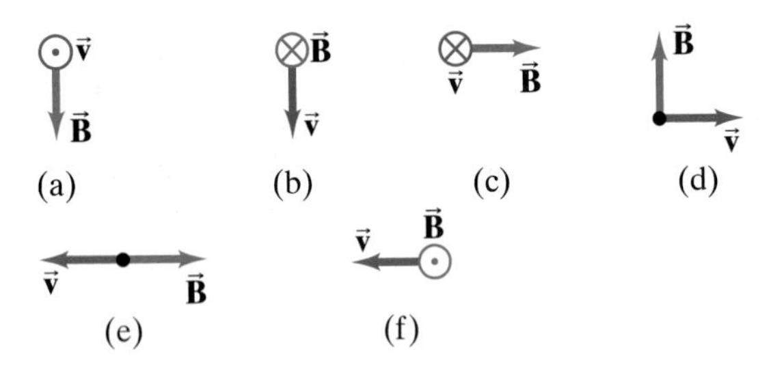

FIGURE 20–51 Problem 11.

12. (I) Determine the direction of $\vec{\mathbf{B}}$ for each case in Fig. 20–52, where $\vec{\mathbf{F}}$ represents the maximum magnetic force on a positively charged particle moving with velocity $\vec{\mathbf{v}}$.

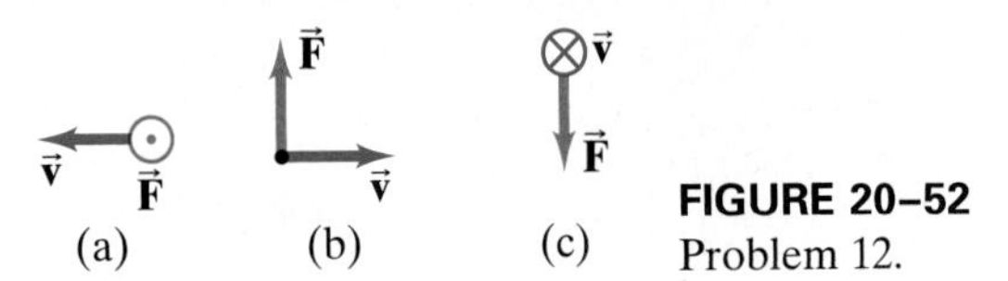

FIGURE 20–52 Problem 12.

13. (I) An electron is projected vertically upward with a speed of 1.70×10^6 m/s into a uniform magnetic field of 0.350 T that is directed horizontally away from the observer. Describe the electron's path in this field.
14. (II) A 5.0-MeV (kinetic energy) proton enters a 0.20-T field, in a plane perpendicular to the field. What is the radius of its path?
15. (II) An electron experiences the greatest force as it travels 2.9×10^6 m/s in a magnetic field when it is moving northward. The force is upward and of magnitude 7.2×10^{-13} N. What are the magnitude and direction of the magnetic field?
16. (II) What is the velocity of a beam of electrons that go undeflected when passing through perpendicular electric and magnetic fields of magnitude 8.8×10^3 V/m and 3.5×10^{-3} T, respectively? What is the radius of the electron orbit if the electric field is turned off?
17. (II) A doubly charged helium atom whose mass is 6.6×10^{-27} kg is accelerated by a voltage of 2100 V. (*a*) What will be its radius of curvature if it moves in a plane perpendicular to a uniform 0.340-T field? (*b*) What is its period of revolution?
18. (II) A proton (mass m_p), a deuteron ($m = 2m_p$, $Q = e$), and an alpha particle ($m = 4m_p$, $Q = 2e$) are accelerated by the same potential difference V and then enter a uniform magnetic field $\vec{\mathbf{B}}$, where they move in circular paths perpendicular to $\vec{\mathbf{B}}$. Determine the radius of the paths for the deuteron and alpha particle in terms of that for the proton.
19. (II) Show that the time T required for a particle of charge q moving with constant speed v to make one circular revolution in a uniform magnetic field $\vec{\mathbf{B}}$ ($\perp \vec{\mathbf{v}}$) is
$$T = \frac{2\pi m}{qB}.$$
[*Hint:* see Example 20–5 and Chapter 5.]
20. (II) A particle of charge q moves in a circular path of radius r in a uniform magnetic field B. Show that its momentum is $p = qBr$.
21. (II) A particle of mass m and charge q moves in a circular path in a magnetic field B. Show that its kinetic energy is proportional to r^2, the square of the radius of curvature of its path.
22. (II) Show that the angular momentum of the particle in Problem 21 is $L = qBr^2$ about the center of the circle.
23. (III) A 3.40-g bullet moves with a speed of 160 m/s perpendicular to the Earth's magnetic field of 5.00×10^{-5} T. If the bullet possesses a net charge of 13.5×10^{-9} C, by what distance will it be deflected from its path due to the Earth's magnetic field after it has traveled 1.00 km?
24. (III) Suppose the Earth's magnetic field at the equator has magnitude 0.40×10^{-4} T and a northerly direction at all points. Estimate the speed a singly ionized uranium ion ($m = 238$ u, $q = e$) would need to circle the Earth 5.0 km above the equator. Can you ignore gravity?
25. (III) A proton moving with speed $v = 2.0 \times 10^5$ m/s in a field-free region abruptly enters an essentially uniform magnetic field $B = 0.850$ T ($\vec{\mathbf{B}} \perp \vec{\mathbf{v}}$). If the proton enters the magnetic field region at a 45° angle as shown in Fig. 20–53, (*a*) at what angle does it leave, and (*b*) at what distance x does it exit the field?

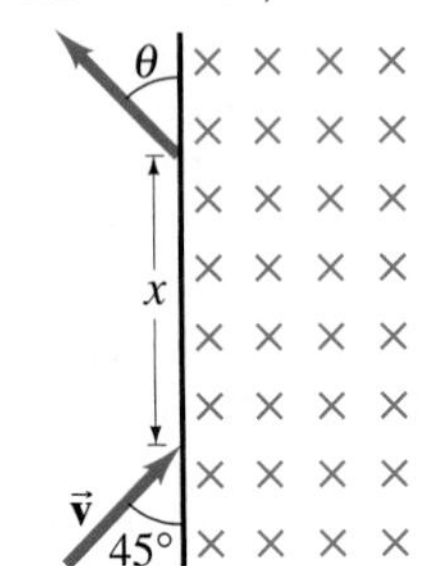

FIGURE 20–53 Problem 25.

20–5 and 20–6 Magnetic Field of Straight Wire, Force Between Two Wires

26. (I) A jumper cable used to start a stalled vehicle carries a 65-A current. How strong is the magnetic field 6.0 cm away from it? Compare to the Earth's magnetic field.
27. (I) If an electric wire is allowed to produce a magnetic field no larger than that of the Earth (0.55×10^{-4} T) at a distance of 25 cm, what is the maximum current the wire can carry?
28. (I) In Fig. 20–54, a long straight wire carries current I out of the page toward you. Indicate, with appropriate arrows, the direction of $\vec{\mathbf{B}}$ at each of the points C, D, and E in the plane of the page.

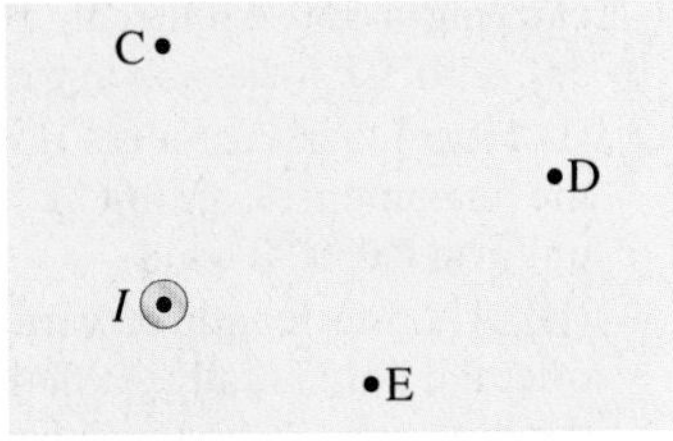

FIGURE 20–54 Problem 28.

29. (I) A vertical straight wire carrying an upward 24-A current exerts an attractive force per unit length of 8.8×10^{-4} N/m on a second parallel wire 7.0 cm away. What current (magnitude and direction) flows in the second wire?
30. (I) Determine the magnitude and direction of the force between two parallel wires 35 m long and 6.0 cm apart, each carrying 25 A in the same direction.

31. (II) An experiment on the Earth's magnetic field is being carried out 1.00 m from an electric cable. What is the maximum allowable current in the cable if the experiment is to be accurate to $\pm 1.0\%$?

32. (II) A power line carries a current of 95 A along the tops of 8.5-m-high poles. What is the magnitude of the magnetic field produced by this wire at the ground? How does this compare with the Earth's field of about $\frac{1}{2}$ G?

33. (II) Two long thin parallel wires 13.0 cm apart carry 25-A currents in the same direction. Determine the magnetic field at point P, 12.0 cm from one wire and 5.0 cm from the other (Fig. 20–55).

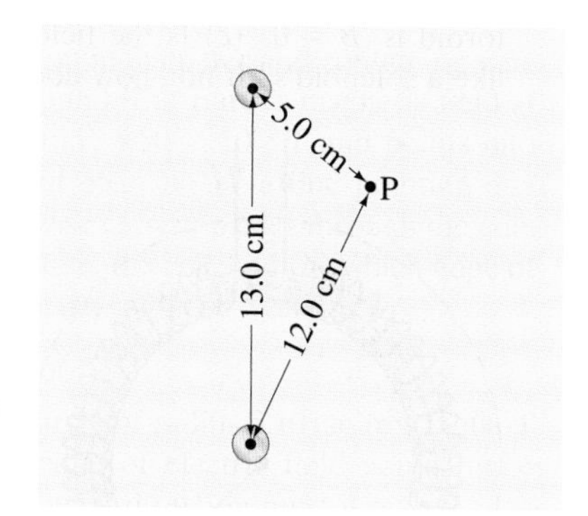

FIGURE 20–55 Problem 33.

34. (II) A horizontal compass is placed 18 cm due south from a straight vertical wire carrying a 35-A current downward. In what direction does the compass needle point at this location? Assume the horizontal component of the Earth's field at this point is 0.45×10^{-4} T and the magnetic declination is 0°.

35. (II) A long horizontal wire carries 22.0 A of current due north. What is the net magnetic field 20.0 cm due west of the wire if the Earth's field there points north but downward, 37° below the horizontal, and has magnitude 5.0×10^{-5} T?

36. (II) A straight stream of protons passes a given point in space at a rate of 1.5×10^9 protons/s. What magnetic field do they produce 2.0 m from the beam?

37. (II) Determine the magnetic field midway between two long straight wires 2.0 cm apart in terms of the current I in one when the other carries 15 A. Assume these currents are (*a*) in the same direction, and (*b*) in opposite directions.

38. (II) A long pair of wires conducts 25.0 A of dc current to, and from, an instrument. If the insulated wires are of negligible diameter but are 2.8 mm apart, what is the magnetic field 10.00 cm from their midpoint, in their plane (Fig. 20–56)? Compare to the magnetic field of the Earth.

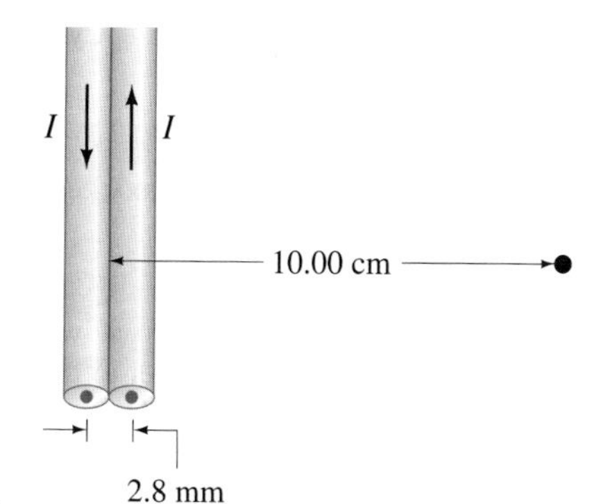

FIGURE 20–56 Problems 38 and 39.

39. (II) A third wire is placed in the plane of the two wires shown in Fig. 20–56, parallel and just to the right. If it carries 25.0 A upward, what force per meter of length does it exert on each of the other two wires? Assume it is 2.8 mm from the nearest wire, center to center.

40. (II) A compass needle points 23° E of N outdoors. However, when it is placed 12.0 cm to the east of a vertical wire inside a building, it points 55° E of N. What are the magnitude and direction of the current in the wire? The Earth's field there is 0.50×10^{-4} T and is horizontal.

41. (II) A rectangular loop of wire lies in the same plane as a straight wire, as shown in Fig. 20–57. There is a current of 2.5 A in both wires. Determine the magnitude and direction of the net force on the loop.

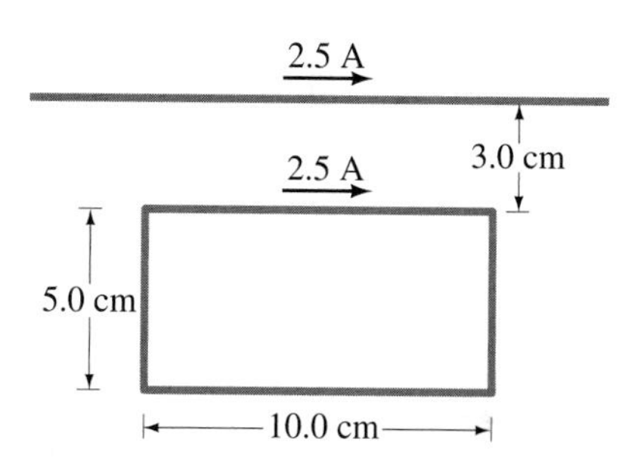

FIGURE 20–57 Problem 41.

42. (II) A long horizontal wire carries a current of 48 A. A second wire, made of 2.5-mm-diameter copper wire and parallel to the first, is kept in suspension magnetically 15 cm below (Fig. 20–58). (*a*) Determine the magnitude and direction of the current in the lower wire. (*b*) Is the lower wire in stable equilibrium? (*c*) Repeat parts (*a*) and (*b*) if the second wire is suspended 15 cm *above* the first due to the latter's field.

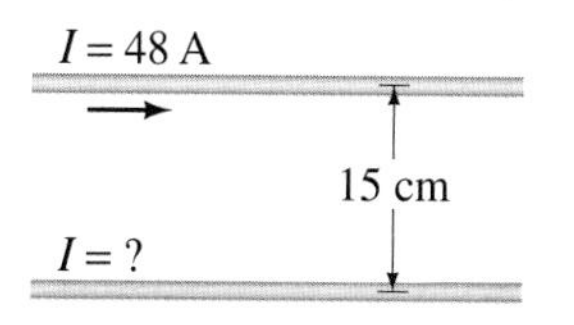

FIGURE 20–58 Problem 42.

43. (II) Two long wires are oriented so that they are perpendicular to each other. At their closest, they are 20.0 cm apart (Fig. 20–59). What is the magnitude of the magnetic field at a point midway between them if the top one carries a current of 20.0 A and the bottom one carries 5.0 A?

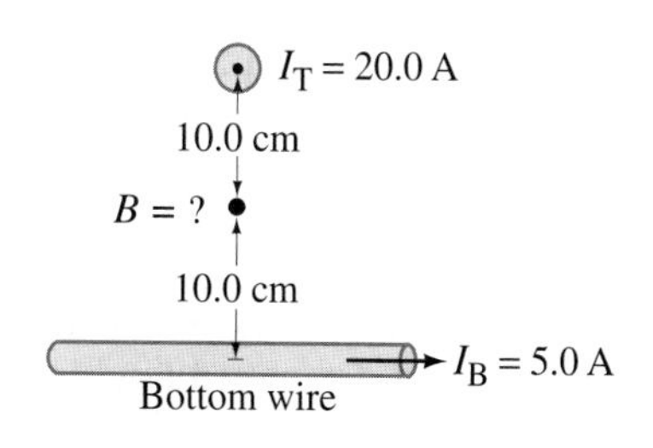

FIGURE 20–59 Problem 43.

75. Two stiff parallel wires a distance l apart in a horizontal plane act as rails to support a light metal rod of mass m (perpendicular to each rail), Fig. 20–66. A magnetic field $\vec{\mathbf{B}}$, directed vertically upward (outward in the diagram), acts throughout. At $t = 0$, wires connected to the rails are connected to a constant current source and a current I begins to flow through the system. Determine the speed of the rod, which starts from rest at $t = 0$, as a function of time (a) assuming no friction between the rod and the rails, and (b) if the coefficient of friction is μ_k. (c) Does the rod move east or west if the current through it heads north?

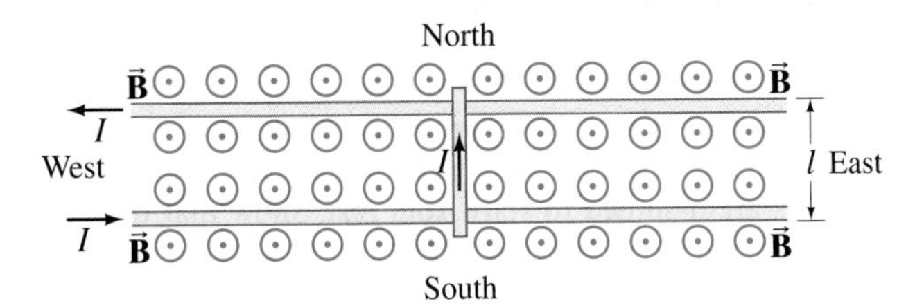

FIGURE 20–66 Looking down on a rod sliding on rails. Problem 75.

76. Estimate the approximate maximum deflection of the electron beam near the center of a TV screen due to the Earth's 5.0×10^{-5} T field. Assume the CRT screen (Section 17–10) is 22 cm from the electron gun, where the electrons are accelerated (a) by 2.0 kV, or (b) by 30 kV. Note that in color TV sets, the CRT beam must be directed accurately to within less than 1 mm in order to strike the correct phosphor. Because the Earth's field is significant here, mu-metal shields are used to reduce the Earth's field in the CRT.

77. The **cyclotron** (Fig. 20–67) is a device used to accelerate elementary particles such as protons to high speeds. Particles starting at point A with some initial velocity travel in circular orbits in the magnetic field B. The particles are accelerated to higher speeds each time they pass through the gap between the metal "dees," where there is an electric field E. (There is no electric field inside the hollow metal dees.) The electric field changes direction each half-cycle, owing to an ac voltage $V = V_0 \sin 2\pi f t$, so that the particles are increased in speed at each passage through the gap. (a) Show that the frequency f of the voltage must be $f = Bq/2\pi m$, where q is the charge on the particles and m their mass. (b) Show that the kinetic energy of the particles increases by $2qV_0$ each revolution, assuming that the gap is small. (c) If the radius of the cyclotron is 2.0 m and the magnetic field strength is 0.50 T, what will be the maximum kinetic energy of accelerated protons in MeV?

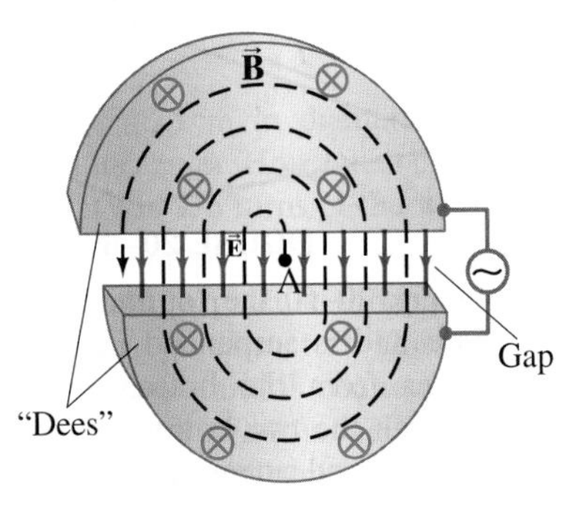

FIGURE 20–67 A cyclotron. Problem 77.

78. Four very long straight parallel wires, located at the corners of a square of side l, carry equal currents I_0 perpendicular to the page as shown in Fig. 20–68. Determine the magnitude and direction of $\vec{\mathbf{B}}$ at the center C of the square.

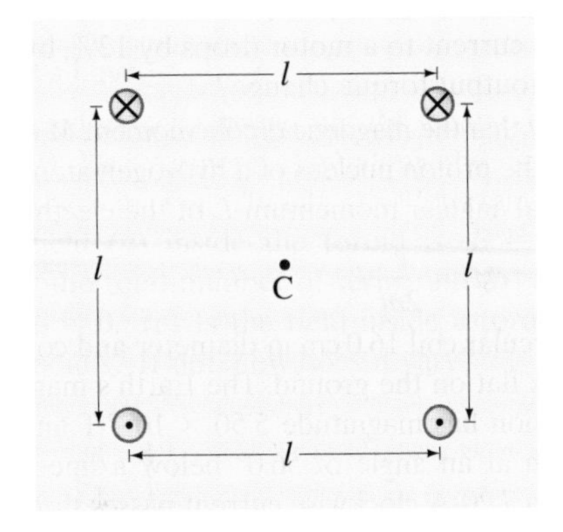

FIGURE 20–68 Problem 78.

79. Magnetic fields are very useful in particle accelerators for "beam steering"; that is, the magnetic fields can be used to change the beam's direction without altering its speed (Fig. 20–69). Show how this works with a beam of protons. What happens to protons that are not moving with the speed that the magnetic field is designed for? If the field extends over a region 5.0 cm wide and has a magnitude of 0.33 T, by approximately what angle will a beam of protons traveling at 1.0×10^7 m/s be bent?

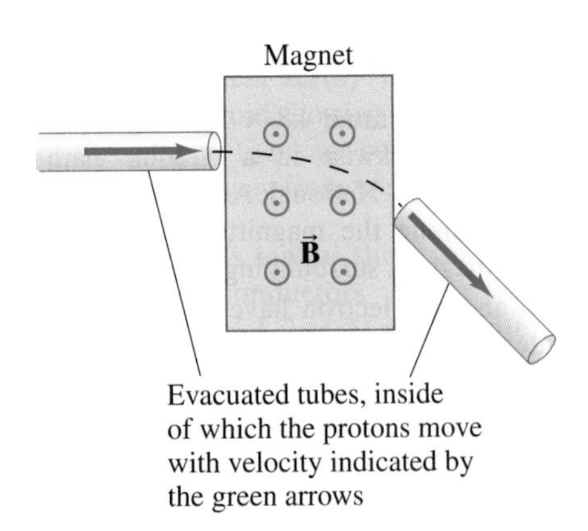

FIGURE 20–69 Problem 79.

80. The magnetic field B at the center of a circular coil of wire carrying a current I (as in Fig. 20–9) is

$$B = \frac{\mu_0 N I}{2r},$$

where N is the number of loops in the coil and r is its radius. Suppose that an electromagnet uses a coil 1.2 m in diameter made from square copper wire 1.6 mm on a side. The power supply produces 120 V at a maximum power output of 4.0 kW. (a) How many turns are needed to run the power supply at maximum power? (b) What is the magnetic field strength at the center of the coil? (c) If you use a greater number of turns and this same power supply (so the voltage remains at 120 V), will a greater magnetic field strength result? Explain.

81. Near the Earth's poles the magnetic field is about 1 G (1×10^{-4} T). Imagine a simple model in which the Earth's field is produced by a single current loop around the equator. Roughly estimate the current this loop would carry. [*Hint:* use the formula given in Problem 80.]

82. You want to get an idea of the magnitude of magnetic fields produced by overhead power lines. You estimate that the two wires are each about 30 m above the ground and are about 3 m apart. The local power company tells you that the lines operate at 10 kV and provide a maximum of 40 MW to the local area. Estimate the maximum magnetic field you might experience walking under these power lines, and compare to the Earth's field. [For an ac current, values are rms, and the magnetic field will be changing.]

83. (*a*) What value of magnetic field would make a beam of electrons, traveling to the right at a speed of 4.8×10^6 m/s, go undeflected through a region where there is a uniform electric field of 10,000 V/m pointing vertically up? (*b*) What is the direction of the magnetic field if it is known to be perpendicular to the electric field? (*c*) What is the frequency of the circular orbit of the electrons if the electric field is turned off?

84. A proton follows a spiral path through a gas in a magnetic field of 0.010 T, perpendicular to the plane of the spiral, as shown in Fig. 20–70. In two successive loops, at points P and Q, the radii are 10.0 mm and 8.5 mm, respectively. Calculate the change in the kinetic energy of the proton as it travels from P to Q.

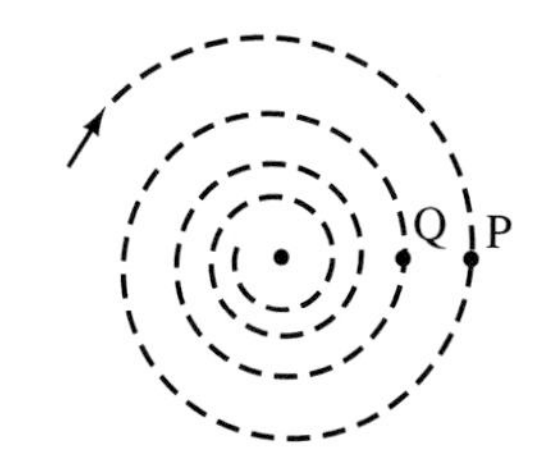

FIGURE 20–70 Problem 84.

85. A 32-cm-long solenoid, 1.8 cm in diameter, is to produce a 0.30-T magnetic field at its center. If the maximum current is 5.7 A, how many turns must the solenoid have?

86. Two long straight aluminum wires, each of diameter 0.50 mm, carry the same current but in opposite directions. They are suspended by 0.50-m-long strings as shown in Fig. 20–71. If the suspension strings make an angle of 3.0° with the vertical, what is the current in the wires?

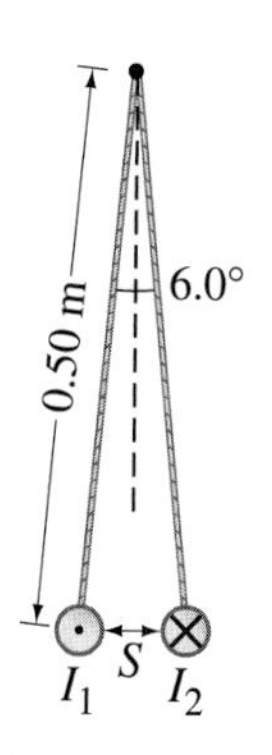

FIGURE 20–71 Problem 86.

87. An electron enters a uniform magnetic field $B = 0.23$ T at a 45° angle to $\vec{\mathbf{B}}$. Determine the radius r and pitch p (distance between loops) of the electron's helical path assuming its speed is 3.0×10^6 m/s. See Fig. 20–72.

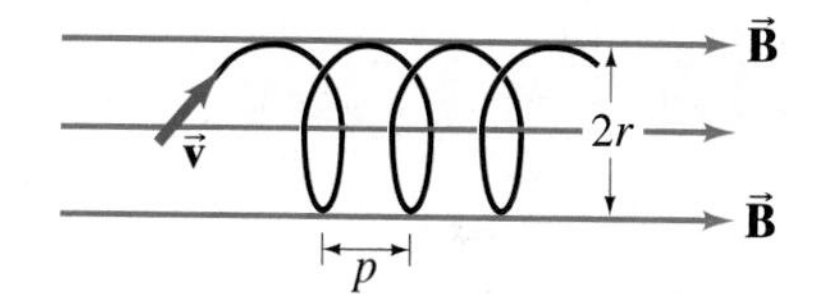

FIGURE 20–72 Problem 87.

Answers to Exercises

A: Near the poles, where the field lines are closer together.
B: Counterclockwise.
C: 0.15 N.
D: Zero.
E: Negative; the direction of the helical path would be reversed.
F: 2.0 cm.

One of the great laws of physics is Faraday's law of induction, which says that a changing magnetic flux produces an induced emf. This photo shows a bar magnet moving inside a coil of wire, and the galvanometer registers an induced current. This phenomenon of electromagnetic induction is the basis for many practical devices, from generators to alternators to transformers, tape recording, and computer memory.

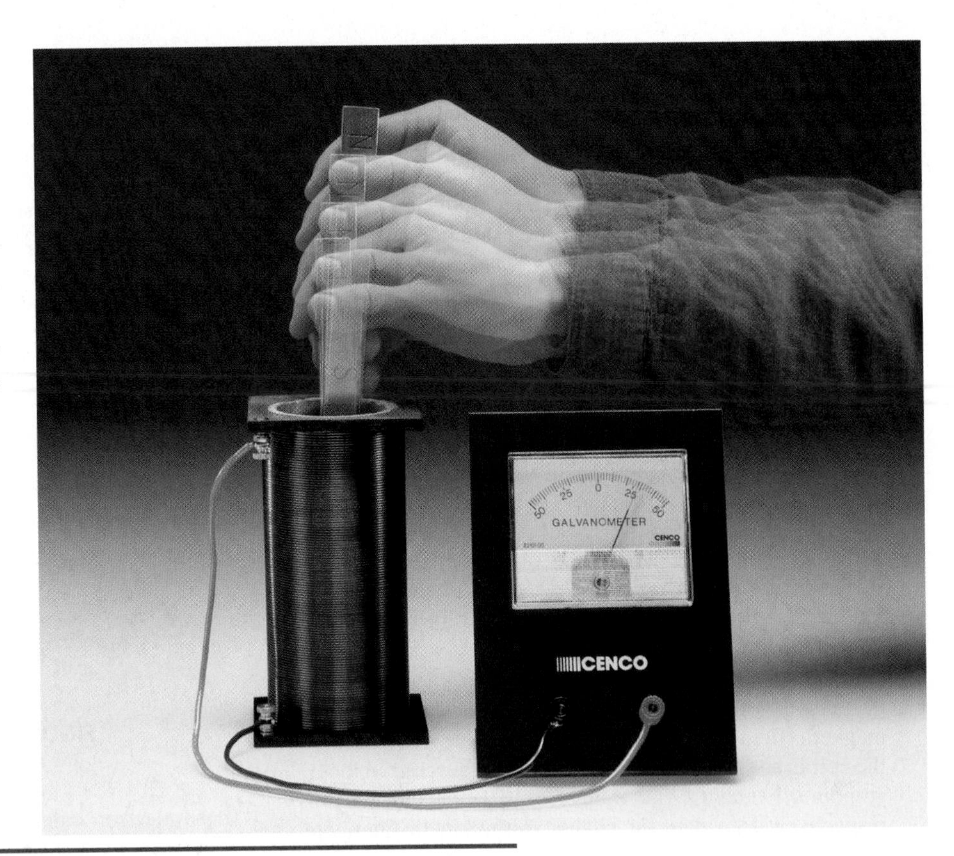

CHAPTER 21

Electromagnetic Induction and Faraday's Law

In Chapter 20, we discussed two ways in which electricity and magnetism are related: (1) an electric current produces a magnetic field; and (2) a magnetic field exerts a force on an electric current or moving electric charge. These discoveries were made in 1820–1821. Scientists then began to wonder: if electric currents produce a magnetic field, is it possible that a magnetic field can produce an electric current? Ten years later the American Joseph Henry (1797–1878) and the Englishman Michael Faraday (1791–1867) independently found that it was possible. Henry actually made the discovery first. But Faraday published his results earlier and investigated the subject in more detail. We now discuss this phenomenon and some of its world-changing applications such as the electric generator.

21–1 Induced EMF

In his attempt to produce an electric current from a magnetic field, Faraday used an apparatus like that shown in Fig. 21–1. A coil of wire, X, was connected to a battery. The current that flowed through X produced a magnetic field that was intensified by the iron core around which the wire was wrapped. Faraday hoped that a strong steady current in X would produce a great enough magnetic

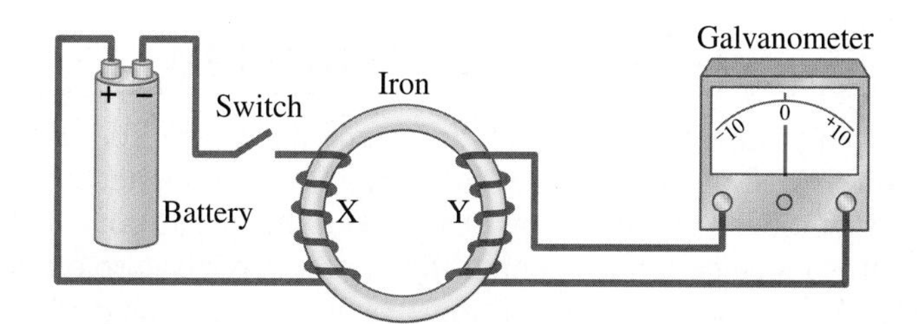

FIGURE 21–1 Faraday's experiment to induce an emf.

field to produce a current in a second coil Y which shared the same iron core. This second circuit, Y, contained a galvanometer to detect any current but contained no battery. He met no success with constant currents. But the long-sought effect was finally observed when Faraday noticed the galvanometer in circuit Y deflect strongly at the moment he closed the switch in circuit X. And the galvanometer deflected strongly in the opposite direction when he opened the switch in X. A constant current in X produced a constant magnetic field which produced *no* current in Y. Only when the current in X was starting or stopping was a current produced in Y.

Faraday concluded that although a constant magnetic field produces no current in a conductor, a *changing* magnetic field can produce an electric current. Such a current is called an **induced current**. When the magnetic field through coil Y changes, a current occurs in Y as if there were a source of emf in circuit Y. We therefore say that

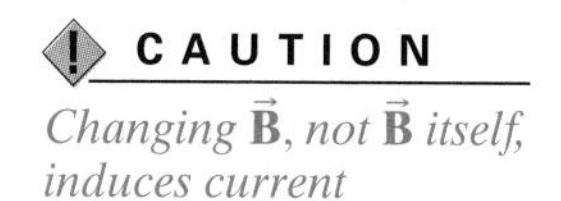

Changing $\vec{\mathbf{B}}$, *not* $\vec{\mathbf{B}}$ *itself, induces current*

a changing magnetic field induces an emf.

Changing $\vec{\mathbf{B}}$ *induces an emf*

Faraday did further experiments on **electromagnetic induction**, as this phenomenon is called. For example, Fig. 21–2 shows that if a magnet is moved quickly into a coil of wire, a current is induced in the wire. If the magnet is quickly removed, a current is induced in the opposite direction ($\vec{\mathbf{B}}$ through the coil decreases). Furthermore, if the magnet is held steady and the coil of wire is moved toward or away from the magnet, again an emf is induced and a current flows. Motion or change is required to induce an emf. It doesn't matter whether the magnet or the coil moves. It is their *relative motion* that counts.

CAUTION

Relative motion—magnet or *coil moving induces current*

FIGURE 21–2 (a) A current is induced when a magnet is moved toward a coil, momentarily increasing the magnetic field through the coil. (b) The induced current is opposite when the magnet is moved away from the coil ($\vec{\mathbf{B}}$ decreases). Note that the galvanometer zero is at the center of the scale and the needle deflects left or right, depending on the direction of the current. In (c), no current is induced if the magnet does not move relative to the coil. It is the relative motion that counts here: the magnet can be held steady and the coil moved, which also induces an emf.

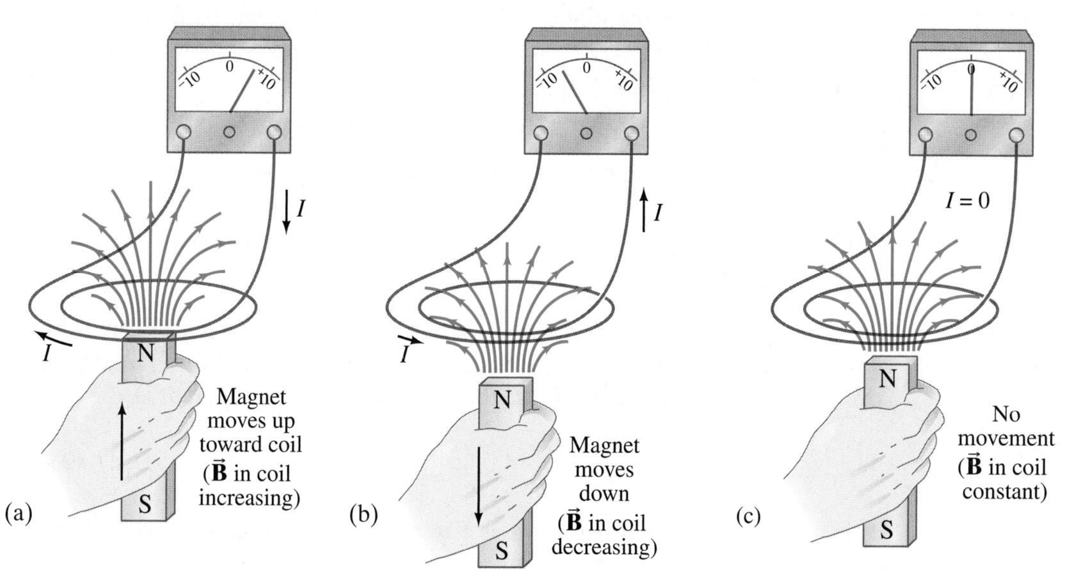

21–2 Faraday's Law of Induction; Lenz's Law

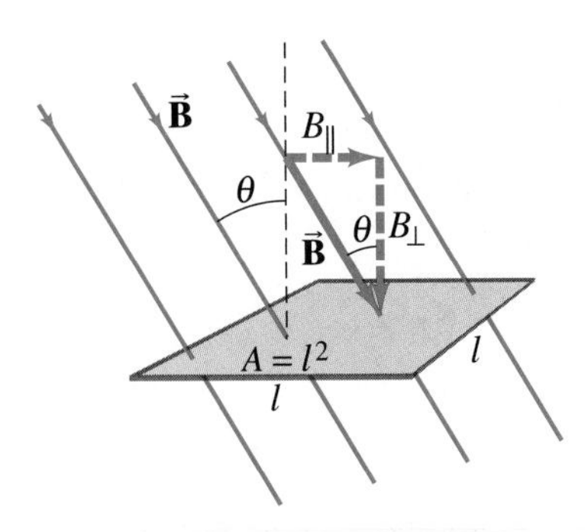

FIGURE 21–3 Determining the flux through a flat loop of wire. This loop is square, of side l and area $A = l^2$.

Faraday investigated quantitatively what factors influence the magnitude of the emf induced. He found first of all that the more rapidly the magnetic field changes, the greater the induced emf in a loop of wire. But the emf is not simply proportional to the rate of change of the magnetic field, $\vec{\mathbf{B}}$; it depends also on the loop's area and angle. That is, the emf is proportional to the rate of change of the **magnetic flux**, Φ_B, through the loop. Magnetic flux for a uniform magnetic field through a loop of area A is defined as

$$\Phi_B = B_\perp A = BA\cos\theta. \qquad [B \text{ uniform}] \quad \textbf{(21–1)}$$

Here $B_\perp$ is the component of the magnetic field $\vec{\mathbf{B}}$ perpendicular to the face of the loop, and θ is the angle between $\vec{\mathbf{B}}$ and a line perpendicular to the face of the loop. These quantities are shown in Fig. 21–3 for a square loop of side l whose area is $A = l^2$. When the face of the loop is parallel to $\vec{\mathbf{B}}$, $\theta = 90°$ and $\Phi_B = 0$. When $\vec{\mathbf{B}}$ is perpendicular to the loop, $\theta = 0°$, and

$$\Phi_B = BA. \qquad [\text{uniform } \vec{\mathbf{B}} \perp \text{loop face}]$$

FIGURE 21–4 Magnetic flux Φ_B is proportional to the number of lines of $\vec{\mathbf{B}}$ that pass through the loop.

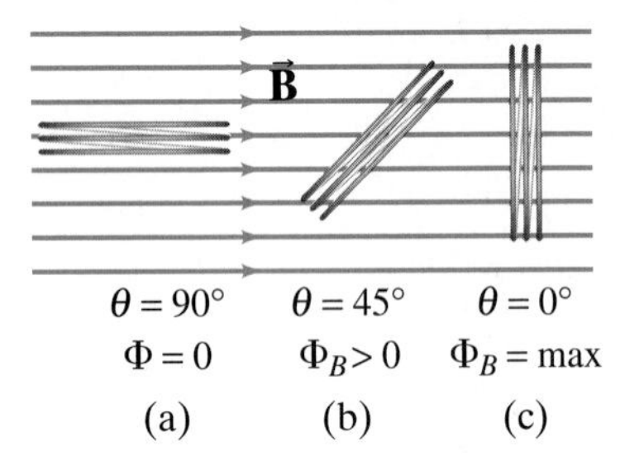

As we saw in Chapter 20, the lines of $\vec{\mathbf{B}}$ (like lines of $\vec{\mathbf{E}}$) can be drawn such that the number of lines per unit area is proportional to the field strength. Then the flux Φ_B can be thought of as being proportional to the *total number of lines passing through the area enclosed by the loop*. This is illustrated in Fig. 21–4, where the loop is viewed from the side (on edge). For $\theta = 90°$, no magnetic field lines pass through the loop and $\Phi_B = 0$, whereas Φ_B is a maximum when $\theta = 0°$. The unit of magnetic flux is the tesla-meter^2; this is called a **weber**: $1\text{ Wb} = 1\text{ T}\cdot\text{m}^2$.

CONCEPTUAL EXAMPLE 21–1 **Determining flux.** A square loop of wire encloses area A_1 as shown in Fig. 21–5. A uniform magnetic field $\vec{\mathbf{B}}$ perpendicular to the loop extends over the area A_2. What is the magnetic flux through the loop A_1?

FIGURE 21–5 Example 21–1.

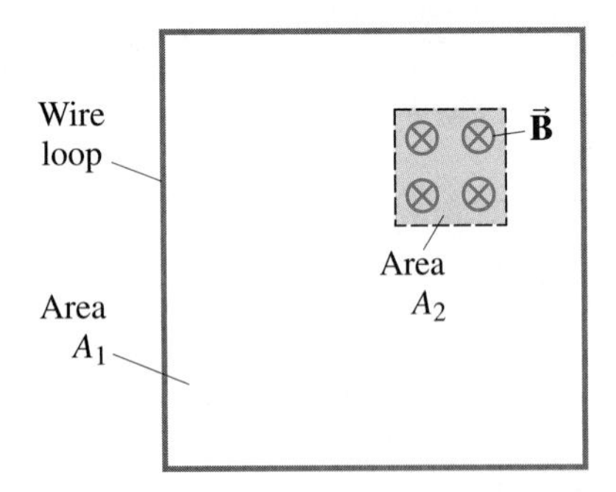

RESPONSE We assume that the magnetic field is zero outside the area A_2. The total magnetic flux through area A_1 is the flux through area A_2, which by Eq. 21–1 for a uniform field is BA_2, plus the flux through the remaining area $(= A_1 - A_2)$, which is zero because $B = 0$. So the total flux is $\Phi_B = BA_2 + 0(A_1 - A_2) = BA_2$. It is *not* equal to BA_1 because $\vec{\mathbf{B}}$ is not uniform over A_1.

EXAMPLE 21–2 **Calculate the flux.** A square loop of wire 10.0 cm on a side is in a 1.25-T magnetic field B. What are the maximum and minimum values of flux that can pass through the loop?

APPROACH The flux is given by Eq. 21–1. It is a maximum for $\theta = 0°$, which occurs when the plane of the loop is perpendicular to $\vec{\mathbf{B}}$. The minimum value occurs when $\theta = 90°$ and the plane of the loop is aligned with $\vec{\mathbf{B}}$.

SOLUTION From Eq. 21–1, the maximum value is

$$\Phi_B = BA\cos\theta = (1.25\text{ T})(0.100\text{ m})(0.100\text{ m})\cos 0° = 0.0125\text{ Wb}.$$

The minimum value is 0 Wb when $\theta = 90°$ and $\cos 90° = 0$.

EXERCISE A Find the flux in Example 21–2 when the perpendicular to the coil makes a 35° angle with $\vec{\mathbf{B}}$.

With our definition of flux, Eq. 21–1, we can now write down the results of Faraday's investigations. If the flux through a loop of wire changes by an amount $\Delta\Phi_B$ over a very brief time interval Δt, the induced emf at this instant is

FARADAY'S LAW OF INDUCTION

$$\mathscr{E} = -\frac{\Delta\Phi_B}{\Delta t}. \qquad [1 \text{ loop}] \quad \textbf{(21–2a)}$$

This fundamental result is known as **Faraday's law of induction**, and it is one of the basic laws of electromagnetism.

If the circuit contains N closely wrapped loops, the emfs induced in each loop add together, so

$$\mathcal{E} = -N\frac{\Delta\Phi_B}{\Delta t}. \qquad [N \text{ loops}] \quad \textbf{(21–2b)}$$

FARADAY'S LAW OF INDUCTION

The minus sign in Eqs. 21–2 is there to remind us in which direction the induced emf acts. Experiments show that

a current produced by an induced emf moves in a direction so that its magnetic field opposes the original change in flux.

Lenz's law

This is known as **Lenz's law**. Be aware that we are now discussing two distinct magnetic fields: (1) the changing magnetic field or flux that induces the current, and (2) the magnetic field produced by the induced current (all currents produce a field). The second field opposes the change in the first.

CAUTION
Distinguish two different magnetic fields

Let us now apply Lenz's law to the relative motion between a magnet and a coil, Fig. 21–2. The changing flux through the coil induces an emf in the coil, producing a current. This induced current produces its own magnetic field. In Fig. 21–2a the distance between the coil and the magnet decreases. The magnet's magnetic field (and number of field lines) through the coil increases, and therefore the flux increases. The magnetic field of the magnet points upward. To oppose the upward increase, the magnetic field inside the coil produced by the induced current needs to point *downward*. Thus, Lenz's law tells us that the current moves as shown (use the right-hand rule). In Fig. 21–2b, the flux *decreases* (because the magnet is moved away and B decreases), so the induced current in the coil produces an *upward* magnetic field through the coil that is "trying" to maintain the status quo. Thus the current in Fig. 21–2b is in the opposite direction from Fig. 21–2a.

It is important to note that an emf is induced whenever there is a change in *flux* through the coil, and we now consider some more possibilities.

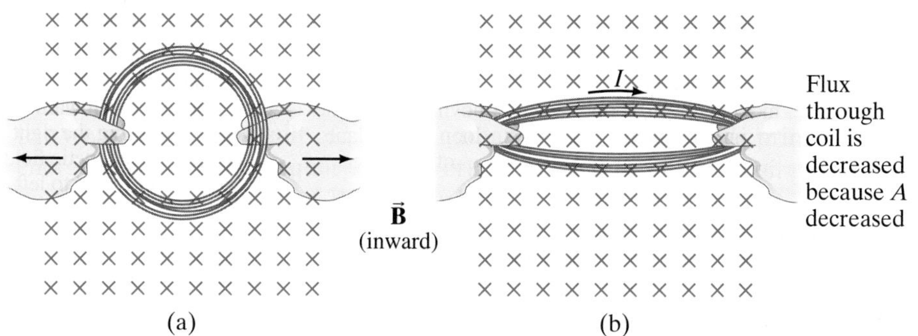

FIGURE 21–6 A current can be induced by changing the area of the coil, even though B doesn't change. In both this case and that of Fig. 21–7, the *flux* through the coil is reduced as we go from (a) to (b). Here the brief induced current acts in the direction shown so as to try to maintain the original flux ($\Phi = BA$) by producing its own magnetic field into the page. That is, as the area A decreases, the current acts to increase B in the original (inward) direction.

Since magnetic flux $\Phi_B = BA\cos\theta$, we see that an emf can be induced in three ways: (1) by a changing magnetic field B; (2) by changing the area A of the loop in the field; or (3) by changing the loop's orientation θ with respect to the field. Figures 21–1 and 21–2 illustrated case 1. Examples of cases 2 and 3 are illustrated in Figs. 21–6 and 21–7, respectively.

Three ways to change the magnetic flux: change B, A, or θ

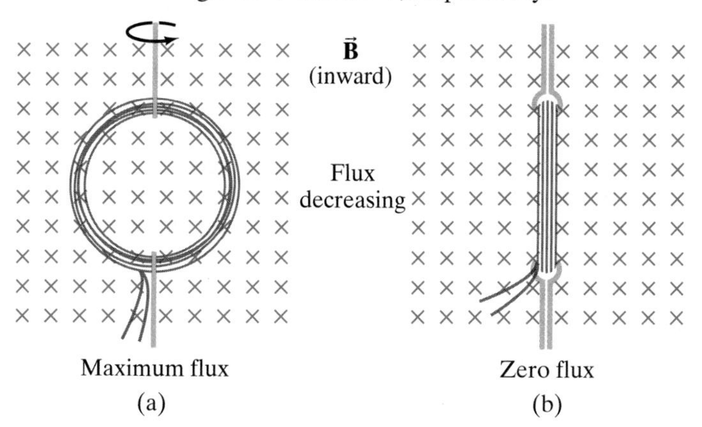

FIGURE 21–7 A current can be induced by rotating a coil in a magnetic field. The flux through the coil changes from (a) to (b) because θ (in Eq. 21–1) went from $0°$ ($\cos\theta = 1$) to $90°$ ($\cos\theta = 0$).

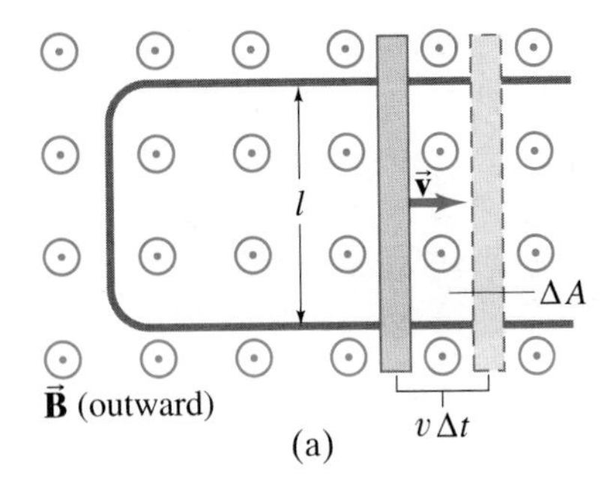

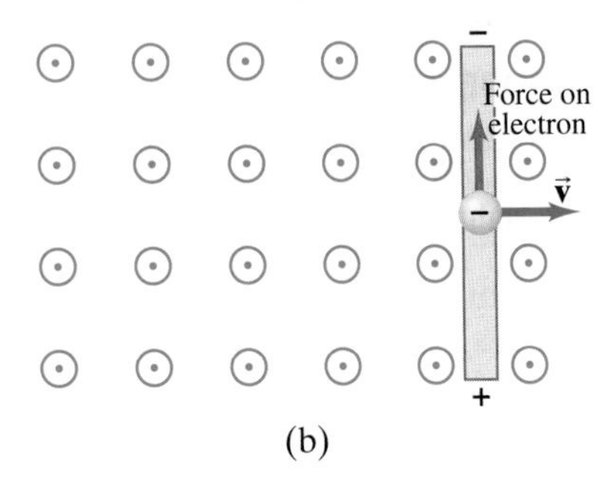

FIGURE 21–12 (a) A conducting rod is moved to the right on a U-shaped conductor in a uniform magnetic field $\vec{\mathbf{B}}$ that points out of the paper. (b) Upward force on an electron in the metal rod (moving to the right) due to $\vec{\mathbf{B}}$ pointing out of page.

21–3 EMF Induced in a Moving Conductor

Another way to induce an emf is shown in Fig. 21–12a, and this situation helps illuminate the nature of the induced emf. Assume that a uniform magnetic field $\vec{\mathbf{B}}$ is perpendicular to the area bounded by the U-shaped conductor and the movable rod resting on it. If the rod is made to move at a speed v, it travels a distance $\Delta x = v\,\Delta t$ in a time Δt. Therefore, the area of the loop increases by an amount $\Delta A = l\,\Delta x = lv\,\Delta t$ in a time Δt. By Faraday's law there is an induced emf $\mathscr{E}$ whose magnitude is given by

$$\mathscr{E} = \frac{\Delta \Phi_B}{\Delta t} = \frac{B\,\Delta A}{\Delta t} = \frac{Blv\,\Delta t}{\Delta t} = Blv. \qquad \textbf{(21–3)}$$

Equation 21–3 is valid as long as B, l, and v are mutually perpendicular. (If they are not, we use only the components of each that are mutually perpendicular.) An emf induced on a conductor moving in a magnetic field is sometimes called *motional emf*.

We can also obtain Eq. 21–3 without using Faraday's law. We saw in Chapter 20 that a charged particle moving perpendicular to a magnetic field B with speed v experiences a force $F = qvB$ (Eq. 20–4). When the rod of Fig. 21–12a moves to the right with speed v, the electrons in the rod also move with this speed. Therefore, since $\vec{\mathbf{v}} \perp \vec{\mathbf{B}}$, each electron feels a force $F = qvB$, which acts up the page as shown in Fig. 21–12b. If the rod was not in contact with the U-shaped conductor, electrons would collect at the upper end of the rod, leaving the lower end positive (see signs in Fig. 21–12b). There must thus be an induced emf. If the rod does slide on the U-shaped conductor (Fig. 21–12a), the electrons will flow into the U. There will then be a clockwise (conventional) current in the loop. To calculate the emf, we determine the work W needed to move a charge q from one end of the rod to the other against this potential difference: W = force × distance = $(qvB)(l)$. The emf equals the work done per unit charge, so $\mathscr{E} = W/q = qvBl/q = Blv$, the same result as from Faraday's law above, Eq. 21–3.

EXERCISE C In what direction will the electrons flow in Fig. 21–12 if the rod moves to the left, decreasing the area of the current loop?

FIGURE 21–13 Example 21–6.

EXAMPLE 21–6 **Does a moving airplane develop a large emf?** An airplane travels 1000 km/h in a region where the Earth's magnetic field is 5.0×10^{-5} T and is nearly vertical (Fig. 21–13). What is the potential difference induced between the wing tips that are 70 m apart?

APPROACH We consider the wings to be a 70-m-long conductor moving through the Earth's magnetic field. We use Eq. 21–3 to get the emf.

SOLUTION Since v = 1000 km/h = 280 m/s, and $\vec{\mathbf{v}} \perp \vec{\mathbf{B}}$, we have

$$\mathscr{E} = Blv = (5.0 \times 10^{-5}\,\mathrm{T})(70\,\mathrm{m})(280\,\mathrm{m/s}) = 1.0\,\mathrm{V}.$$

NOTE Not much to worry about.

Blood flow measurement

FIGURE 21–14 Measurement of blood velocity from the induced emf. Example 21–7.

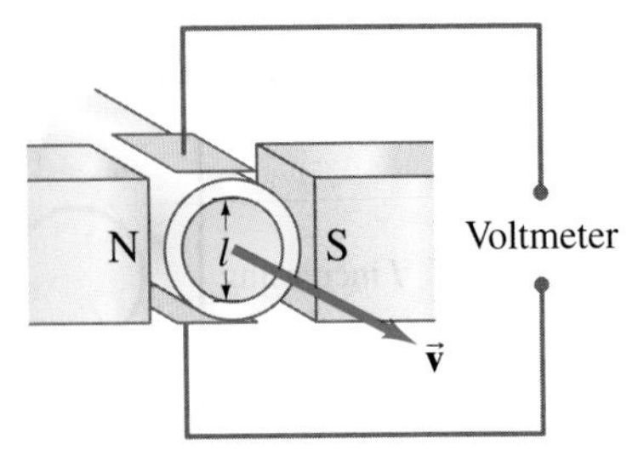

EXAMPLE 21–7 **Electromagnetic blood-flow measurement.** The rate of blood flow in our body's vessels can be measured using the apparatus shown in Fig. 21–14, since blood contains charged ions. Suppose that the blood vessel is 2.0 mm in diameter, the magnetic field is 0.080 T, and the measured emf is 0.10 mV. What is the flow velocity of the blood?

APPROACH The magnetic field $\vec{\mathbf{B}}$ points horizontally from left to right (N pole toward S pole). The induced emf acts over the width l = 2.0 mm of the blood vessel (Fig. 21–14), perpendicular to $\vec{\mathbf{B}}$ and $\vec{\mathbf{v}}$, just as in Fig. 21–12. We can then use Eq. 21–3 to get v.

SOLUTION We solve for v in Eq. 21–3:

$$v = \frac{\mathscr{E}}{Bl} = \frac{(1.0 \times 10^{-4}\,\mathrm{V})}{(0.080\,\mathrm{T})(2.0 \times 10^{-3}\,\mathrm{m})} = 0.63\,\mathrm{m/s}.$$

NOTE In actual practice, an alternating current is used to produce an alternating magnetic field. The induced emf is then alternating.

Additional Example

EXAMPLE 21–8 **Force on the rod.** To make the rod of Fig. 21–12a move to the right at speed v, you need to apply a force on the rod to the right. (*a*) Explain and determine the magnitude of the required force. (*b*) What external power is needed to move the rod? (Do not confuse this force on the rod with force on the electrons shown in Fig. 21–12b.)

APPROACH When the rod moves to the right, electrons flow upward in the rod according to right-hand-rule-3 (p. 562). So the conventional current is downward in the rod. We can see this also from Lenz's law: the outward magnetic flux through the loop is increasing, so the induced current must oppose the increase. Thus the current is clockwise so as to produce a magnetic field into the page (right-hand-rule-1). The magnetic force on the moving rod is $F = IlB$ for a constant B (Eq. 20–2). Right-hand-rule-2 tells us this magnetic force is to the left, and is thus a "drag force" opposing our effort to move the rod to the right.

SOLUTION (*a*) The magnitude of the external force, to the right, needs to balance the magnetic force $F = IlB$. The current $I = \mathscr{E}/R = Blv/R$ (see Eq. 21–3), and the resistance R is that of the whole circuit: the rod and the U-shaped conductor. The force F required to move the rod is thus

$$F = IlB = \left(\frac{Blv}{R}\right)lB = \frac{B^2l^2}{R}v.$$

If B, l, and R are constant, then a constant speed v is produced by a constant force. Constant R implies that parallel rails have negligible resistance.

(*b*) The external power needed to move the rod for constant R is

$$P_{\text{ext}} = Fv = \frac{B^2l^2v^2}{R}.$$

The power dissipated in the resistance is $P = I^2R$. With $I = \mathscr{E}/R = Blv/R$,

$$P_{\text{R}} = I^2R = \frac{B^2l^2v^2}{R},$$

so the power input equals that dissipated in the resistance at any moment.

21–4 Changing Magnetic Flux Produces an Electric Field

We have seen that a changing magnetic flux induces an emf; there also is an induced current. This implies there is an electric field in a wire, causing the electrons to start moving. Indeed, this and other results suggest the important conclusion that

a changing magnetic field induces an electric field.

This applies not only to wires and other conductors, but is a general result that applies to any region in space: an electric field will be induced at any point in space where there is a changing magnetic field.

We can get a simple formula for E in terms of B for the case of electrons in a moving conductor, as in Fig. 21–12. The electrons feel a force (upwards in Fig. 21–12b); and if we put ourselves in the reference frame of the conductor, this force accelerating the electrons implies that there is an electric field in the conductor. Electric field is defined as the force per unit charge, $E = F/q$, where $F = qvB$ (Eq. 20–4). Thus the effective field E in the rod must be

$$E = \frac{F}{q} = \frac{qvB}{q} = vB. \qquad \textbf{(21–4)}$$

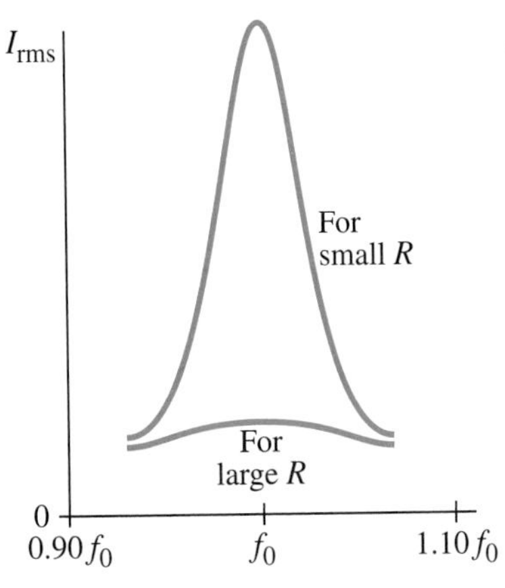

FIGURE 21–42 Current in an LRC circuit as a function of frequency, showing resonance peak at $f = f_0 = (1/2\pi)\sqrt{1/LC}$.

* 21–14 Resonance in AC Circuits

The rms current in an LRC series circuit is given by (see Eqs. 21–14, 21–15, 21–11b, and 21–12b):

$$I_{rms} = \frac{V_{rms}}{Z} = \frac{V_{rms}}{\sqrt{R^2 + \left(2\pi f L - \frac{1}{2\pi f C}\right)^2}}. \quad \textbf{(21–18)}$$

Because the reactance of inductors and capacitors depends on the frequency f of the source, the current in an LRC circuit depends on frequency. From Eq. 21–18 we see that the current will be maximum at a frequency that satisfies

$$2\pi f L - \frac{1}{2\pi f C} = 0.$$

We solve this for f, and call the solution f_0:

Resonant frequency

$$f_0 = \frac{1}{2\pi}\sqrt{\frac{1}{LC}}. \quad \textbf{(21–19)}$$

When $f = f_0$, the circuit is in resonance, and f_0 is the **resonant frequency** of the circuit. At this frequency, $X_C = X_L$, so the impedance is purely resistive. A graph of I_{rms} versus f is shown in Fig. 21–42 for particular values of R, L, and C. For smaller R compared to X_L and X_C, the resonance peak will be higher and sharper.

LC circuit

When R is very small, we speak of an ***LC* circuit**. The energy in an LC circuit oscillates, at frequency f_0, between the inductor and the capacitor, with some being dissipated in R (some resistance is unavoidable). This is called an ***LC* oscillation** or an **electromagnetic oscillation**. Not only does the charge oscillate back and forth, but so does the energy, which oscillates between being stored in the electric field of the capacitor and in the magnetic field of the inductor.

EM oscillations

Electric resonance is used in many circuits. Radio and TV sets, for example, use resonant circuits for tuning in a station. Many frequencies reach the circuit from the antenna, but a significant current flows only for frequencies at or near the resonant frequency. Either L or C is variable so that different stations can be tuned in (more on this in Chapter 22).

Summary

The **magnetic flux** passing through a loop is equal to the product of the area of the loop times the perpendicular component of the magnetic field strength:

$$\Phi_B = B_\perp A = BA\cos\theta. \quad \textbf{(21–1)}$$

If the magnetic flux through a coil of wire changes in time, an emf is induced in the coil. The magnitude of the induced emf equals the time rate of change of the magnetic flux through the loop times the number N of loops in the coil:

$$\mathscr{E} = -N\frac{\Delta\Phi_B}{\Delta t}. \quad \textbf{(21–2b)}$$

This is **Faraday's law of induction**.

The induced emf can produce a current whose magnetic field opposes the original change in flux (**Lenz's law**).

Faraday's law also tells us that a changing magnetic field produces an electric field; and that a straight wire of length l moving with speed v perpendicular to a magnetic field of strength B has an emf induced between its ends equal to

$$\mathscr{E} = Blv. \quad \textbf{(21–3)}$$

An electric **generator** changes mechanical energy into electrical energy. Its operation is based on Faraday's law: a coil of wire is made to rotate uniformly by mechanical means in a magnetic field, and the changing flux through the coil induces a sinusoidal current, which is the output of the generator.

[*A motor, which operates in the reverse of a generator, acts like a generator in that a **back emf**, or **counter emf**, is induced in its rotating coil. Because this back emf opposes the input voltage, it can act to limit the current in a motor coil. Similarly, a generator acts somewhat like a motor in that a **counter torque** acts on its rotating coil.]

A **transformer**, which is a device to change the magnitude of an ac voltage, consists of a primary coil and a secondary coil. The changing flux due to an ac voltage in the primary coil induces an ac voltage in the secondary coil. In a 100% efficient transformer, the ratio of output to input voltages (V_S/V_P) equals the ratio of the number of turns N_S in the secondary to the number N_P in the primary:

$$\frac{V_S}{V_P} = \frac{N_S}{N_P}. \quad \textbf{(21–6)}$$

The ratio of secondary to primary current is in the inverse ratio of turns:

$$\frac{I_S}{I_P} = \frac{N_P}{N_S}. \quad \textbf{(21–7)}$$

Microphones, ground fault circuit interrupters, seismographs, and read/write heads for computer drives and tape recorders are applications of electromagnetic induction.

[*A changing current in a coil of wire will produce a changing magnetic field that induces an emf in a second coil placed nearby. The **mutual inductance**, M, is defined by

$$\mathscr{E}_2 = -M\frac{\Delta I_1}{\Delta t}. \quad \textbf{(21–8)]}$$

[*Within a single coil, the changing B due to a changing current induces an opposing emf, $\mathscr{E}$, so a coil has a **self-inductance** L defined by

$$\mathscr{E} = -L\frac{\Delta I}{\Delta t}. \quad \textbf{(21–9)]}$$

[*The energy stored in an inductance L carrying current I is given by $U = \frac{1}{2}LI^2$. This energy can be thought of as being stored in the magnetic field of the inductor. The energy density u in any magnetic field B is given by

$$u = \frac{1}{2}\frac{B^2}{\mu_0}. \quad \textbf{(21–10)]}$$

[*When an inductance L and resistor R are connected in series to a source of emf, V, the current rises as

$$I = \frac{V}{R}\left(1 - e^{-t/\tau}\right),$$

where $\tau = L/R$ is the time constant. If the battery is suddenly switched out of the LR circuit, the current drops exponentially, $I = I_{\text{max}}e^{-t/\tau}$.]

[*Inductive and capacitive **reactance**, X, defined as for resistors, is the proportionality constant between voltage and current (either the rms or peak values). Across an inductor,

$$V = IX_L, \quad \textbf{(21–11a)}$$

and across a capacitor,

$$V = IX_C. \quad \textbf{(21–12a)}$$

The reactance of an inductor increases with frequency

$$X_L = 2\pi fL, \quad \textbf{(21–11b)}$$

whereas the reactance of a capacitor decreases with frequency f,

$$X_C = \frac{1}{2\pi fC}. \quad \textbf{(21–12b)}$$

The current through a resistor is always in phase with the voltage across it, but in an inductor, the current lags the voltage by 90°, and in a capacitor the current leads the voltage by 90°.]

[*In an LRC series circuit, the total **impedance** Z is defined by the equivalent of $V = IR$ for resistance, namely,

$$V_0 = I_0Z \quad \text{or} \quad V_{\text{rms}} = I_{\text{rms}}Z; \quad \textbf{(21–14)}$$

Z is given by

$$Z = \sqrt{R^2 + (X_L - X_C)^2}. \quad \textbf{(21–15a)]}$$

[*An LRC series circuit **resonates** at a frequency given by

$$f_0 = \frac{1}{2\pi}\sqrt{\frac{1}{LC}}. \quad \textbf{(21–19)}$$

The rms current in the circuit is largest when the applied voltage has a frequency equal to f_0.]

Questions

1. What would be the advantage, in Faraday's experiments (Fig. 21–1), of using coils with many turns?

2. What is the difference between magnetic flux and magnetic field?

3. Suppose you are holding a circular ring of wire and suddenly thrust a magnet, south pole first, away from you toward the center of the circle. Is a current induced in the wire? Is a current induced when the magnet is held steady within the ring? Is a current induced when you withdraw the magnet? In each case, if your answer is yes, specify the direction.

4. Two loops of wire are moving in the vicinity of a very long straight wire carrying a steady current as shown in Fig. 21–43. Find the direction of the induced current in each loop.

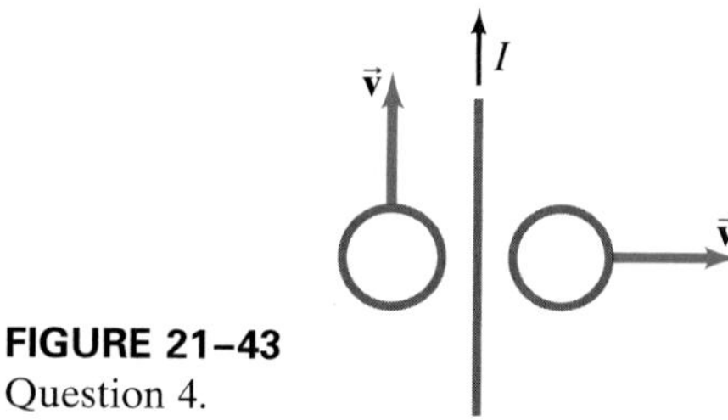

FIGURE 21–43 Question 4.

5. Suppose you are looking along a line through the centers of two circular (but separate) wire loops, one behind the other. A battery is suddenly connected to the front loop, establishing a clockwise current. (*a*) Will a current be induced in the second loop? (*b*) If so, when does this current start? (*c*) When does it stop? (*d*) In what direction is this current? (*e*) Is there a force between the two loops? (*f*) If so, in what direction?

6. In Fig. 21–44, determine the direction of the induced current in resistor R_A when (*a*) coil B is moved toward coil A, (*b*) when coil B is moved away from A, (*c*) when the resistance R_B is increased.

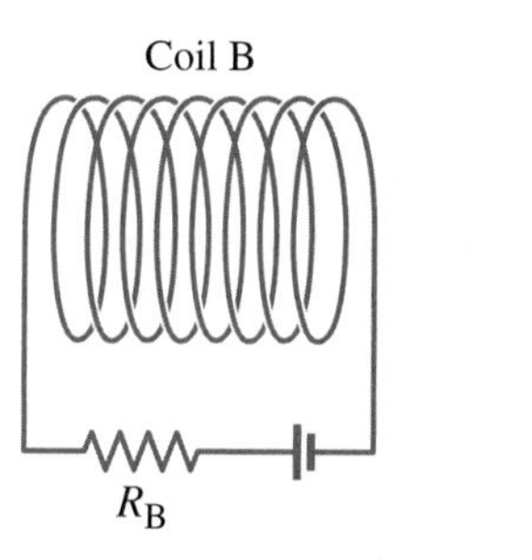

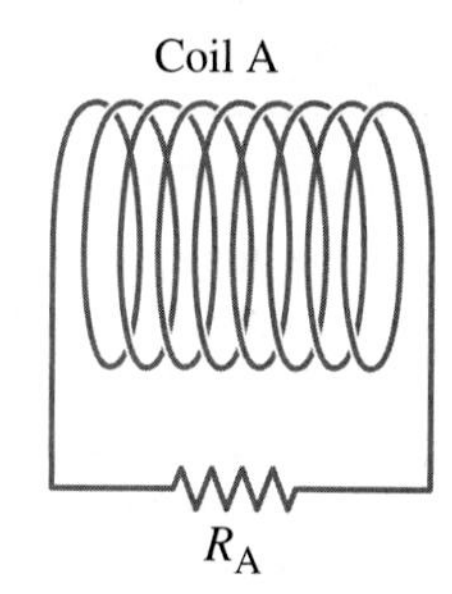

FIGURE 21–44 Question 6.

7. In situations where a small signal must travel over a distance, a "shielded cable" is used in which the signal wire is surrounded by an insulator and then enclosed by a cylindrical conductor carrying the return current. Why is a "shield" necessary?

8. What is the advantage of placing the two insulated electric wires carrying ac close together or even twisted about each other?

***9.** Explain why, exactly, the lights may dim briefly when a refrigerator motor starts up. When an electric heater is turned on, the lights may stay dimmed as long as the heater is on. Explain the difference.

FIGURE 23–5 When you look in a mirror, you see an image of yourself and objects around you. You don't see yourself as others see you, because left and right appear reversed in the image.

When you look straight into a mirror, you see what appears to be yourself as well as various objects around and behind you, Fig. 23–5. Your face and the other objects look as if they are in front of you, beyond the mirror; but they aren't. What you see in the mirror is an **image** of the objects, including yourself, that are in front of the mirror.

A "plane" mirror is one with a smooth flat reflecting surface. Figure 23–6 shows how an image is formed by a plane mirror according to the ray model. We are viewing the mirror, on edge, in the diagram of Fig. 23–6, and the rays are shown reflecting from the front surface. (Good mirrors are generally made by putting a highly reflective metallic coating on one surface of a very flat piece of glass.) Rays from two different points on an object (a bottle) are shown in Fig. 23–6: two rays are shown leaving from a point on the top of the bottle, and two more from a point on the bottom. Rays leave each point on the object going in many directions, but only those that enclose the bundle of rays that enter the eye from each of the two points are shown. Each set of diverging rays that enter the eye *appear* to come from a single point (called the image point) behind the mirror, as shown by the dashed lines. That is, our eyes and brain interpret any rays that enter an eye as having traveled straight-line paths. The point from which each bundle of rays seems to come is one point on the image. For each point on the object, there is a corresponding image point.

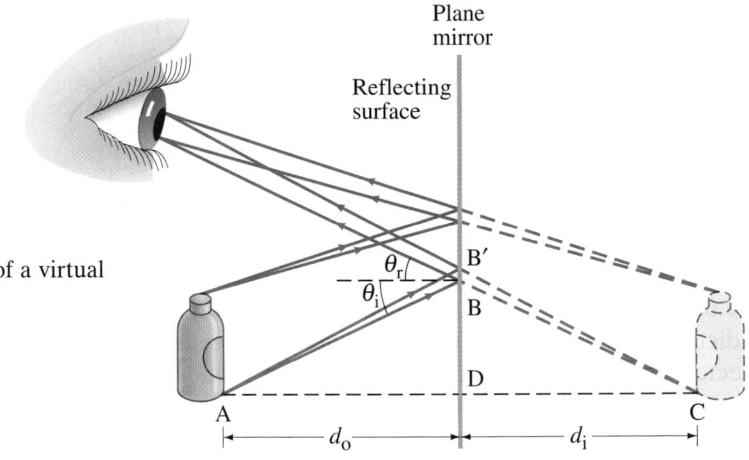

FIGURE 23–6 Formation of a virtual image by a plane mirror.

Let us concentrate on the two rays that leave point A on the object in Fig. 23–6, and strike the mirror at points B and B′. We use geometry now, for the rays at B. The angles ADB and CDB are right angles; and because of the law of reflection, $\theta_i = \theta_r$ at point B. Therefore, by geometry, angles ABD and CBD are also equal. The two triangles ABD and CBD are thus congruent, and the length AD = CD. That is, the image appears as far behind the mirror as the object is in front. The **image distance**, d_i (distance from mirror to image, Fig. 23–6), equals the **object distance**, d_o (distance from object to mirror). From the geometry, we also see that the height of the image is the same as that of the object.

Image distance = object distance (plane mirror)

The light rays do not actually pass through the image location itself in Fig. 23–6. (Note where the red lines are dashed to show they are our projections, not rays.) The image would not appear on paper or film placed at the location of the image. Therefore, it is called a **virtual image**. This is to distinguish it from a **real image** in which the light does pass through the image and which therefore could appear on paper or film placed at the image position. Our eyes can see both real and virtual images, as long as the diverging rays enter our pupils. We will see that curved mirrors and lenses can form real images, as well as virtual. A movie projector lens, for example, produces a real image that is visible on the screen.

Real and virtual images

EXAMPLE 23–1 **How tall must a full-length mirror be?** A woman 1.60 m tall stands in front of a vertical plane mirror. What is the minimum height of the mirror, and how high must its lower edge be above the floor, if she is to be able to see her whole body? (Assume her eyes are 10 cm below the top of her head.)

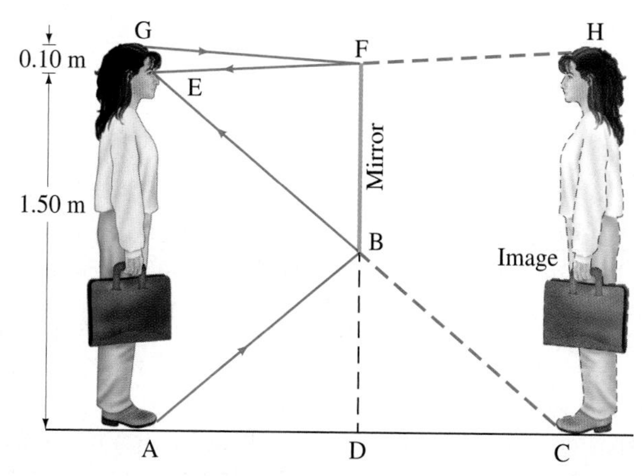

FIGURE 23–7 Seeing oneself in a mirror. Example 23–1.

APPROACH For her to see her whole body, light rays from the top of her head and from the bottom of her foot must reflect from the mirror and enter her eye: see Fig. 23–7. We don't show two rays diverging from each point as we did in Fig. 23–6, where we wanted to find where the image is. Now that we know the image is the same distance behind a plane mirror as the object is in front, we only need to show one ray leaving point G (top of head) and one ray leaving point A (her toe), and then use simple geometry.

SOLUTION First consider the ray that leaves her foot at A, reflects at B, and enters the eye at E. The mirror needs to extend no lower than B. Because the angle of reflection equals the angle of incidence, the height BD is half of the height AE. Because $AE = 1.60\,\text{m} - 0.10\,\text{m} = 1.50\,\text{m}$, then $BD = 0.75\,\text{m}$. Similarly, if the woman is to see the top of her head, the top edge of the mirror only needs to reach point F, which is 5 cm below the top of her head (half of $GE = 10\,\text{cm}$). Thus, $DF = 1.55\,\text{m}$, and the mirror need have a vertical height of only $(1.55\,\text{m} - 0.75\,\text{m}) = 0.80\,\text{m}$. And the mirror's bottom edge must be 0.75 m above the floor.

NOTE We see that a mirror need be only half as tall as a person for that person to see all of himself or herself.

PHYSICS APPLIED

How tall a mirror do you need to see a reflection of your entire self?

EXERCISE A Does the result of Example 23–1 depend on the person's distance from the mirror? (Try it and see, it's fun.)

23–3 Formation of Images by Spherical Mirrors

Reflecting surfaces do not have to be flat. The most common *curved* mirrors are *spherical*, which means they form a section of a sphere. A spherical mirror is called **convex** if the reflection takes place on the outer surface of the spherical shape so that the center of the mirror surface bulges out toward the viewer (Fig. 23–8a). A mirror is called **concave** if the reflecting surface is on the inner surface of the sphere so that the center of the mirror sinks away from the viewer (like a "cave"), Fig. 23–8b. Concave mirrors are used as shaving or cosmetic mirrors (Fig. 23–9a), and convex mirrors are sometimes used on cars and trucks (rearview mirrors) and in shops (to watch for thieves), because they take in a wide field of view (Fig. 23–9b).

FIGURE 23–8 Mirrors with convex and concave spherical surfaces. Note that $\theta_r = \theta_i$ for each ray.

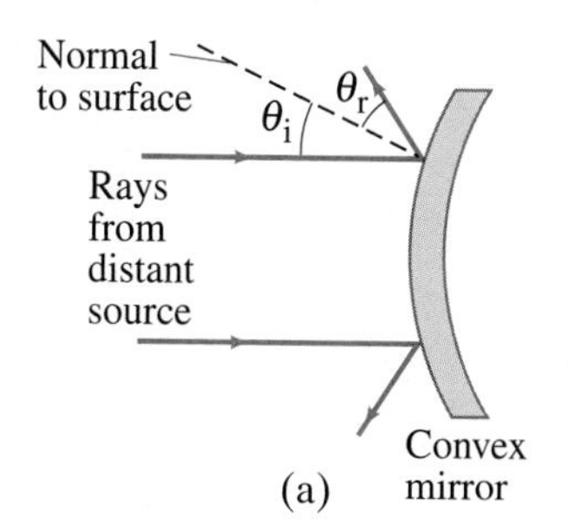

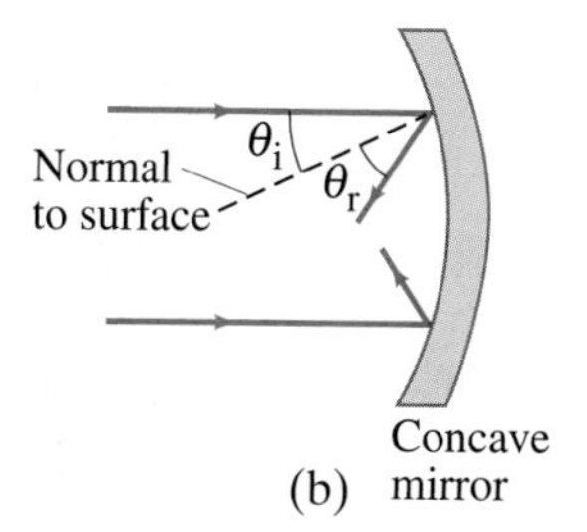

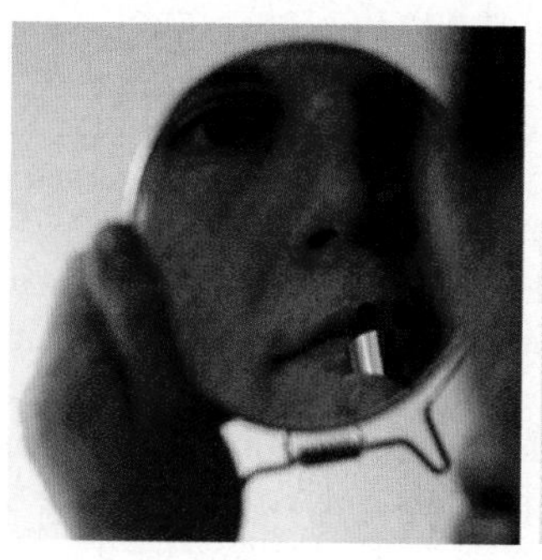
(a)

(b)

FIGURE 23–9 (a) A concave cosmetic mirror gives a magnified image. (b) A convex mirror in a store reduces image size and so includes a wide field of view.

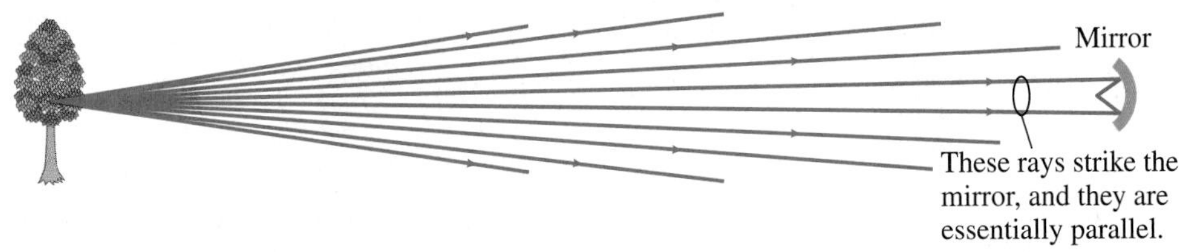

FIGURE 23–10 If the object's distance is large compared to the size of the mirror (or lens), the rays are nearly parallel. They are parallel for an object at infinity (∞).

Focal Point and Focal Length

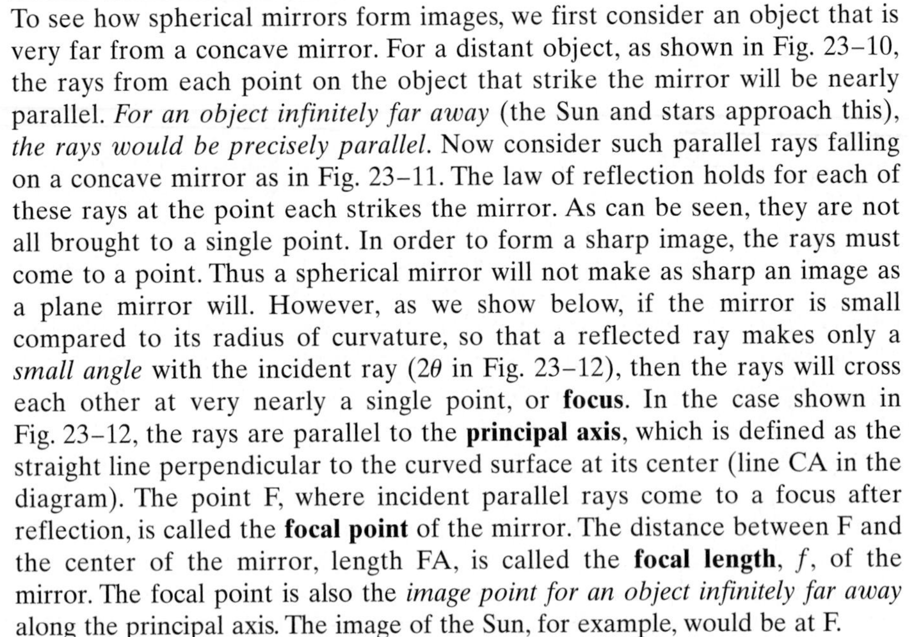

To see how spherical mirrors form images, we first consider an object that is very far from a concave mirror. For a distant object, as shown in Fig. 23–10, the rays from each point on the object that strike the mirror will be nearly parallel. *For an object infinitely far away* (the Sun and stars approach this), *the rays would be precisely parallel.* Now consider such parallel rays falling on a concave mirror as in Fig. 23–11. The law of reflection holds for each of these rays at the point each strikes the mirror. As can be seen, they are not all brought to a single point. In order to form a sharp image, the rays must come to a point. Thus a spherical mirror will not make as sharp an image as a plane mirror will. However, as we show below, if the mirror is small compared to its radius of curvature, so that a reflected ray makes only a *small angle* with the incident ray (2θ in Fig. 23–12), then the rays will cross each other at very nearly a single point, or **focus**. In the case shown in Fig. 23–12, the rays are parallel to the **principal axis**, which is defined as the straight line perpendicular to the curved surface at its center (line CA in the diagram). The point F, where incident parallel rays come to a focus after reflection, is called the **focal point** of the mirror. The distance between F and the center of the mirror, length FA, is called the **focal length**, f, of the mirror. The focal point is also the *image point for an object infinitely far away* along the principal axis. The image of the Sun, for example, would be at F.

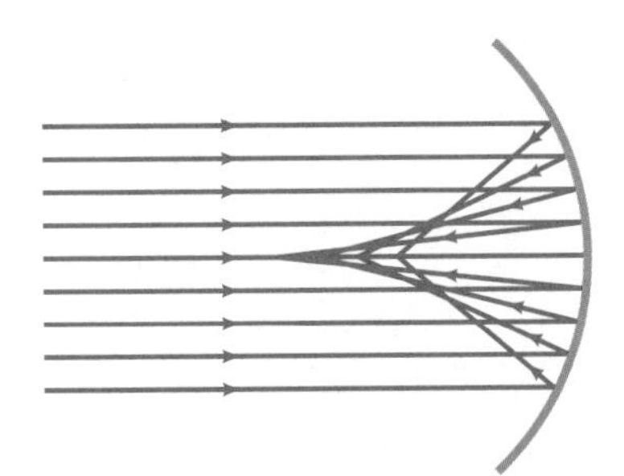

FIGURE 23–11 Parallel rays striking a concave spherical mirror do not focus at precisely a single point. (This "defect" is referred to as "spherical aberration.")

Now we will show, for a mirror whose reflecting surface is small compared to its radius of curvature, that the rays very nearly meet at a common point, F, and we will also calculate the focal length f. In this approximation, we consider only rays that make a small angle with the principal axis; such rays are called **paraxial rays**, and their angles are exaggerated in Fig. 23–12 to make the labels clear. First we consider a ray that strikes the mirror at B in Fig. 23–12. The point C is the center of curvature of the mirror (the center of the sphere of which the mirror is a part). So the dashed line CB is equal to r, the radius of curvature, and CB is normal to the mirror's surface at B. The incoming ray that hits the mirror at B makes an angle θ with this normal, and hence the reflected ray, BF, also makes an angle θ with the normal (law of reflection). Note that angle BCF is also θ as shown. The triangle CBF is isosceles because two of its angles are equal. Thus we have length $\mathrm{CF} = \mathrm{BF}$. We assume the mirror surface is small compared to the mirror's radius of curvature, so the angles are small, and the length FB is nearly equal to length FA. In this approximation, $\mathrm{FA} = \mathrm{FC}$. But $\mathrm{FA} = f$, the focal length, and $\mathrm{CA} = 2 \times \mathrm{FA} = r$. Thus the focal length is half the radius of curvature:

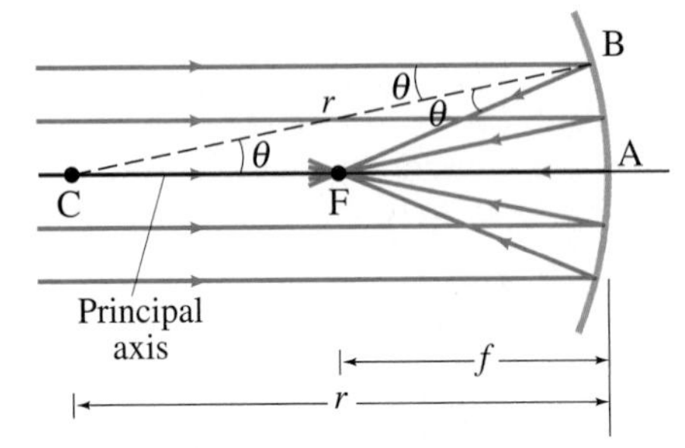

FIGURE 23–12 Rays parallel to the principal axis of a concave spherical mirror come to a focus at F, the focal point, as long as the mirror is small in width as compared to its radius of curvature, r, so that the rays are "paraxial"—that is, make only small angles with the axis.

Focal length of mirror

$$f = \frac{r}{2}. \tag{23–1}$$

We assumed only that the angle θ was small, so this result applies for all other incident paraxial rays. Thus all paraxial rays pass through the same point F.

Since it is only approximately true that the rays come to a perfect focus at F, the more curved the mirror, the worse the approximation (Fig. 23–11) and the more blurred the image. This "defect" of spherical mirrors is called **spherical aberration**; we will discuss it more with regard to lenses in Chapter 25. A *parabolic* reflector, on the other hand, will reflect the rays to a perfect focus. However, because parabolic shapes are much harder to make and thus much more expensive, spherical mirrors are used for most purposes. (Many astronomical telescopes use parabolic reflectors.) We consider here only spherical mirrors and we will assume that they are small compared to their radius of curvature so that the image is sharp and Eq. 23–1 holds.

Parabolic mirror

Image Formation—Ray Diagrams

We saw that for an object at infinity, the image is located at the focal point of a concave spherical mirror, where $f = r/2$. But where does the image lie for an object not at infinity? First consider the object shown as an arrow in Fig. 23–13, which is placed between F and C at point O (O for object). Let us determine where the image will be for a given point O′ at the top of the object. To do this we can draw several rays and make sure these reflect from the mirror such that the angle of reflection equals the angle of incidence. Many rays could be drawn leaving any point on an object, but determining the image position is simplified if we deal with three particularly simple rays. These are the rays labeled 1, 2, and 3 in Fig. 23–13 and we draw them leaving object point O′ as follows:

➡ RAY DIAGRAM

Finding the image position for a curved mirror

Ray 1 is drawn parallel to the axis; therefore after reflection it must pass along a line through F (as we saw in Fig. 23–12, and drawn here in Fig. 23–13a).

Ray 2 leaves O′ and is made to pass through F; therefore it must reflect so it is parallel to the axis (Fig. 23–13b).

Ray 3 passes through C, the center of curvature; it is along a radius of the spherical surface and is perpendicular to the mirror, so it is reflected back on itself (Fig. 23–13c).

All three rays leave a single point O′ on the object. After reflection from a (small) mirror, the point at which these rays cross is the image point I′. All other rays from the same object point will also pass through this image point. To find the image point for any object point, only these three types of rays need to be drawn. Only two of these rays are needed, but the third serves as a check.

Image point is where reflected rays intersect

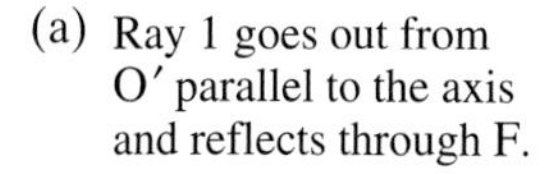
(a) Ray 1 goes out from O′ parallel to the axis and reflects through F.

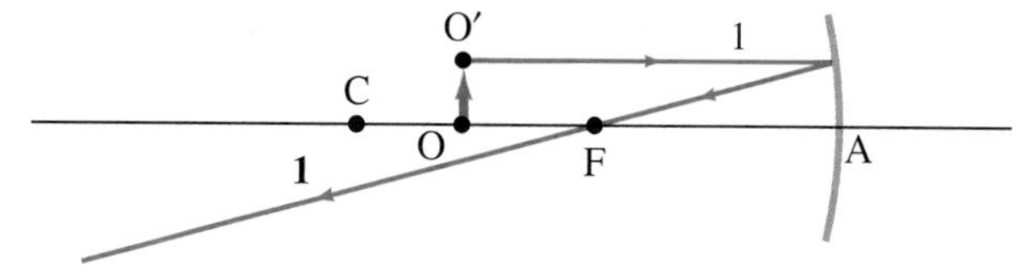

(b) Ray 2 goes through F and then reflects back parallel to the axis.

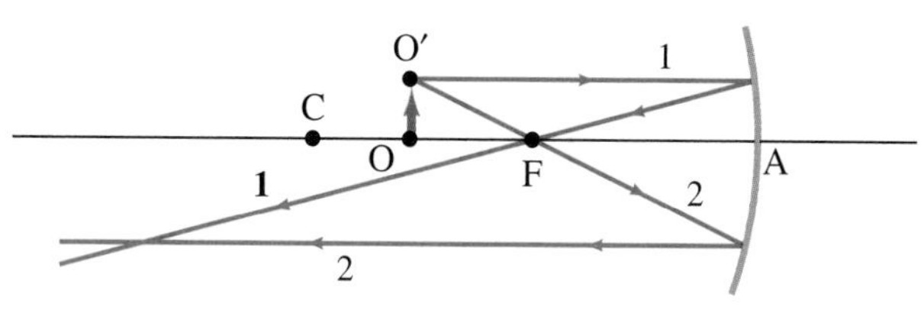

(c) Ray 3 is chosen perpendicular to mirror, and so must reflect back on itself and go through C (center of curvature).

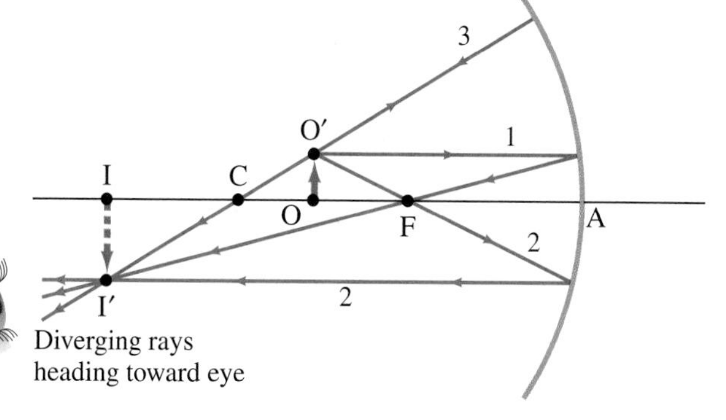

FIGURE 23–13 Rays leave point O′ on the object (an arrow). Shown are the three most useful rays for determining where the image I′ is formed. [Note that our mirror is not small compared to f, so our diagram will not give the precise position of the image.]

We have shown the image point in Fig. 23–13 only for a single point on the object. Other points on the object are imaged nearby, so a complete image of the object is formed, as shown by the dashed arrow in Fig. 23–13c. Because the light actually passes through the image itself, this is a *real image* that will appear on a piece of paper or film placed there. This can be compared to the virtual image formed by a plane mirror (the light does not actually pass through that image, Fig. 23–6).

Real image

The image in Fig. 23–13 can be seen by the eye when the eye is placed to the left of the image so that some of the rays diverging from each point on the image (as point I′) can enter the eye as shown in Fig. 23–13c. (See also Figs. 23–1 and 23–6.)

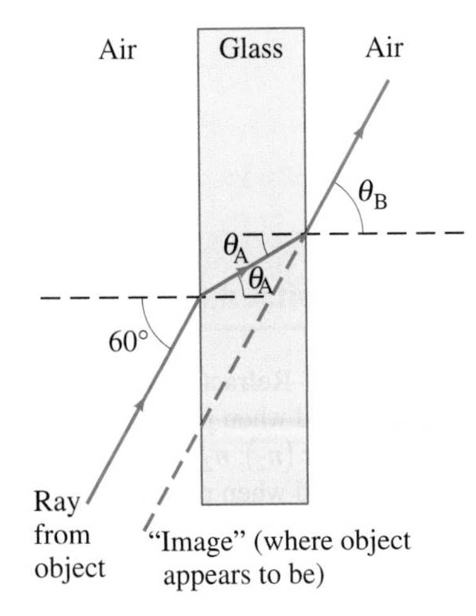

FIGURE 23–22 Light passing through a piece of glass (Example 23–6).

EXERCISE C Light passes from a medium with $n = 1.3$ into a medium with $n = 1.5$. Is the light bent toward or away from the perpendicular to the interface?

EXAMPLE 23–6 Refraction through flat glass. Light traveling in air strikes a flat piece of uniformly thick glass at an incident angle of 60°, as shown in Fig. 23–22. If the index of refraction of the glass is 1.50, (*a*) what is the angle of refraction θ_A in the glass; (*b*) what is the angle θ_B at which the ray emerges from the glass?

APPROACH We apply Snell's law at the first surface, where the light enters the glass, and again at the second surface where it leaves the glass and enters the air.

SOLUTION (*a*) The incident ray is in air, so $n_1 = 1.00$ and $n_2 = 1.50$. Applying Snell's law where the light enters the glass $(\theta_1 = 60°)$ gives

$$\sin \theta_A = \frac{1.00}{1.50} \sin 60° = 0.577,$$

so $\theta_A = 35.2°$.

(*b*) Since the faces of the glass are parallel, the incident angle at the second surface is just θ_A (simple geometry), so $\sin \theta_A = 0.577$. At this second interface, $n_1 = 1.50$ and $n_2 = 1.00$. Thus the ray re-enters the air at an angle $\theta_B\ (= \theta_2)$ given by

$$\sin \theta_B = \frac{1.50}{1.00} \sin \theta_A = 0.866,$$

and $\theta_B = 60°$. The direction of a light ray is thus unchanged by passing through a flat piece of glass of uniform thickness.

NOTE It should be clear that this works for any angle of incidence. The ray is displaced slightly to one side, however. You can observe this by looking through a piece of glass (near its edge) at some object and then moving your head to the side slightly so that you see the object directly. It "jumps."

FIGURE 23–23 Example 23–7.

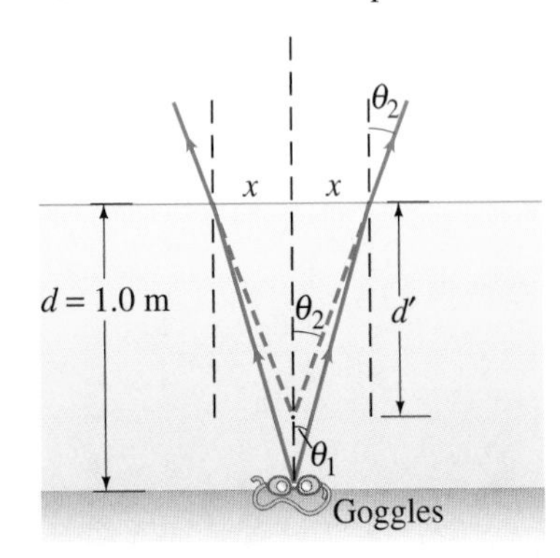

EXAMPLE 23–7 Apparent depth of a pool. A swimmer has dropped her goggles to the bottom of a pool at the shallow end, marked as 1.0 m deep. But the goggles don't look that deep. Why? How deep do the goggles appear to be when you look straight down into the water?

APPROACH We draw a ray diagram showing two rays going upward from a point on the goggles at a small angle, and being refracted at the water's (flat) surface. This is shown in Fig. 23–23, and the dashed lines show why the water seems less deep than it actually is. The two rays traveling upward from the goggles are refracted *away* from the normal as they exit the water, and so appear to be diverging from a point above the goggles (dashed lines).

SOLUTION To calculate the apparent depth d' (Fig. 23–23), given a real depth $d = 1.0$ m, we use Snell's law with $n_1 = 1.33$ for water and $n_2 = 1$ for air:

$$\sin \theta_2 = n_1 \sin \theta_1 .$$

We are considering only small angles, so $\sin \theta \approx \tan \theta \approx \theta$, with θ in radians. So Snell's law becomes

$$\theta_2 \approx n_1 \theta_1 .$$

From Fig. 23–23, we see that

$$\theta_2 \approx \tan \theta_2 = \frac{x}{d'} \qquad \text{and} \qquad \theta_1 \approx \tan \theta_1 = \frac{x}{d}.$$

Putting these into Snell's law, $\theta_2 \approx n_1 \theta_1$, we get

$$\frac{x}{d'} \approx n_1 \frac{x}{d}$$

or

$$d' \approx \frac{d}{n_1} = \frac{1.0\,\text{m}}{1.33} = 0.75\,\text{m}.$$

The pool seems only three-fourths as deep as it actually is.

23–6 Total Internal Reflection; Fiber Optics

When light passes from one material into a second material where the index of refraction is less (say, from water into air), the light bends away from the normal, as for rays I and J in Fig. 23–24. At a particular incident angle, the angle of refraction will be 90°, and the refracted ray would skim the surface (ray K) in this case. The incident angle at which this occurs is called the **critical angle**, θ_C. From Snell's law, θ_C is given by

$$\sin\theta_C = \frac{n_2}{n_1}\sin 90° = \frac{n_2}{n_1}. \qquad \textbf{(23–6)}$$

Critical angle

For any incident angle less than θ_C, there will be a refracted ray, although part of the light will also be reflected at the boundary. However, for incident angles greater than θ_C, Snell's law would tell us that $\sin\theta_2$ is greater than 1.00. Yet the sine of an angle can never be greater than 1.00. In this case there is no refracted ray at all, and *all of the light is reflected*, as for ray L in Fig. 23–24. This effect is called **total internal reflection**. But note that total internal reflection can occur only when light strikes a boundary where the medium beyond has a lower index of refraction.

Total internal reflection (occurs only if refractive index is smaller beyond boundary)

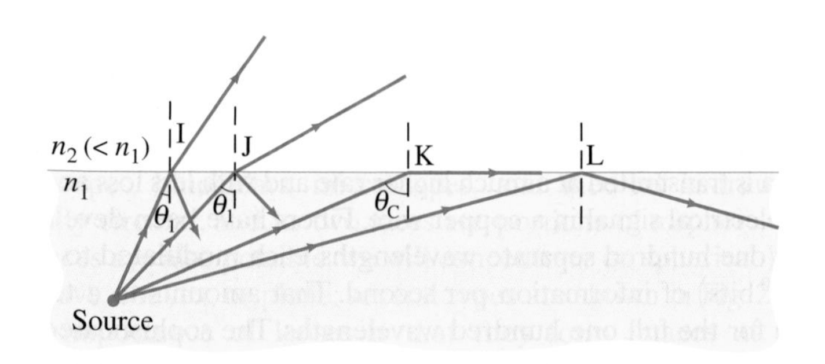

FIGURE 23–24 Since $n_2 < n_1$, light rays are totally internally reflected if the incident angle $\theta_1 > \theta_C$, as for ray L. If $\theta_1 < \theta_C$, as for rays I and J, only a part of the light is reflected, and the rest is refracted.

CONCEPTUAL EXAMPLE 23–8 **View up from under water.** Describe what a person would see who looked up at the world from beneath the perfectly smooth surface of a lake or swimming pool.

RESPONSE For an air–water interface, the critical angle is given by

$$\sin\theta_C = \frac{1.00}{1.33} = 0.750.$$

Therefore, $\theta_C = 49°$. Thus the person would see the outside world compressed into a circle whose edge makes a 49° angle with the vertical. Beyond this angle, the person would see reflections from the sides and bottom of the lake or pool (Fig. 23–25).

EXERCISE D Light traveling in air strikes a glass surface with $n = 1.48$. For what range of angles will total internal reflection occur?

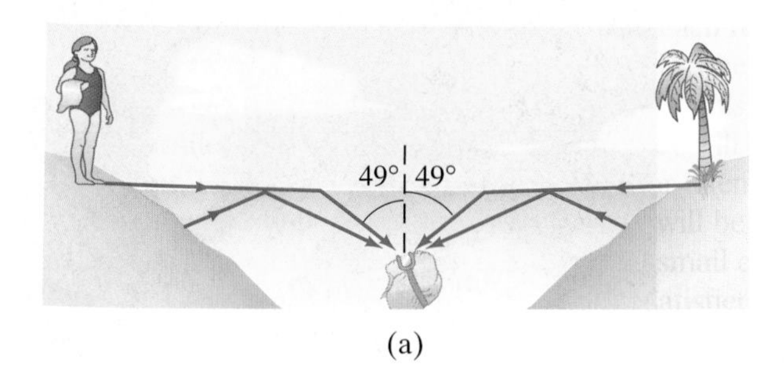

(a)

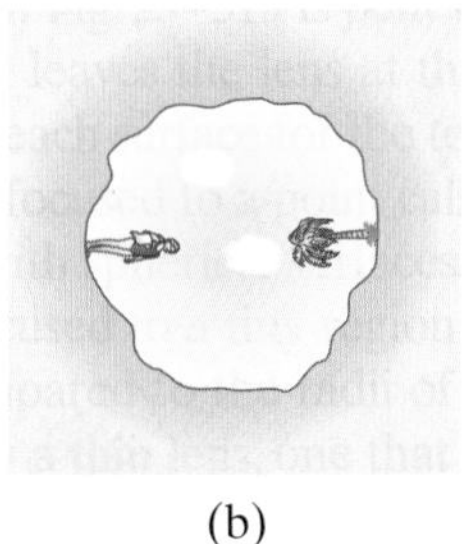

(b)

FIGURE 23–25 (a) Light rays, and (b) view looking upward from beneath the water (the surface of the water must be very smooth). Example 23–8.

24–3 Interference—Young's Double-Slit Experiment

In 1801, the Englishman Thomas Young (1773–1829) obtained convincing evidence for the wave nature of light and was even able to measure wavelengths for visible light. Figure 24–5a shows a schematic diagram of Young's famous double-slit experiment. Light from a single source (Young used the Sun) falls on a screen containing two closely spaced slits S_1 and S_2. If light consists of tiny particles, we might expect to see two bright lines on a screen placed behind the slits as in (b). But instead a series of bright lines are seen, as in (c). Young was able to explain this result as a **wave-interference** phenomenon. To see this, imagine plane waves of light of a single wavelength—called **monochromatic**, meaning "one color"—falling on the two slits as shown in Fig. 24–6. Because of diffraction, the waves leaving the two small slits spread out as shown. This is equivalent to the interference pattern produced when two rocks are thrown into a lake (Fig. 11–37), or when sound from two loudspeakers interferes (Fig. 12–16). Recall Section 11–12 on wave interference.

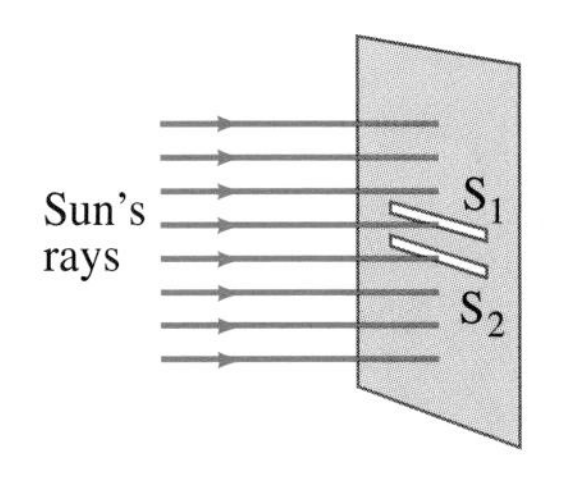

(a)

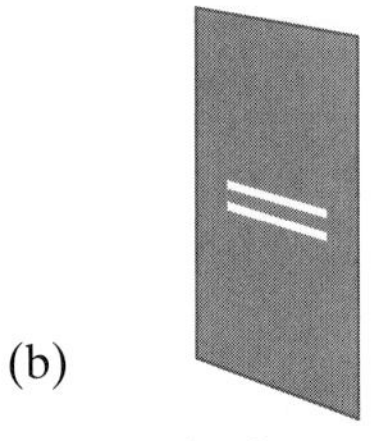

(b)

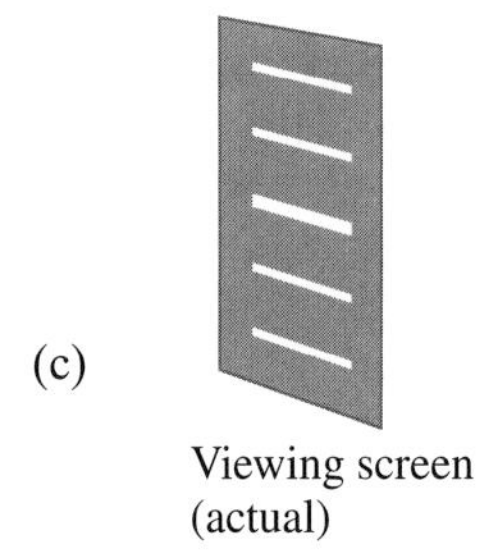

(c)

FIGURE 24–5 (a) Young's double-slit experiment. (b) If light consists of particles, we would expect to see two bright lines on the screen behind the slits. (c) In fact, many lines are observed.

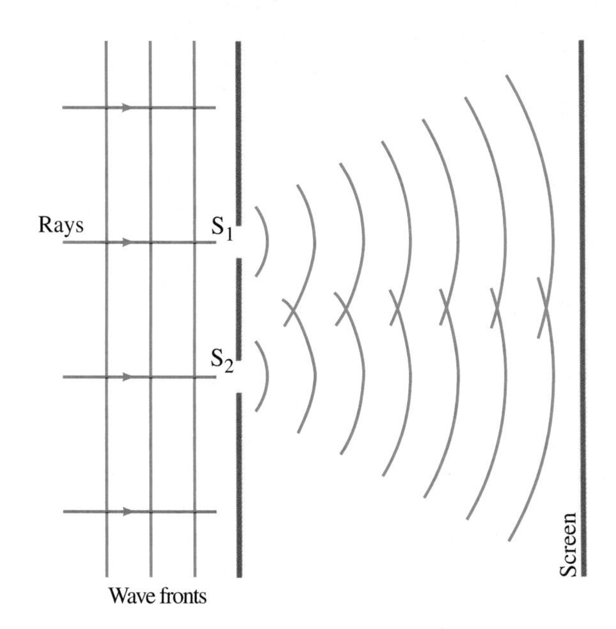

FIGURE 24–6 If light is a wave, light passing through one of two slits should interfere with light passing through the other slit.

To see how an interference pattern is produced on the screen, we make use of Fig. 24–7. Waves of wavelength λ are shown entering the slits S_1 and S_2, which are a distance d apart. The waves spread out in all directions after passing through the slits, but they are shown only for three different angles θ. In Fig. 24–7a, the waves reaching the center of the screen are shown ($\theta = 0°$). The waves from the two slits travel the same distance, so they are in phase: a crest of one wave arrives at the same time as a crest of the other wave. Hence the amplitudes of the two waves add to form a larger amplitude as shown in Fig. 24–8a. This is **constructive interference**, and there is a bright area at the center of the screen. Constructive interference also occurs when the paths of the two rays differ by one wavelength (or any whole number of wavelengths), as shown in Fig. 24–7b; also here there will be brightness on the screen. But if one ray travels an extra distance of one-half wavelength (or $\frac{3}{2}\lambda$, $\frac{5}{2}\lambda$, and so on), the two waves are exactly out of phase when they reach the screen: the crests of one wave arrive at the same time as the troughs of the other wave, and so they add to produce zero amplitude (Fig. 24–8b). This is **destructive interference**, and the screen is dark, Fig. 24–7c. Thus, there will be a series of bright and dark lines (or **fringes**) on the viewing screen.

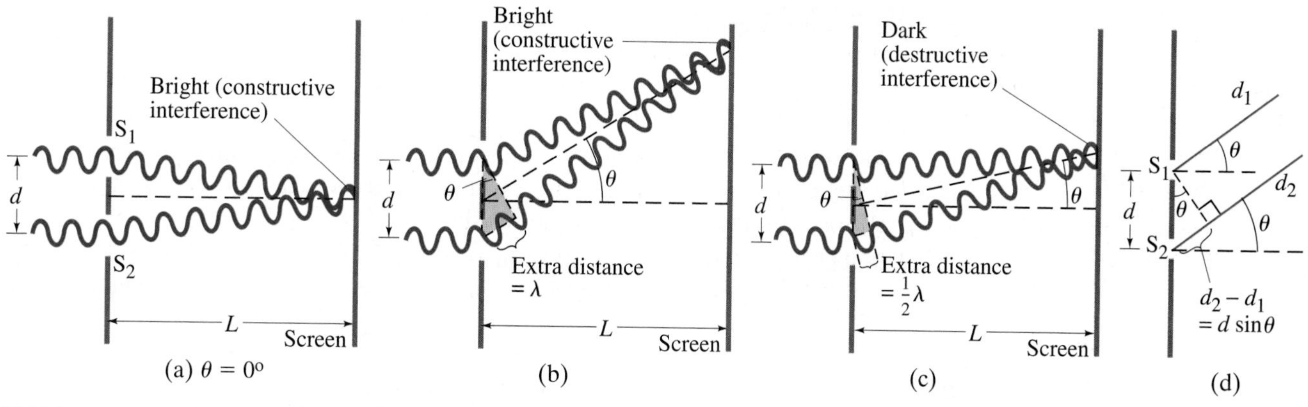

FIGURE 24–7 How the wave theory explains the pattern of lines seen in the double-slit experiment. (a) At the center of the screen the waves from each slit travel the same distance and are in phase. (b) At this angle θ, the lower wave travels an extra distance of one whole wavelength, and the waves are in phase; note from the shaded triangle that the path difference equals $d \sin\theta$. (c) For this angle θ, the lower wave travels an extra distance equal to one-half wavelength, so the two waves arrive at the screen fully out of phase. (d) A more detailed diagram showing the geometry for parts (b) and (c).

To determine exactly where the bright lines fall, first note that Fig. 24–7 is somewhat exaggerated; in real situations, the distance d between the slits is very small compared to the distance L to the screen. The rays from each slit for each case will therefore be essentially parallel, and θ is the angle they make with the horizontal as shown in Fig. 24–7d. From the shaded right triangles shown in Figs. 24–7b and c, we can see that the extra distance traveled by the lower ray is $d \sin\theta$ (seen more clearly in Fig. 24–7d). **Constructive interference** will occur, and a bright fringe will appear on the screen, when the *path difference*, $d \sin\theta$, equals a whole number of wavelengths:

$$d \sin\theta = m\lambda, \qquad m = 0, 1, 2, \cdots. \qquad \left[\begin{array}{c}\text{constructive}\\\text{interference}\\\text{(bright)}\end{array}\right] \quad \textbf{(24–2a)}$$

The value of m is called the **order** of the interference fringe. The first order ($m = 1$), for example, is the first fringe on each side of the central fringe (which is at $\theta = 0$, $m = 0$). Destructive interference occurs when the path difference $d \sin\theta$ is $\frac{1}{2}\lambda$, $\frac{3}{2}\lambda$, and so on:

$$d \sin\theta = \left(m + \tfrac{1}{2}\right)\lambda, \qquad m = 0, 1, 2, \cdots. \qquad \left[\begin{array}{c}\text{destructive}\\\text{interference}\\\text{(dark)}\end{array}\right] \quad \textbf{(24–2b)}$$

The bright fringes are peaks or maxima of light intensity, the dark fringes are minima. The intensity of the bright fringes is greatest for the central fringe ($m = 0$) and decreases for higher orders, as shown in Fig. 24–9. How much the intensity decreases with increasing order depends on the width of the two slits.

FIGURE 24–8 Two traveling waves are waves shown undergoing (a) constructive interference, (b) destructive interference. (See also Section 11–12.)

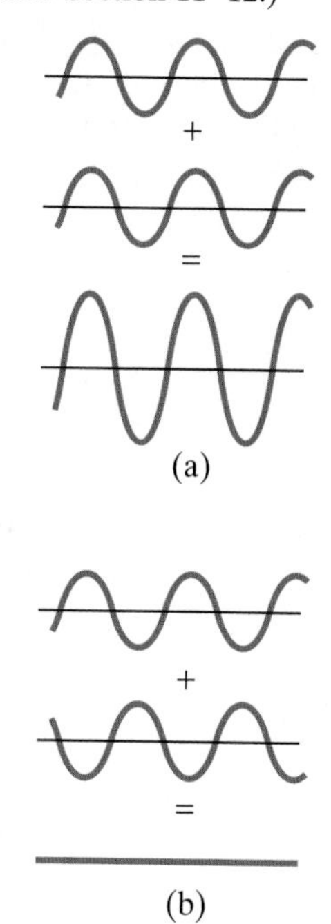

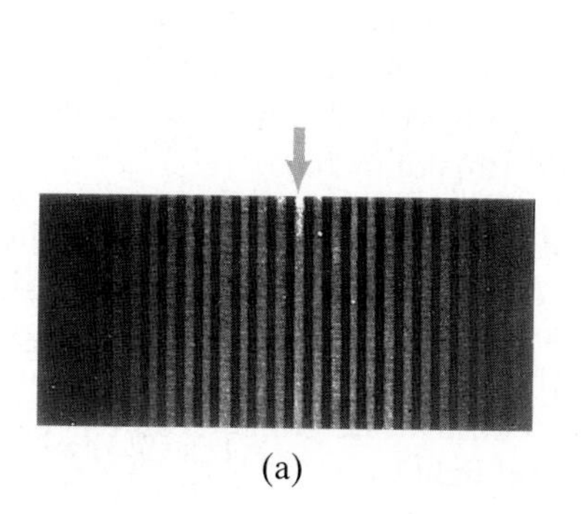

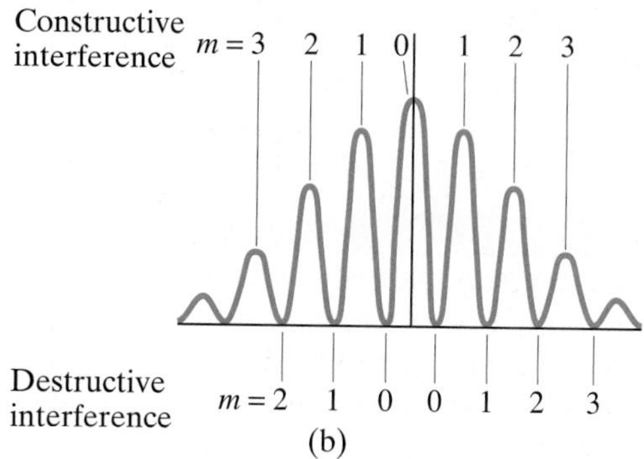

FIGURE 24–9 (a) Interference fringes produced by a double-slit experiment and detected by photographic film placed on the viewing screen. The arrow marks the central fringe. (b) Graph of the intensity of light in the interference pattern. Also shown are values of m for Eq. 24–2a (constructive interference) and Eq. 24–2b (destructive interference).

EXAMPLE 24–1 **Line spacing for double-slit interference.** A screen containing two slits 0.100 mm apart is 1.20 m from the viewing screen. Light of wavelength $\lambda = 500\text{ nm}$ falls on the slits from a distant source. Approximately how far apart will adjacent bright interference fringes be on the screen?

APPROACH The angular position of bright (constructive interference) fringes is found using Eq. 24–2a. The distance between the first two fringes (say) can be found using right triangles as shown in Fig. 24–10.

SOLUTION Given $d = 0.100\text{ mm} = 1.00 \times 10^{-4}\text{ m}$, $\lambda = 500 \times 10^{-9}\text{ m}$, and $L = 1.20\text{ m}$, the first-order fringe ($m = 1$) occurs at an angle θ given by

$$\sin\theta_1 = \frac{m\lambda}{d} = \frac{(1)(500 \times 10^{-9}\text{ m})}{1.00 \times 10^{-4}\text{ m}} = 5.00 \times 10^{-3}.$$

CAUTION

Use the approximation $\theta \approx \tan\theta$ or $\theta \approx \sin\theta$ only if θ is small and in radians

This is a very small angle, so we can take $\sin\theta \approx \theta$, with θ in radians. The first-order fringe will occur a distance x_1 above the center of the screen (see Fig. 24–10), given by $x_1/L = \tan\theta_1 \approx \theta_1$, so

$$x_1 \approx L\theta_1 = (1.20\text{ m})(5.00 \times 10^{-3}) = 6.00\text{ mm}.$$

The second-order fringe ($m = 2$) will occur at

$$x_2 \approx L\theta_2 = L\frac{2\lambda}{d} = 12.0\text{ mm}$$

above the center, and so on. Thus the lower order fringes are 6.00 mm apart.

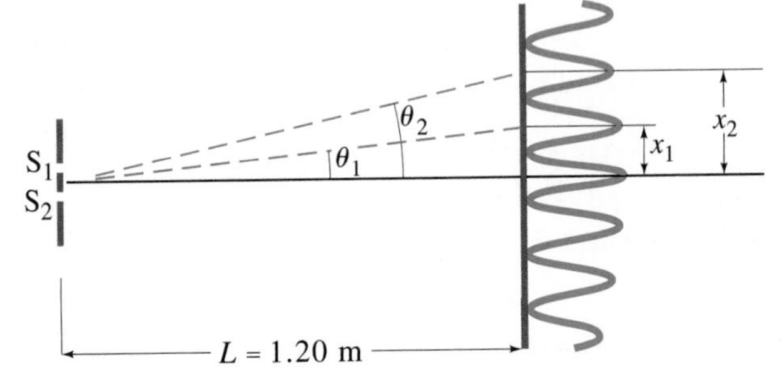

FIGURE 24–10 Examples 24–1 and 24–2. For small angles, the interference fringes occur at distance $x = \theta L$ above the center fringe ($m = 0$); θ_1 and x_1 are for the first-order fringe ($m = 1$), θ_2 and x_2 are for $m = 2$.

CONCEPTUAL EXAMPLE 24–2 **Changing the wavelength.** (*a*) What happens to the interference pattern shown in Fig. 24–10, Example 24–1, if the incident light (500 nm) is replaced by light of wavelength 700 nm? (*b*) What happens instead if the wavelength stays at 500 nm but the slits are moved farther apart?

RESPONSE (*a*) When λ increases in Eq. 24–2a but d stays the same, then the angle θ for bright fringes increases and the interference pattern spreads out. (*b*) Increasing the slit spacing d reduces θ for each order, so the lines are closer together.

Wavelength (or frequency) determines color

From Eqs. 24–2 we can see that, except for the zeroth-order fringe at the center, the position of the fringes depends on wavelength. Consequently, when white light falls on the two slits, as Young found in his experiments, the central fringe is white, but the first- (and higher-) order fringes contain a spectrum of colors like a rainbow; θ was found to be smallest for violet light and largest for red (Fig. 24–11). By measuring the position of these fringes, Young was the first to determine the wavelengths of visible light (using Eqs. 24–2). In doing so, he showed that what distinguishes different colors physically is their wavelength (or frequency), an idea put forward earlier by Grimaldi in 1665.

FIGURE 24–11 First-order fringes are a full spectrum, like a rainbow. Also Example 24–3.

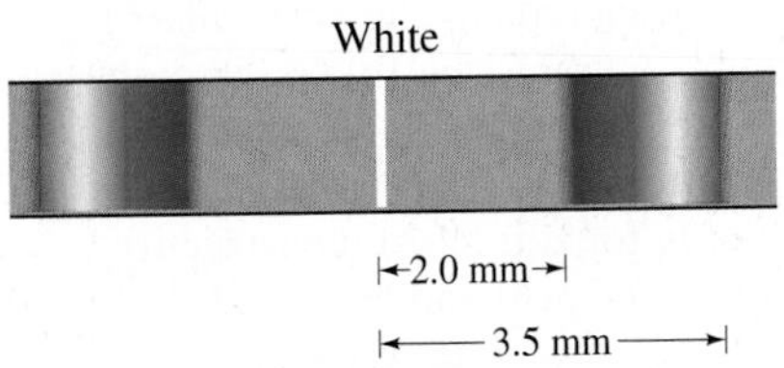

White

2.0 mm

3.5 mm

EXAMPLE 24–3 **Wavelengths from double-slit interference.** White light passes through two slits 0.50 mm apart, and an interference pattern is observed on a screen 2.5 m away. The first-order fringe resembles a rainbow with violet and red light at opposite ends. The violet light falls about 2.0 mm and the red 3.5 mm from the center of the central white fringe (Fig. 24–11). Estimate the wavelengths for the violet and red light.

APPROACH We find the angles for violet and red light from the distances given and the diagram of Fig. 24–10. Then we use Eq. 24–2a to obtain the wavelengths. Because 3.5 mm is much less than 2.5 m, we can use the small-angle approximation.

SOLUTION We use Eq. 24–2a with $m = 1$ and $\sin\theta \approx \tan\theta \approx \theta$. Then for violet light, $x = 2.0\,\text{mm}$, so (see also Fig. 24–10)

$$\lambda = \frac{d\sin\theta}{m} \approx \frac{d\theta}{m} \approx \frac{d}{m}\frac{x}{L} = \left(\frac{5.0 \times 10^{-4}\,\text{m}}{1}\right)\left(\frac{2.0 \times 10^{-3}\,\text{m}}{2.5\,\text{m}}\right) = 4.0 \times 10^{-7}\,\text{m},$$

or 400 nm. For red light, $x = 3.5\,\text{mm}$, so

$$\lambda = \frac{d}{m}\frac{x}{L} = \left(\frac{5.0 \times 10^{-4}\,\text{m}}{1}\right)\left(\frac{3.5 \times 10^{-3}\,\text{m}}{2.5\,\text{m}}\right) = 7.0 \times 10^{-7}\,\text{m} = 700\,\text{nm}.$$

EXERCISE A For the setup in Example 24–3, how far from the central white fringe is the first-order fringe for green light $\lambda = 500\,\text{nm}$?

Coherent Light

The two slits in Fig. 24–7 act as if they were two sources of radiation. They are called **coherent sources** because the waves leaving them have the same wavelength and frequency, and bear the same phase relationship to each other at all times. This happens because the waves come from a single source to the left of the two slits in Fig. 24–7. An interference pattern is observed only when the sources are coherent. If two tiny lightbulbs replaced the two slits, an interference pattern would not be seen. The light emitted by one lightbulb would have a random phase with respect to the second bulb, and the screen would be more or less uniformly illuminated. Two such sources, whose output waves bear no fixed phase relationship to each other, are called **incoherent sources**.

Coherent and incoherent sources

Interference patterns occur only if sources are coherent

24–4 The Visible Spectrum and Dispersion

The two most obvious properties of light are readily describable in terms of the wave theory of light: intensity (or brightness) and color. The **intensity** of light is the energy it carries per unit area per unit time, and is related to the square of the amplitude of the wave, just as for any wave (see Section 11–10, or Eqs. 22–7 and 22–8). The **color** of the light is related to the frequency f or wavelength λ of the light. (Recall $\lambda f = c = 3.0 \times 10^8\,\text{m/s}$, Eq. 22–4.) Visible light—that to which our eyes are sensitive—consists of frequencies from $4 \times 10^{14}\,\text{Hz}$ to $7.5 \times 10^{14}\,\text{Hz}$, corresponding to wavelengths in air of about 400 nm to 750 nm.† This is known as the **visible spectrum**, and within it lie the different colors from violet to red, as shown in Fig. 24–12. Light with wavelength shorter than 400 nm is called **ultraviolet** (UV), and light with wavelength longer than 750 nm is called **infrared** (IR).‡ Although human eyes are not sensitive to UV or IR, some types of photographic film and other detectors do respond to them.

†Sometimes the angstrom (Å) unit is used when referring to light: $1\,\text{Å} = 1 \times 10^{-10}\,\text{m}$. Visible light has wavelengths in air of 4000 Å to 7500 Å.

‡The complete electromagnetic spectrum is illustrated in Fig. 22–8.

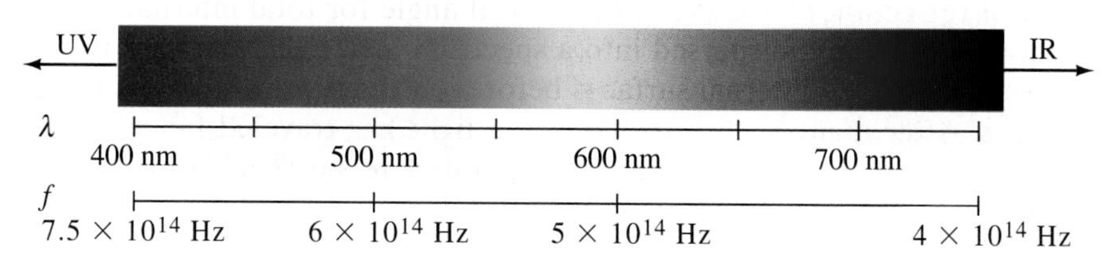

FIGURE 24–12 The spectrum of visible light, showing the range of frequencies and wavelengths (in air) for the various colors.

To see how a diffraction pattern arises, we will analyze the important case of monochromatic light passing through a narrow slit. We will assume that parallel rays (plane waves) of light fall on the slit of width D, and pass through to a viewing screen very far away.† As we know from studying water waves and from Huygens' principle, the waves passing through the slit spread out in all directions. We will now examine how the waves passing through different parts of the slit interfere with each other.

Parallel rays of monochromatic light pass through the narrow slit as shown in Fig. 24–20a. The light falls on a screen which is assumed to be very far away, so the rays heading for any point are very nearly parallel before they meet at the screen. First we consider rays that pass straight through as in Fig. 24–20a. They are all in phase, so there will be a central bright spot on the screen. In Fig. 24–20b, we consider rays moving at an angle θ such that the ray from the top of the slit travels exactly one wavelength farther than the ray from the bottom edge to reach the screen. The ray passing through the very center of the slit will travel one-half wavelength farther than the ray at the bottom of the slit. These two rays will be exactly out of phase with one another and so will destructively interfere when they overlap at the screen. Similarly, a ray slightly above the bottom one will cancel a ray that is the same distance above the central one. Indeed, each ray passing through the lower half of the

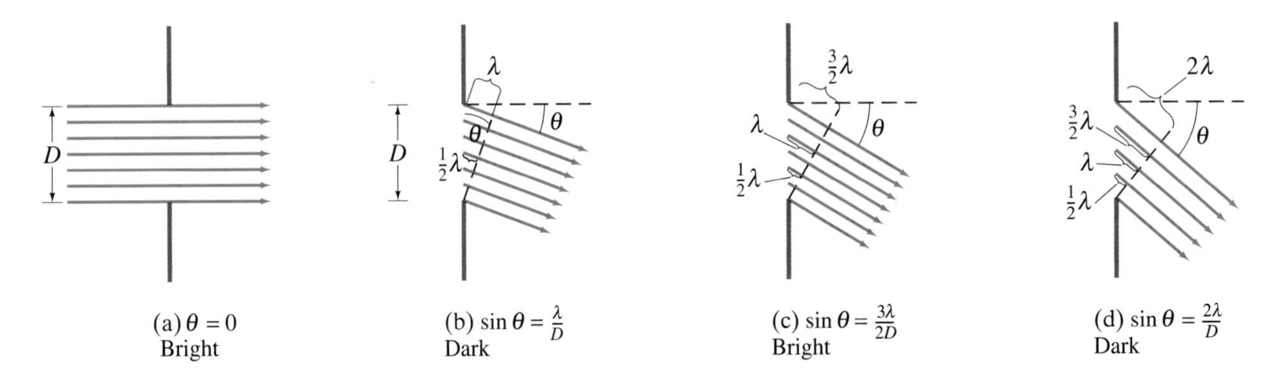

FIGURE 24–20 Analysis of diffraction pattern formed by light passing through a narrow slit.

slit will cancel with a corresponding ray passing through the upper half. Thus, all the rays destructively interfere in pairs, and so the light intensity will be zero on the viewing screen at this angle. The angle θ at which this takes place can be seen from Fig. 24–20b to occur when $\lambda = D\sin\theta$, so

Diffraction equation (angular half width of central spot)

$$\sin\theta = \frac{\lambda}{D}. \qquad \text{[first minimum]} \quad \textbf{(24–3a)}$$

The light intensity is a maximum at $\theta = 0°$ and decreases to a minimum (intensity = zero) at the angle θ given by Eq. 24–3a.

Now consider a larger angle θ such that the top ray travels $\frac{3}{2}\lambda$ farther than the bottom ray, as in Fig. 24–20c. In this case, the rays from the bottom third of the slit will cancel in pairs with those in the middle third because they will be $\lambda/2$ out of phase. However, light from the top third of the slit will still reach the screen, so there will be a bright spot centered near $\sin\theta \approx 3\lambda/2D$, but it will not be nearly as bright as the central spot at $\theta = 0°$. For an even larger angle θ such that the top ray travels 2λ farther than the bottom ray, Fig. 24–20d, rays from the bottom quarter of the slit will cancel with those in the quarter just above it because the path lengths differ by $\lambda/2$. And the rays through the quarter of the slit just above center will cancel with those through the top quarter. At this angle there will again be a minimum of zero intensity in the diffraction

†If the viewing screen is not far away, lenses can be used to make the rays parallel.

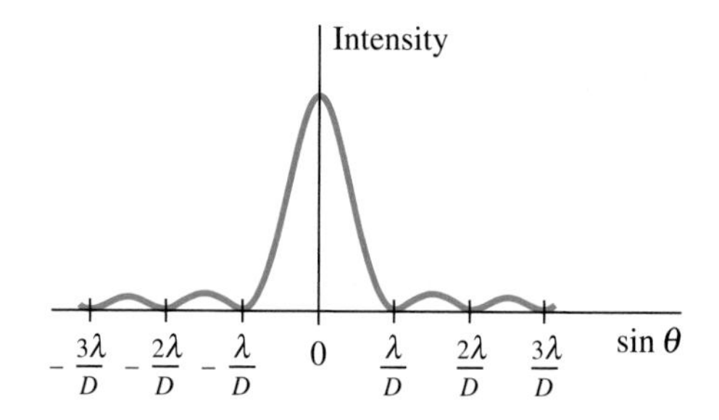

FIGURE 24–21 Intensity in the diffraction pattern of a single slit as a function of $\sin\theta$. Note that the central maximum is not only much higher than the maxima to each side, but it is also twice as wide ($2\lambda/D$ wide) as any of the others (only λ/D wide each).

pattern. A plot of the intensity as a function of angle is shown in Fig. 24–21. This corresponds well with the photo of Fig. 24–19c. Notice that minima (zero intensity) occur at

$$D\sin\theta = m\lambda, \qquad m = 1, 2, 3, \cdots, \qquad \text{[minima]} \quad \textbf{(24–3b)}$$

Single-slit diffraction minima

but *not* at $m = 0$ where there is the strongest maximum. Between the minima, smaller intensity maxima occur at approximately (not exactly) $m \approx \frac{3}{2}, \frac{5}{2}, \cdots$.

Note that the *minima* for a diffraction pattern, Eq. 24–3b, satisfy a criterion that looks very similar to that for the *maxima* (bright spots) for double-slit interference, Eq. 24–2a. Also note that D is a single slit width, whereas d in Eq. 24–2 is the distance between two slits.

Don't confuse Eqs. 24–2 for interference with Eqs. 24–3 for diffraction; note the differences

EXAMPLE 24–4 Single-slit diffraction maximum. Light of wavelength 750 nm passes through a slit 1.0×10^{-3} mm wide. How wide is the central maximum (*a*) in degrees, and (*b*) in centimeters, on a screen 20 cm away?

APPROACH The width of the central maximum goes from the first minimum on one side to the first minimum on the other side. We use Eq. 24–3a to find the angular position of the first single-slit diffraction minimum.

SOLUTION (*a*) The first minimum occurs at

$$\sin\theta = \frac{\lambda}{D} = \frac{7.5 \times 10^{-7}\,\text{m}}{1 \times 10^{-6}\,\text{m}} = 0.75.$$

So $\theta = 49°$. This is the angle between the center and the first minimum, Fig. 24–22. The angle subtended by the whole central maximum, between the minima above and below the center, is twice this, or 98°.

(*b*) The width of the central maximum is $2x$, where $\tan\theta = x/20\,\text{cm}$. So $2x = 2(20\,\text{cm})(\tan 49°) = 46\,\text{cm}$.

NOTE A large width of the screen will be illuminated, but it will not normally be very bright since the amount of light that passes through such a small slit will be small and it is spread over a large area. Note also that we *cannot* use the small-angle approximation here ($\theta \approx \sin\theta \approx \tan\theta$) because θ is large.

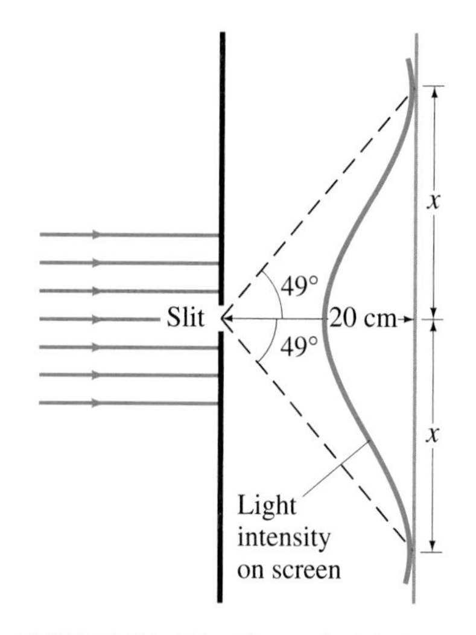

FIGURE 24–22 Example 24–4.

EXERCISE B In Example 24–4, red light ($\lambda = 750\,\text{nm}$) was used. If instead yellow light ($\lambda = 550\,\text{nm}$) had been used, would the central maximum be wider or narrower?

CONCEPTUAL EXAMPLE 24–5 Diffraction spreads. Light shines through a rectangular hole that is narrower in the vertical direction than the horizontal, Fig. 24–23. (*a*) Would you expect the diffraction pattern to be more spread out in the vertical direction or in the horizontal direction? (*b*) Should a rectangular loudspeaker horn at a stadium be high and narrow, or wide and flat?

RESPONSE (*a*) From Eq. 24–3a we can see that if we make the slit (width D) narrower, the pattern spreads out more. This is consistent with our study of waves in Chapter 11. The diffraction through the rectangular hole will be wider vertically, since the opening is smaller in that direction.

(*b*) For a loudspeaker, the sound pattern desired is one spread out horizontally, so the horn should be tall and narrow (rotate Fig. 24–23 by 90°).

FIGURE 24–23 Example 24–5.

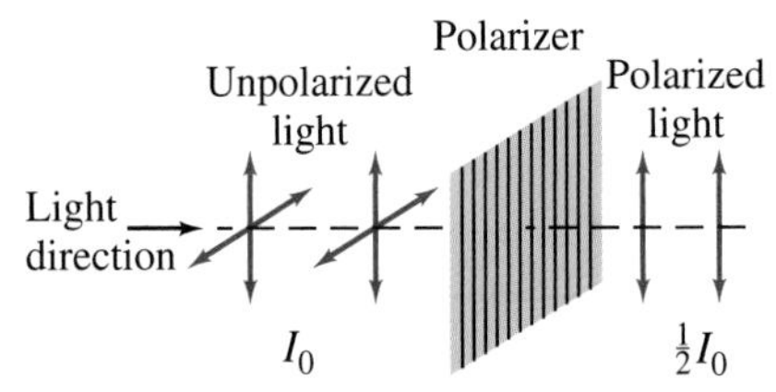

FIGURE 24–42 Unpolarized light has equal intensity vertical and horizontal components. After passing through a polarizer, one of these components is eliminated. The intensity of the light is reduced to half.

Unpolarized light consists of light with random directions of polarization. Each of these polarization directions can be resolved into components along two mutually perpendicular directions. On average, an unpolarized beam can be thought of as two plane-polarized beams of equal magnitude perpendicular to one another. When unpolarized light passes through a polarizer, one component is eliminated. So the intensity of the light passing through is reduced by half since half the light is eliminated: $I = \frac{1}{2}I_0$ (Fig. 24–42).

When two Polaroids are *crossed*—that is, their polarizing axes are perpendicular to one another—unpolarized light can be entirely stopped. As shown in Fig. 24–43, unpolarized light is made plane-polarized by the first Polaroid (the polarizer).

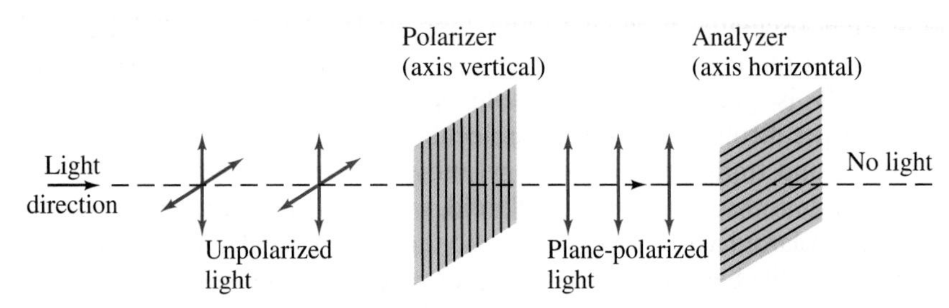

FIGURE 24–43 Crossed Polaroids completely eliminate light.

The second Polaroid, the analyzer, then eliminates this component since its transmission axis is perpendicular to the first. You can try this with Polaroid sunglasses (Fig. 24–44). Note that Polaroid sunglasses eliminate 50% of unpolarized light because of their polarizing property; they absorb even more because they are colored.

FIGURE 24–44 Crossed Polaroids. When the two polarized sunglass lenses overlap, with axes perpendicular, almost no light passes through.

EXAMPLE 24–11 **Two Polaroids at 60°.** Unpolarized light passes through two Polaroids; the axis of one is vertical and that of the other is at 60° to the vertical. Describe the orientation and intensity of the transmitted light.

APPROACH Half of the unpolarized light is absorbed by the first Polaroid, and emerges plane polarized. When that light passes through the second Polaroid, the intensity is further reduced according to Eq. 24–5, and the plane of polarization becomes along the axis of the second Polaroid.

SOLUTION The first Polaroid eliminates half the light, so the intensity is reduced by half: $I_1 = \frac{1}{2}I_0$. The light reaching the second polarizer is vertically polarized and so is reduced in intensity (Eq. 24–5) to

$$I_2 = I_1(\cos 60^\circ)^2 = \tfrac{1}{4}I_1.$$

Thus, $I_2 = \frac{1}{8}I_0$. The transmitted light has an intensity one-eighth that of the original and is plane-polarized at a 60° angle to the vertical.

CONCEPTUAL EXAMPLE 24–12 **Three Polaroids.** We saw in Fig. 24–43 that when unpolarized light falls on two crossed Polaroids (axes at 90°), no light passes through. What happens if a third Polaroid, with axis at 45° to each of the other two, is placed between them (Fig. 24–45a)?

RESPONSE We start just as in Example 24–11 and recall again that light emerging from each Polaroid is polarized parallel to that Polaroid's axis. Thus the angle in Eq. 24–5 is that between the transmission axes of each pair of Polaroids taken in turn. The first Polaroid changes the unpolarized light to plane-polarized and reduces the intensity from I_0 to $I_1 = \frac{1}{2}I_0$. The second polarizer further reduces the intensity by $(\cos 45^\circ)^2$, Eq. 24–5:

$$I_2 = I_1(\cos 45^\circ)^2 = \tfrac{1}{2}I_1 = \tfrac{1}{4}I_0.$$

The light leaving the second polarizer is plane polarized at 45° (Fig. 24–45b) relative to the third polarizer, so the third one reduces the intensity to

$$I_3 = I_2(\cos 45^\circ)^2 = \tfrac{1}{2}I_2,$$

or $I_3 = \frac{1}{8}I_0$. Thus $\frac{1}{8}$ of the original intensity gets transmitted.

NOTE If we don't insert the 45° Polaroid, zero intensity results (Fig. 24–43).

FIGURE 24–45 Example 24–12.

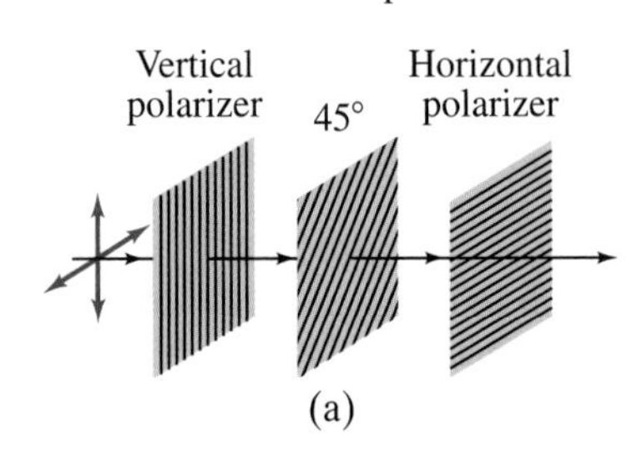

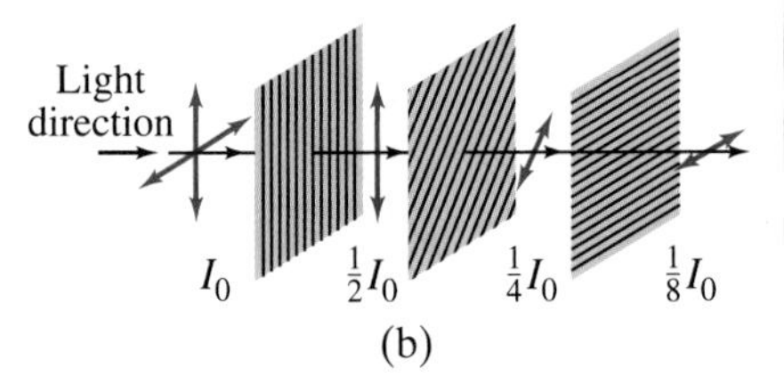

EXERCISE E How much light would pass through if the 45° Polarizer in Example 24–12 was placed not between the other two polarizers but (a) before the vertical (first) polarizer, or (b) after the horizontal polarizer?

Polarization by Reflection

Another means of producing polarized light from unpolarized light is by reflection. When light strikes a nonmetallic surface at any angle other than perpendicular, the reflected beam is polarized preferentially in the plane parallel to the surface, Fig. 24–46. In other words, the component with polarization in the plane perpendicular to the surface is preferentially transmitted or absorbed. You can check this by rotating Polaroid sunglasses while looking through them at a flat surface of a lake or road. Since most outdoor surfaces are horizontal, Polaroid sunglasses are made with their axes vertical to eliminate the more strongly reflected horizontal component, and thus reduce glare. People who go fishing wear Polaroids to eliminate reflected glare from the surface of a lake or stream and thus see beneath the water more clearly (Fig. 24–47).

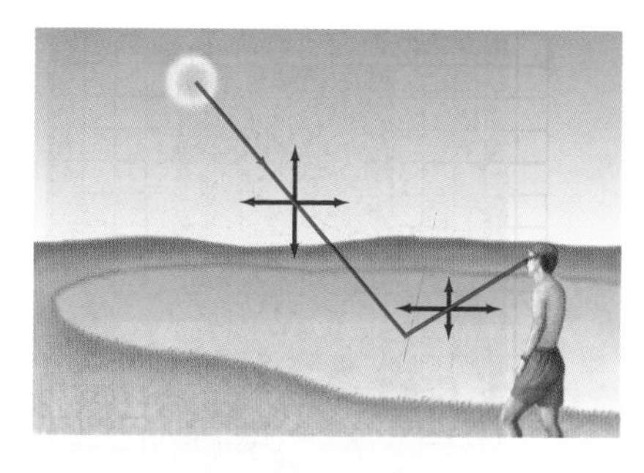

FIGURE 24–46 Light reflected from a nonmetallic surface, such as the smooth surface of water in a lake, is partially polarized parallel to the surface.

(a)

(b)

FIGURE 24–47 Photographs of a river, (a) allowing all light into the camera lens, and (b) using a polarizer. The polarizer is adjusted to absorb most of the (polarized) light reflected from the water's surface, allowing the dimmer light from the bottom of the river, and any fish lying there, to be seen more readily.

The amount of polarization in the reflected beam depends on the angle, varying from no polarization at normal incidence to 100% polarization at an angle known as the **polarizing angle**, θ_p.† This angle is related to the index of refraction of the two materials on either side of the boundary by the equation

$$\tan\theta_p = \frac{n_2}{n_1}, \tag{24–6a}$$

where n_1 is the index of refraction of the material in which the beam is traveling, and n_2 is that of the medium beyond the reflecting boundary. If the beam is traveling in air, $n_1 = 1$, and Eq. 24–6a becomes

$$\tan\theta_p = n. \tag{24–6b}$$

The polarizing angle θ_p is also called **Brewster's angle**, and Eqs. 24–6 *Brewster's law*, after the Scottish physicist David Brewster (1781–1868), who worked it out experimentally in 1812. Equations 24–6 can be derived from the electromagnetic wave theory of light. It is interesting that at Brewster's angle, the reflected ray and the transmitted (refracted) ray make a 90° angle to each other; that is, $\theta_p + \theta_r = 90°$, where θ_r is the refraction angle (Fig. 24–48). This can be seen by substituting Eq. 24–6a, $n_2 = n_1 \tan\theta_p = n_1 \sin\theta_p/\cos\theta_p$, into Snell's law, $n_1 \sin\theta_p = n_2 \sin\theta_r$, and get $\cos\theta_p = \sin\theta_r$ which can only hold if $\theta_p = 90° - \theta_r$.

FIGURE 24–48 At θ_p the reflected light is plane-polarized parallel to the surface, and $\theta_p + \theta_r = 90°$, where θ_r is the refraction angle. (The large dots represent vibrations perpendicular to the page.)

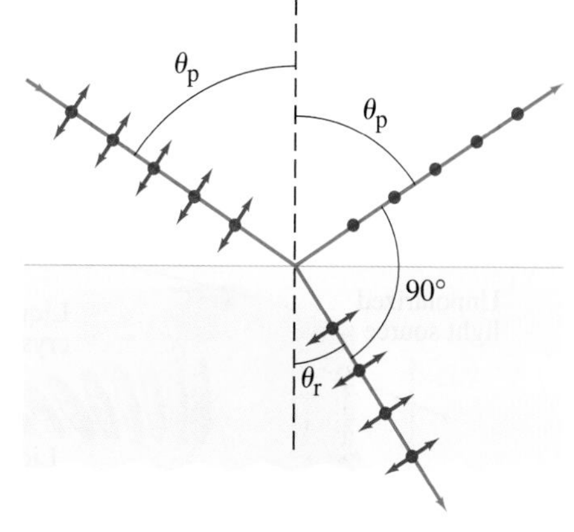

EXAMPLE 24–13 Polarizing angle. (*a*) At what incident angle is sunlight reflected from a lake plane-polarized? (*b*) What is the refraction angle?

APPROACH The polarizing angle at the surface is Brewster's angle, Eq. 24–6b. We find the angle of refraction from Snell's law.

SOLUTION (*a*) We use Eq. 24–6b with $n = 1.33$, so $\tan\theta_p = 1.33$ giving $\theta_p = 53.1°$.
(*b*) From Snell's law, $\sin\theta_r = \sin\theta_p/n = \sin 53.1°/1.33 = 0.601$ giving $\theta_r = 36.9°$.

NOTE $\theta_p + \theta_r = 53.1° + 36.9° = 90.0°$, as expected.

† Only a fraction of the incident light is reflected at the surface of a transparent medium. Although this reflected light is 100% polarized (if $\theta = \theta_p$), the remainder of the light, which is transmitted into the new medium, is only partially polarized.

23. When a compact disk (CD) is held at an angle in white light, the reflected light is a full spectrum (Fig. 24–56). Explain. What would you expect to see if monochromatic light was used?

FIGURE 24–56 Question 23.

24. Why are Newton's rings (Fig. 24–31) closer together farther from the center?

25. Some coated lenses appear greenish yellow when seen by reflected light. What wavelengths do you suppose the coating is designed to transmit completely?

26. A drop of oil on a pond appears bright at its edges, where its thickness is much less than the wavelengths of visible light. What can you say about the index of refraction of the oil?

27. What does polarization tell us about the nature of light?

28. Explain the advantage of polarized sunglasses over normal tinted sunglasses.

29. How can you tell if a pair of sunglasses is polarizing or not?

30. Two polarized sheets rotated at an angle of 90° with respect to each other will not let any light through. Three polarized sheets, each rotated at an angle of 45° with respect to each other, will let some light through. What will happen to unpolarized light if you align four polarized sheets, each rotated at an angle of 30° with respect to the one in front of it?

* 31. What would be the color of the sky if the Earth had no atmosphere?

* 32. If the Earth's atmosphere were 50 times denser than it is, would sunlight still be white, or would it be some other color?

Problems

24–3 Double-Slit Interference

1. (I) Monochromatic light falling on two slits 0.016 mm apart produces the fifth-order fringe at an 8.8° angle. What is the wavelength of the light used?

2. (I) The third-order fringe of 610 nm light is observed at an angle of 18° when the light falls on two narrow slits. How far apart are the slits?

3. (II) Monochromatic light falls on two very narrow slits 0.048 mm apart. Successive fringes on a screen 5.00 m away are 6.5 cm apart near the center of the pattern. Determine the wavelength and frequency of the light.

4. (II) A parallel beam of light from a He-Ne laser, with a wavelength 656 nm, falls on two very narrow slits 0.060 mm apart. How far apart are the fringes in the center of the pattern on a screen 3.6 m away?

5. (II) Light of wavelength 680 nm falls on two slits and produces an interference pattern in which the fourth-order fringe is 38 mm from the central fringe on a screen 2.0 m away. What is the separation of the two slits?

6. (II) If 720-nm and 660-nm light passes through two slits 0.58 mm apart, how far apart are the second-order fringes for these two wavelengths on a screen 1.0 m away?

7. (II) In a double-slit experiment, it is found that blue light of wavelength 460 nm gives a second-order maximum at a certain location on the screen. What wavelength of visible light would have a minimum at the same location?

8. (II) Water waves having parallel crests 2.5 cm apart pass through two openings 5.0 cm apart in a board. At a point 2.0 m beyond the board, at what angle relative to the "straight-through" direction would there be little or no wave action?

9. (II) Suppose a thin piece of glass is placed in front of the lower slit in Fig. 24–7 so that the two waves enter the slits 180° out of phase (Fig. 24–57). Describe in detail the interference pattern on the screen.

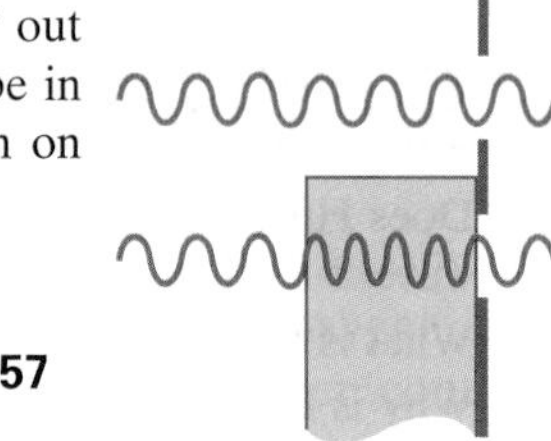

FIGURE 24–57 Problem 9.

10. (II) In a double-slit experiment, the third-order maximum for light of wavelength 500 nm is located 12 mm from the central bright spot on a screen 1.6 m from the slits. Light of wavelength 650 nm is then projected through the same slits. How far from the central bright spot will the second-order maximum of this light be located?

11. (II) Two narrow slits separated by 1.0 mm are illuminated by 544 nm light. Find the distance between adjacent bright fringes on a screen 5.0 m from the slits.

12. (III) Light of wavelength 480 nm in air falls on two slits 6.00×10^{-2} mm apart. The slits are immersed in water, as is a viewing screen 40.0 cm away. How far apart are the fringes on the screen?

13. (III) A very thin sheet of plastic ($n = 1.60$) covers one slit of a double-slit apparatus illuminated by 640-nm light. The center point on the screen, instead of being a maximum, is dark. What is the (minimum) thickness of the plastic?

24–4 Dispersion

14. (I) By what percent, approximately, does the speed of red light (700 nm) exceed that of violet light (400 nm) in silicate flint glass? (See Fig. 24–14.)

15. (II) A light beam strikes a piece of glass at a 60.00° incident angle. The beam contains two wavelengths, 450.0 nm and 700.0 nm, for which the index of refraction of the glass is 1.4820 and 1.4742, respectively. What is the angle between the two refracted beams?

16. (III) A parallel beam of light containing two wavelengths, $\lambda_1 = 450\,\text{nm}$ and $\lambda_2 = 650\,\text{nm}$, enters the silicate flint glass of an equilateral prism as shown in Fig. 24–58. At what angle does each beam leave the prism (give angle with normal to the face)?

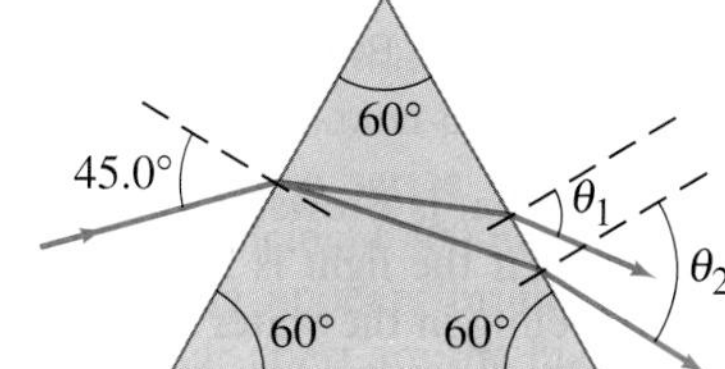

FIGURE 24–58 Problems 16 and 87.

24–5 Single-Slit Diffraction

17. (I) If 580-nm light falls on a slit 0.0440 mm wide, what is the full angular width of the central diffraction peak?
18. (I) Monochromatic light falls on a slit that is 2.60×10^{-3} mm wide. If the angle between the first dark fringes on either side of the central maximum is 35.0° (dark fringe to dark fringe), what is the wavelength of the light used?
19. (II) Light of wavelength 520 nm falls on a slit that is 3.20×10^{-3} mm wide. Estimate how far the first brightish diffraction fringe is from the strong central maximum if the screen is 10.0 m away.
20. (II) A single slit 1.0 mm wide is illuminated by 450-nm light. What is the width of the central maximum (in cm) in the diffraction pattern on a screen 5.0 m away?
21. (II) Monochromatic light of wavelength 653 nm falls on a slit. If the angle between the first bright fringes on either side of the central maximum is 32°, estimate the slit width.
22. (II) How wide is the central diffraction peak on a screen 2.30 m behind a 0.0348-mm-wide slit illuminated by 589-nm light?
23. (II) When blue light of wavelength 440 nm falls on a single slit, the first dark bands on either side of center are separated by 55.0°. Determine the width of the slit.
24. (II) When violet light of wavelength 415 nm falls on a single slit, it creates a central diffraction peak that is 9.20 cm wide on a screen that is 2.55 m away. How wide is the slit?
25. (II) If a slit diffracts 650-nm light so that the diffraction maximum is 4.0 cm wide on a screen 1.50 m away, what will be the width of the diffraction maximum for light of wavelength 420 nm?
26. (II) For a given wavelength λ, what is the maximum slit width for which there will be no diffraction minima?

24–6 and 24–7 Gratings

27. (I) At what angle will 560-nm light produce a second-order maximum when falling on a grating whose slits are 1.45×10^{-3} cm apart?
28. (I) A 3500-line/cm grating produces a third-order fringe at a 28.0° angle. What wavelength of light is being used?
29. (II) How many lines per centimeter does a grating have if the third-order occurs at an 18.0° angle for 630-nm light?
30. (II) A grating has 8300 lines/cm. How many complete spectral orders can be seen (400 nm to 700 nm) when it is illuminated by white light?
31. (II) The first-order line of 589-nm light falling on a diffraction grating is observed at a 15.5° angle. How far apart are the slits? At what angle will the third order be observed?
32. (II) A diffraction grating has 6.0×10^5 lines/m. Find the angular spread in the second-order spectrum between red light of wavelength 7.0×10^{-7} m and blue light of wavelength 4.5×10^{-7} m.
33. (II) Light falling normally on a 9700-line/cm grating is revealed to contain three lines in the first-order spectrum at angles of 31.2°, 36.4°, and 47.5°. What wavelengths are these?
34. (II) What is the highest spectral order that can be seen if a grating with 6000 lines per cm is illuminated with 633-nm laser light? Assume normal incidence.
35. (II) Two (and only two) full spectral orders can be seen on either side of the central maximum when white light is sent through a diffraction grating. What is the maximum number of lines per cm for the grating?
36. (II) White light containing wavelengths from 410 nm to 750 nm falls on a grating with 8500 lines/cm. How wide is the first-order spectrum on a screen 2.30 m away?
37. (II) A He-Ne gas laser which produces monochromatic light of a known wavelength $\lambda = 6.328 \times 10^{-7}$ m is used to calibrate a reflection grating in a spectroscope. The first-order diffraction line is found at an angle of 21.5° to the incident beam. How many lines per meter are there on the grating?
38. (II) Two first-order spectrum lines are measured by a 9500-line/cm spectroscope at angles, on each side of center, of +26°38′, +41°08′ and −26°48′, −41°19′. What are the wavelengths?

24–8 Thin-Film Interference

39. (I) If a soap bubble is 120 nm thick, what wavelength is most strongly reflected at the center of the outer surface when illuminated normally by white light? Assume that $n = 1.34$.
40. (I) How far apart are the dark fringes in Example 24–8 if the glass plates are each 26.5 cm long?
41. (II) What is the smallest thickness of a soap film ($n = 1.42$) that would appear black if illuminated with 480-nm light? Assume there is air on both sides of the soap film.
42. (II) A lens appears greenish yellow ($\lambda = 570$ nm is strongest) when white light reflects from it. What minimum thickness of coating ($n = 1.25$) do you think is used on such a glass ($n = 1.52$) lens, and why?
43. (II) A total of 31 bright and 31 dark Newton's rings (not counting the dark spot at the center) are observed when 550-nm light falls normally on a planoconvex lens resting on a flat glass surface (Fig. 24–31). How much thicker is the center than the edges?
44. (II) A fine metal foil separates one end of two pieces of optically flat glass, as in Fig. 24–33. When light of wavelength 670 nm is incident normally, 28 dark lines are observed (with one at each end). How thick is the foil?
45. (II) How thick (minimum) should the air layer be between two flat glass surfaces if the glass is to appear bright when 450-nm light is incident normally? What if the glass is to appear dark?
46. (II) A piece of material, suspected of being a stolen diamond ($n = 2.42$), is submerged in oil of refractive index 1.43 and illuminated by unpolarized light. It is found that the reflected light is completely polarized at an angle of 59°. Is it diamond?
47. (III) A thin film of alcohol ($n = 1.36$) lies on a flat glass plate ($n = 1.51$). When monochromatic light, whose wavelength can be changed, is incident normally, the reflected light is a minimum for $\lambda = 512$ nm and a maximum for $\lambda = 640$ nm. What is the minimum thickness of the film?

CONCEPTUAL EXAMPLE 25–2 **Shutter speed.** To improve the depth of field, you "stop down" your camera lens by two f-stops from $f/4$ to $f/8$. What should you do to the shutter speed to maintain the same exposure?

RESPONSE The amount of light admitted by the lens is proportional to the area of the lens opening. Reducing the lens opening by two f-stops reduces the diameter by a factor of 2, and the area by a factor of 4. To maintain the same exposure, the shutter must be open four times as long. If the shutter speed had been $\frac{1}{500}$ s, you would have to increase it to $\frac{1}{125}$ s.

Picture Sharpness

The sharpness of a picture depends not only on accurate focusing, but also on the graininess of the film, or the number of pixels for a digital camera. Fine-grained films are "slower," meaning they require longer exposures for a given light level.

The quality of the lens strongly affects the image quality, and we discuss lens resolution and diffraction effects in Sections 25–6 and 25–7. The sharpness, or *resolution*, of a lens is often given as so many lines per millimeter, measured by photographing a standard set of parallel lines on fine-grain film. The minimum spacing of distinguishable lines gives the resolution; 50 lines/mm is reasonable, 100 lines/mm is very good.

Pixels and resolution

EXAMPLE 25–3 **Pixels and resolution.** A high-quality 6-MP (6-megapixel) digital camera offers a maximum resolution of 2000×3000 pixels on a 16-mm $\times$ 24-mm CCD sensor. How sharp should the lens be to make use of this resolution?

APPROACH We find the number of pixels per millimeter and require the lens to be at least that good.

SOLUTION We can either take the image height (2000 pixels in 16 mm) or the width (3000 pixels in 24 mm):

$$\frac{3000 \text{ pixels}}{24 \text{ mm}} = 125 \text{ pixels/mm}.$$

We would want the lens to be able to resolve at least 125 lines/mm as well. If it can't, we could use fewer pixels and less memory.

NOTE Increasing lens resolution is a tougher problem today than is squeezing more pixels on a CCD.

When is a photo sharp?

EXAMPLE 25–4 **Blown-up photograph.** An enlarged photograph looks sharp at normal viewing distances if the dots or lines are resolved to about 10 dots/mm. Would an 8×10-inch enlargement of a photo taken by the camera in Example 25–3 seem sharp? To what maximum size could you enlarge this 2000×3000-pixel image?

APPROACH We assume the image is 2000×3000 pixels on a 16×24-mm CCD as in Example 25–3, or 125 pixels/mm. We make an enlarged photo $8 \times 10 \text{ in.} = 20 \text{ cm} \times 25 \text{ cm}$.

SOLUTION The short side of the CCD is $16 \text{ mm} = 1.6 \text{ cm}$ long, and that side of the photograph is 8 inches or 20 cm. Thus the enlargement is by a factor of $20 \text{ cm}/1.6 \text{ cm} = 12.5\times$ (or $25 \text{ cm}/2.4 \text{ cm} \approx 10\times$). To fill the 8×10-in. paper, we assume the enlargement is $12.5\times$. The pixels are thus enlarged $12.5\times$; so the pixel count of 125/mm on the CCD becomes 10/mm on the print, which is just about the maximum possible for a sharp photograph. If you feel 7 dots/mm is good enough, you can enlarge to maybe 11×14 inches.

Telephotos and Wide-angles

Camera lenses are categorized into normal, telephoto, and wide angle, according to focal length and film size. A **normal lens** covers the film with a field of view that corresponds approximately to that of normal vision. A normal lens for 35-mm film has a focal length in the vicinity of 50 mm.† **Telephoto lenses** act like telescopes to magnify images. They have longer focal lengths than a normal lens: as we saw in Chapter 23 (Eq. 23–9), the height of the image for a given object distance is proportional to the image distance, and the image distance will be greater for a lens with longer focal length. For distant objects, the image height is very nearly proportional to the focal length. Thus a 200-mm telephoto lens for use with a 35-mm camera gives a 4× magnification over the normal 50-mm lens. A **wide-angle lens** has a shorter focal length than normal: a wider field of view is included, and objects appear smaller. A **zoom lens** is one whose focal length can be changed so that you seem to zoom up to, or away from, the subject as you change the focal length.

Telephoto and wide-angle lenses

Digital cameras may have an "optical zoom" meaning the lens can change focal length and maintain resolution. But an "electronic" or "digital zoom" just enlarges the dots (pixels) with loss of sharpness.

Different types of viewing systems are common in cameras today. In many cameras, you view through a small window just above the lens as in Fig. 25–1. In a **single-lens reflex** camera (SLR), you actually view through the lens with the use of prisms and mirrors (Fig. 25–7). A mirror hangs at a 45° angle behind the lens and flips up out of the way just before the shutter opens. SLRs have the great advantage that you can see almost exactly what you will get on film. This is also true of the LCD display on a digital camera if it is carefully constructed.

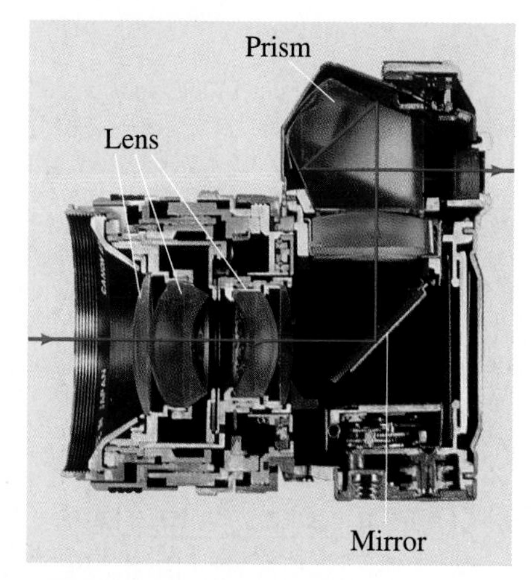

FIGURE 25–7 Single-lens reflex (SLR) camera, showing how the image is viewed through the lens with the help of a movable mirror and prism.

25–2 The Human Eye; Corrective Lenses

The eye

The human eye resembles a camera in its basic structure (Fig. 25–8), but is far more sophisticated. The interior of the eye is filled with a transparent gel-like substance called the *vitreous humor* with index of refraction $n = 1.337$. Light enters this enclosed volume through the cornea and lens. Between the cornea and lens is a watery fluid, the aqueous humor (*aqua* is "water" in Latin) with $n = 1.336$. A diaphragm, called the **iris** (the colored part of your eye) adjusts automatically to control the amount of light entering the eye, similar to a camera. The hole in the iris through which light passes (the **pupil**) is black because no light is reflected from it (it's a hole), and very little light is reflected back out from the interior of the eye. The **retina**, which plays the role of the film or sensor in a camera, is on the curved rear surface of the eye. The retina consists of a complex array of nerves and receptors known as *rods* and *cones* which act to change light energy into electrical signals that travel along the nerves. The reconstruction of the image from all these tiny receptors is done mainly in the brain, although some analysis may also be done in the complex interconnected nerve network at the retina itself. At the center of the retina is a small area called the **fovea**, about 0.25 mm in diameter, where the cones are very closely packed and the sharpest image and best color discrimination are found.

Anatomy of the eye

FIGURE 25–8 Diagram of a human eye.

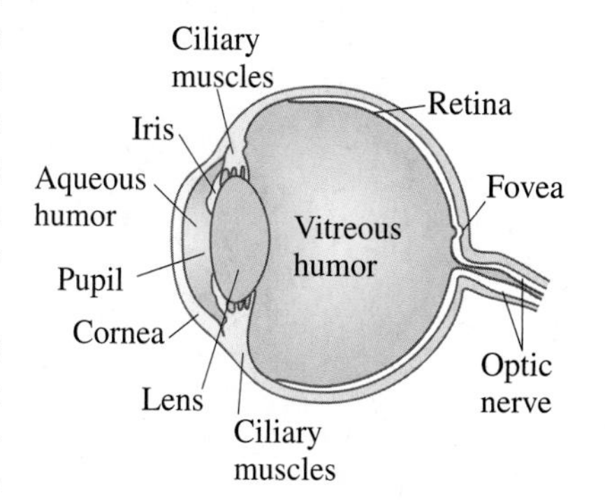

Unlike a camera, the eye contains no shutter. The equivalent operation is carried out by the nervous system, which analyzes the signals to form images at the rate of about 30 per second. This can be compared to motion picture or television cameras, which operate by taking a series of still pictures at a rate of 24 (movies) or 30 (U.S. television) per second. Their rapid projection on the screen gives the appearance of motion.

Focusing the eye

The lens of the eye ($n = 1.386$ to 1.406) does little of the bending of the light rays. Most of the refraction is done at the front surface of the **cornea** ($n = 1.376$) at its interface with air ($n = 1.0$). The lens acts as a fine adjustment for focusing at different distances. This is accomplished by the ciliary muscles (Fig. 25–8), which change the curvature of the lens so that its focal length is changed.

† A "35-mm camera" uses film that is 35 mm wide; that 35 mm is not to be confused with a focal length.

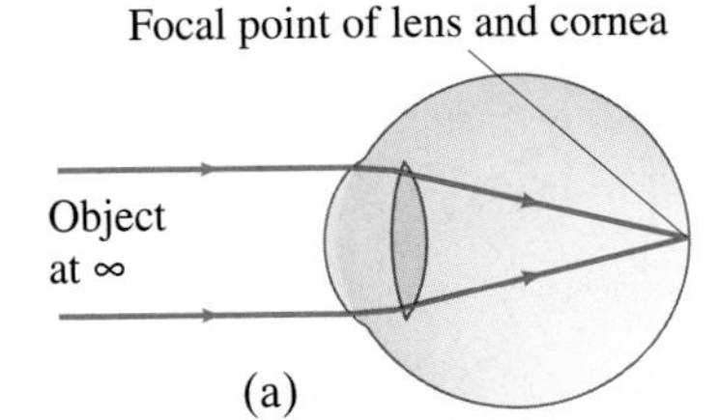

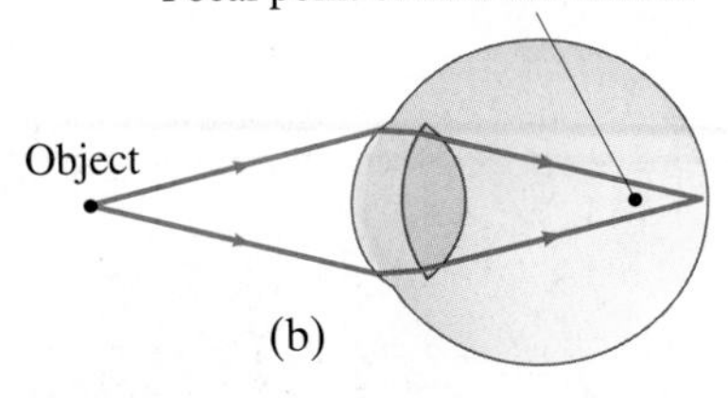

FIGURE 25–9 Accommodation by a normal eye: (a) lens relaxed, focused at infinity; (b) lens thickened, focused on a nearby object.

To focus on a distant object, the ciliary muscles of the eye are relaxed and the lens is thin, as shown in Fig. 25–9a, and parallel rays focus at the focal point (on the retina). To focus on a nearby object, the muscles contract, causing the center of the lens to thicken, Fig. 25–9b, thus shortening the focal length so that images of nearby objects can be focused on the retina, behind the new focal point. This focusing adjustment is called **accommodation**.

The closest distance at which the eye can focus clearly is called the **near point** of the eye. For young adults it is typically 25 cm, although younger children can often focus on objects as close as 10 cm. As people grow older, the ability to accommodate is reduced and the near point increases. A given person's **far point** is the farthest distance at which an object can be seen clearly. For some purposes it is useful to speak of a **normal eye** (a sort of average over the population), defined as an eye having a near point of 25 cm and a far point of infinity. To check your own near point, place this book close to your eye and slowly move it away until the type is sharp.

The "normal" eye is sort of an ideal. Many people have eyes that do not accommodate within the "normal" range of 25 cm to infinity, or have some other defect. Two common defects are nearsightedness and farsightedness. Both can be corrected to a large extent with lenses—either eyeglasses or contact lenses.

Corrective lenses

Nearsightedness

In **nearsightedness**, or *myopia*, the eye can focus only on nearby objects. The far point is not infinity but some shorter distance, so distant objects are not seen clearly. It is usually caused by an eyeball that is too long, although sometimes it is the curvature of the cornea that is too great. In either case, images of distant objects are focused in front of the retina. A diverging lens, because it causes parallel rays to diverge, allows the rays to be focused at the retina (Fig. 25–10a) and thus corrects this defect.

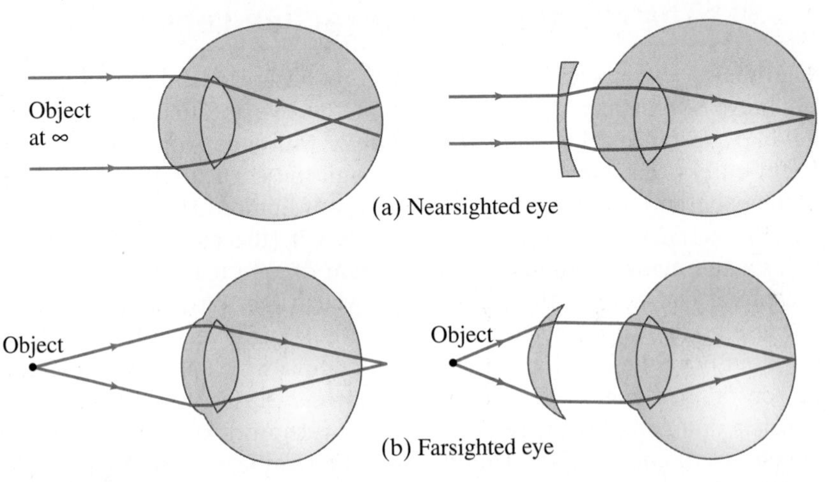

FIGURE 25–10 Correcting eye defects with lenses: (a) a nearsighted eye, which cannot focus clearly on distant objects, can be corrected by use of a diverging lens; (b) a farsighted eye, which cannot focus clearly on nearby objects, can be corrected by use of a converging lens.

Farsightedness

In **farsightedness**, or *hyperopia*, the eye cannot focus on nearby objects. Although distant objects are usually seen clearly, the near point is somewhat greater than the "normal" 25 cm, which makes reading difficult. This defect is caused by an eyeball that is too short or (less often) by a cornea that is not sufficiently curved. It is corrected by a converging lens, Fig. 25–10b. Similar to hyperopia is *presbyopia*, which refers to the lessening ability of the eye to accommodate as one ages, and the near point moves out. Converging lenses also compensate for this.

Astigmatism

Astigmatism is usually caused by an out-of-round cornea or lens so that point objects are focused as short lines, which blurs the image. It is as if the cornea were spherical with a cylindrical section superimposed. As shown in Fig. 25–11, a cylindrical lens focuses a point into a line parallel to its axis. An astigmatic eye may focus rays in one plane, such as the vertical plane, at a shorter distance than it does for rays in a horizontal plane. Astigmatism is corrected with the use of a compensating cylindrical lens. Lenses for eyes that are nearsighted or farsighted as well as astigmatic are ground with superimposed spherical and cylindrical surfaces, so that the radius of curvature of the correcting lens is different in different planes.

FIGURE 25–11 A cylindrical lens forms a line image of a point object because it is converging in one plane only.

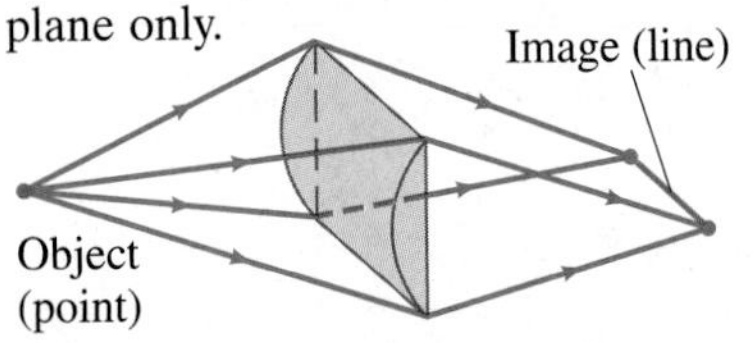

EXAMPLE 25–5 Farsighted eye. Sue is farsighted with a near point of 100 cm. Reading glasses must have what lens power so that she can read a newspaper at a distance of 25 cm? Assume the lens is very close to the eye.

APPROACH When the object is placed 25 cm from the lens, we want the image to be 100 cm away on the *same* side of the lens (so the eye can focus it), and so the image is virtual, Fig. 25–12, and $d_i = -100\text{ cm}$ will be negative. We use the thin lens equation (Eq. 23–8) to determine the needed focal length. Optometrists' prescriptions specify the power ($P = 1/f$, Eq. 23–7) given in diopters $(1\text{ D} = 1\text{ m}^{-1})$.

SOLUTION Given that $d_o = 25\text{ cm}$ and $d_i = -100\text{ cm}$, the thin lens equation gives

$$\frac{1}{f} = \frac{1}{d_o} + \frac{1}{d_i} = \frac{1}{25\text{ cm}} + \frac{1}{-100\text{ cm}} = \frac{4-1}{100\text{ cm}} = \frac{1}{33\text{ cm}}.$$

So $f = 33\text{ cm} = 0.33\text{ m}$. The power P of the lens is $P = 1/f = +3.0\text{ D}$. The plus sign indicates that it is a converging lens.

NOTE We chose the image position to be where the eye can actually focus. The lens needs to put the image there, given the desired placement of the object (newspaper).

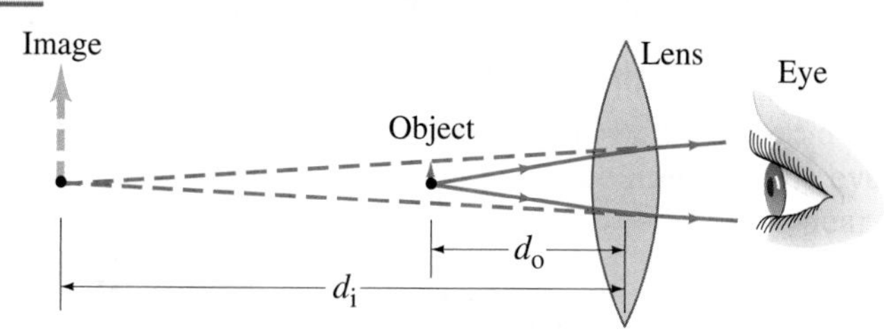

FIGURE 25–12 Lens of reading glasses (Example 25–5).

EXAMPLE 25–6 Nearsighted eye. A nearsighted eye has near and far points of 12 cm and 17 cm, respectively. (*a*) What lens power is needed for this person to see distant objects clearly, and (*b*) what then will be the near point? Assume that the lens is 2.0 cm from the eye (typical for eyeglasses).

APPROACH For a distant object $(d_o = \infty)$, the lens must put the image at the far point of the eye as shown in Fig. 25–13a, 17 cm in front of the eye. We can use the thin lens equation to find the focal length of the lens, and from this its lens power. The new near point (as shown in Fig. 25–13b) can be calculated for the lens by again using the thin lens equation.

SOLUTION (*a*) For an object at infinity $(d_o = \infty)$, the image must be in front of the lens 17 cm from the eye or $(17\text{ cm} - 2\text{ cm}) = 15\text{ cm}$ from the lens; hence $d_i = -15\text{ cm}$. We use the thin lens equation to solve for the focal length of the needed lens:

$$\frac{1}{f} = \frac{1}{d_o} + \frac{1}{d_i} = \frac{1}{\infty} + \frac{1}{-15\text{ cm}} = -\frac{1}{15\text{ cm}}.$$

So $f = -15\text{ cm} = -0.15\text{ m}$ or $P = 1/f = -6.7\text{ D}$. The minus sign indicates that it must be a diverging lens for the myopic eye.

(*b*) The near point when glasses are worn is where an object is placed (d_o) so that the lens forms an image at the "near point of the naked eye," namely 12 cm from the eye. That image point is $(12\text{ cm} - 2\text{ cm}) = 10\text{ cm}$ in front of the lens, so $d_i = -0.10\text{ m}$ and the thin lens equation gives

$$\frac{1}{d_o} = \frac{1}{f} - \frac{1}{d_i} = -\frac{1}{0.15\text{ m}} + \frac{1}{0.10\text{ m}} = \frac{-2+3}{0.30\text{ m}} = \frac{1}{0.30\text{ m}}.$$

So $d_o = 30\text{ cm}$, which means the near point when the person is wearing glasses is 30 cm in front of the lens.

FIGURE 25–13 Example 25–6.

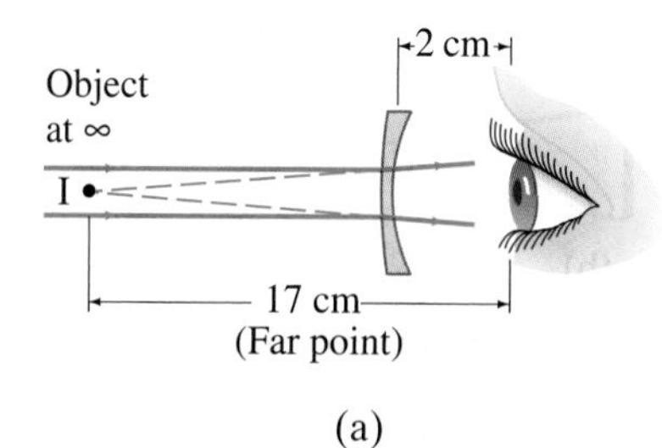

(a)

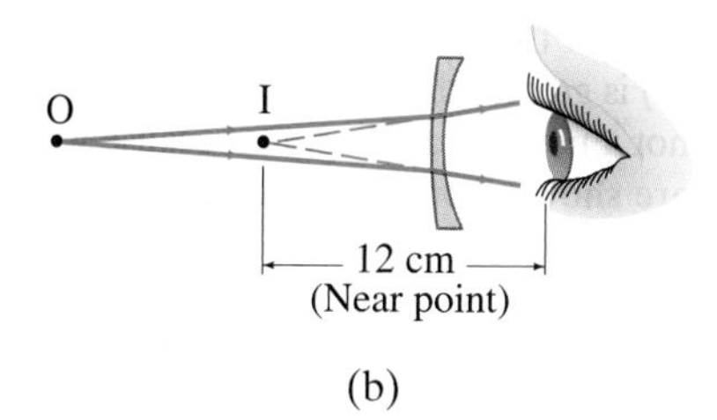

(b)

Contact lenses

Suppose contact lenses are used to correct the eye in Example 25–6. Since contacts are placed directly on the cornea, we would not subtract out the 2.0 cm for the image distances. That is, for distant objects $d_i = f = -17\text{ cm}$, so $P = 1/f = -5.9\text{ D}$. The new near point would be 41 cm. Thus we see that a

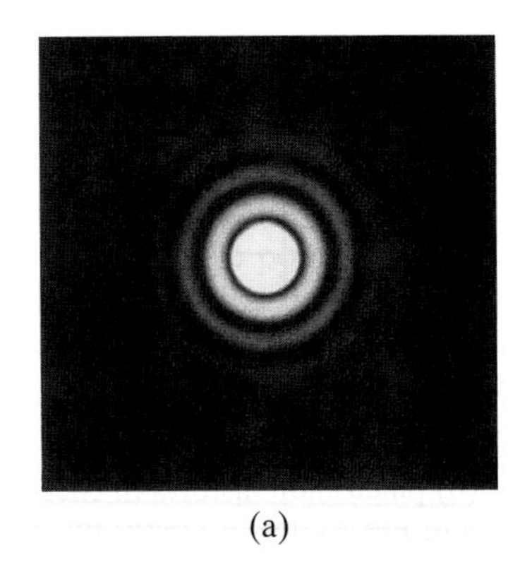

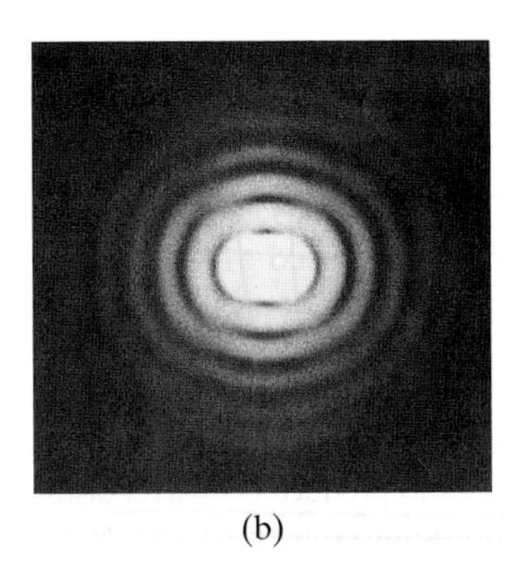

(a) (b)

FIGURE 25–28 Photographs of images (greatly magnified) formed by a lens, showing the diffraction pattern of an image for: (a) a single point object; (b) two point objects whose images are barely resolved.

In the analysis that follows, we assume that the lens is free of aberrations, so we can concentrate on diffraction effects and how much they limit the resolution of a lens. In Fig. 24–21 we saw that the diffraction pattern produced by light passing through a rectangular slit has a central maximum in which most of the light falls. This central peak falls to a minimum on either side of its center at an angle θ given by

$$\sin\theta = \frac{\lambda}{D}$$

(this is Eq. 24–3a), where D is the slit width and λ the wavelength of light used. θ is the angular half-width of the central maximum, and for small angles can be written

$$\theta \approx \sin\theta = \frac{\lambda}{D}.$$

There are also low-intensity fringes beyond. For a lens, or any circular hole, the image of a point object will consist of a *circular* central peak (called the *diffraction spot* or *Airy disk*) surrounded by faint circular fringes, as shown in Fig. 25–28a. The central maximum has an angular half width given by

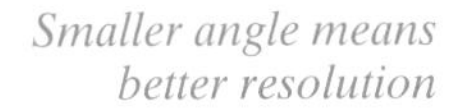
Smaller angle means better resolution

$$\theta = \frac{1.22\lambda}{D},$$

where D is the diameter of the circular opening.

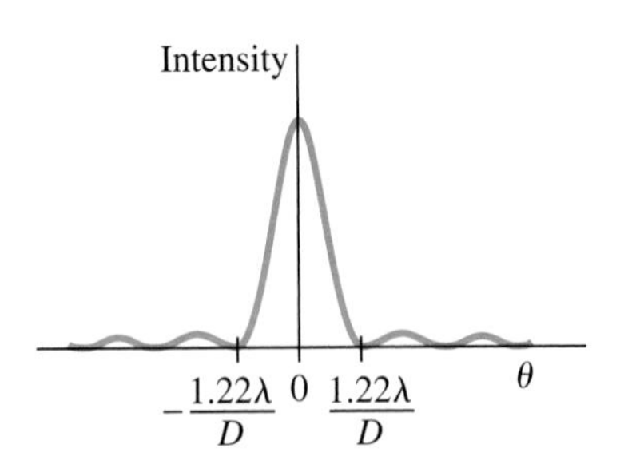

FIGURE 25–29 Intensity of light across the diffraction pattern of a circular hole.

This formula differs from that for a slit (Eq. 24–3) by the factor 1.22. This factor appears because the width of a circular hole is not uniform (like a rectangular slit) but varies from its diameter D to zero. A mathematical analysis shows that the "average" width is $D/1.22$. Hence we get the equation above rather than Eq. 24–3. The intensity of light in the diffraction pattern of light from a point source passing through a circular opening is shown in Fig. 25–29. The image for a non-point source is a superposition of such patterns. For most purposes we need consider only the central spot, since the concentric rings are so much dimmer.

If two point objects are very close, the diffraction patterns of their images will overlap as shown in Fig. 25–28b. As the objects are moved closer, a separation is reached where you can't tell if there are two overlapping images or a single image. The separation at which this happens may be judged differently by different observers. However, a generally accepted criterion is that proposed by Lord Rayleigh (1842–1919). This **Rayleigh criterion** states that *two images are just resolvable when the center of the diffraction disk of one image is directly over the first minimum in the diffraction pattern of the other*. This is shown in Fig. 25–30. Since the first minimum is at an angle $\theta = 1.22\lambda/D$ from the central maximum, Fig. 25–30 shows that two objects can be considered *just resolvable* if they are separated by at least the angle θ given by

Rayleigh criterion (resolution limit)

$$\theta = \frac{1.22\lambda}{D}. \qquad [\theta \text{ in radians}] \quad \textbf{(25–7)}$$

This is the limit on resolution set by the wave nature of light due to diffraction. A smaller angle means better resolution: you can make out closer objects. We see from Eq. 25–7 that using a shorter wavelength λ can increase resolution.

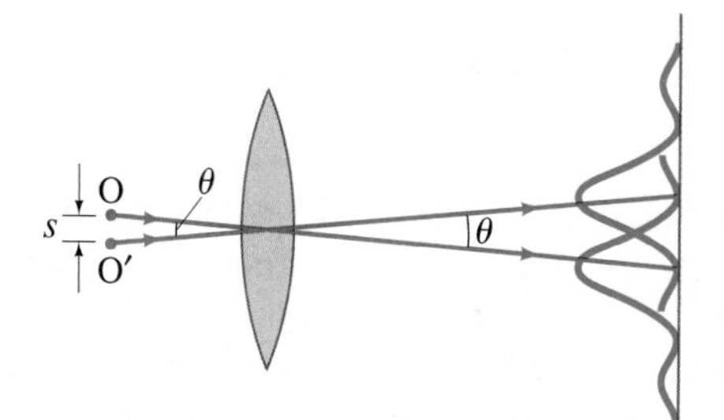

FIGURE 25–30 The *Rayleigh criterion.* Two images are just resolvable when the center of the diffraction peak of one is directly over the first minimum in the diffraction pattern of the other. The two point objects O and O′ subtend an angle θ at the lens; only one ray (it passes through the center of the lens) is drawn for each object, to indicate the center of the diffraction pattern of its image.

EXAMPLE 25–10 Hubble Space Telescope. The Hubble Space Telescope (HST) is a reflecting telescope that was placed in orbit above the Earth's atmosphere, so its resolution would not be limited by turbulence in the atmosphere (Fig. 25–31). Its objective diameter is 2.4 m. For visible light, say $\lambda = 550\,\text{nm}$, estimate the improvement in resolution the Hubble offers over Earth-bound telescopes, which are limited in resolution by movement of the Earth's atmosphere to about half an arc second. (Each degree is divided into 60 minutes each containing 60 seconds, so $1° = 3600$ arc seconds.)

APPROACH Angular resolution for the Hubble is given (in radians) by Eq. 25–7. The resolution for Earth telescopes is given, and we first convert it to radians so we can compare.

SOLUTION Earth-bound telescopes are limited to an angular resolution of

$$\theta = \tfrac{1}{2}\left(\frac{1}{3600}\right)^{\circ}\left(\frac{2\pi\,\text{rad}}{360°}\right) = 2.4 \times 10^{-6}\,\text{rad}.$$

The Hubble, on the other hand, is limited by diffraction (Eq. 25–7) which for $\lambda = 550\,\text{nm}$ is

$$\theta = \frac{1.22\lambda}{D} = \frac{1.22(550 \times 10^{-9}\,\text{m})}{2.4\,\text{m}} = 2.8 \times 10^{-7}\,\text{rad},$$

thus giving almost ten times better resolution $(2.4\times10^{-6}\,\text{rad}/2.8\times10^{-7}\,\text{rad} \approx 9\times)$.

NOTE The Hubble can also observe radiation in the near ultraviolet (wavelengths as small as 115 nm) and infrared (wavelengths as long as 1 mm), which are ranges of the spectrum blocked by the atmosphere. The sensor is a CCD, as in a camera (see Section 25–1), with a pixel count of 16 MP.

FIGURE 25–31 Hubble Space Telescope, with Earth in the background. The flat orange panels are solar cells that collect energy from the Sun.

EXAMPLE 25–11 ESTIMATE Eye resolution. You are in an airplane at an altitude of 10,000 m. If you look down at the ground, estimate the minimum separation s between objects that you could distinguish. Consider only diffraction, and assume your pupil is about 3.0 mm in diameter and $\lambda = 550\,\text{nm}$.

PHYSICS APPLIED
How well the eye can see

APPROACH We use the Rayleigh criterion, Eq. 25–7, to estimate θ. The separation s of objects equals their distance away, $L = 10^4\,\text{m}$, times θ (in radians) as θ is small, so $s = L\theta$.

SOLUTION In Eq. 25–7, we set $D = 3.0\,\text{mm}$ for the opening of the eye:

$$s = L\theta = L\frac{1.22\lambda}{D}$$
$$= \frac{(10^4\,\text{m})(1.22)(550 \times 10^{-9}\,\text{m})}{3.0 \times 10^{-3}\,\text{m}} = 2.2\,\text{m}.$$

EXERCISE D Someone claims a spy satellite camera can see 3-cm-high newspaper headlines from an altitude of 100 km. If diffraction were the only limitation ($\lambda = 550\,\text{nm}$), use Eq. 25–7 to determine what diameter lens the camera would have.

25–8 Resolution of Telescopes and Microscopes; the λ Limit

You might think that a microscope or telescope could be designed to produce any desired magnification, depending on the choice of focal lengths and quality of the lenses. But this is not possible, because of diffraction. An increase in magnification above a certain point merely results in magnification of the diffraction patterns. This can be highly misleading since we might think we are seeing details of an object when we are really seeing details of the diffraction pattern. To examine this problem, we apply the Rayleigh criterion: two objects (or two nearby points on one object) are just resolvable if they are separated by an angle θ (Fig. 25–30) given by Eq. 25–7:

$$\theta = \frac{1.22\lambda}{D}.$$

This formula is valid for either a microscope or a telescope, where D is the diameter of the objective lens. For a telescope, the resolution is specified by stating θ as given by this equation.†

For a microscope, it is more convenient to specify the actual distance, s, between two points that are just barely resolvable: see Fig. 25–30. Since objects are normally placed near the focal point of the microscope objective, the angle subtended by two objects is $\theta = s/f$, or $s = f\theta$. If we combine this with Eq. 25–7, we obtain for the **resolving power (RP)** of a microscope

Resolving power

$$\text{RP} = s = f\theta = \frac{1.22\lambda f}{D}, \qquad \textbf{(25–8)}$$

where f is the objective lens' focal length (not frequency). This distance s is called the resolving power of the lens because it is the minimum separation of two object points that can just be resolved, assuming the highest quality lens since this limit is imposed by the wave nature of light. A smaller RP means better resolution, better detail.

FIGURE 25–32 The 300-meter radiotelescope in Arecibo, Puerto Rico, uses radio waves (Fig. 22–8) instead of visible light.

EXAMPLE 25–12 Telescope resolution (radio wave vs. visible light). What is the theoretical minimum angular separation of two stars that can just be resolved by (*a*) the 200-inch telescope on Palomar Mountain (Fig. 25–21c); and (*b*) the Arecibo radiotelescope (Fig. 25–32), whose diameter is 300 m and whose radius of curvature is also 300 m. Assume $\lambda = 550\,\text{nm}$ for the visible-light telescope in part (*a*), and $\lambda = 4\,\text{cm}$ (the shortest wavelength at which the radiotelescope has been operated) in part (*b*).

APPROACH We apply the Rayleigh criterion (Eq. 25–7) for each telescope.

SOLUTION (*a*) Since $D = 200\,\text{in.} = 5.1\,\text{m}$, we have from Eq. 25–7 that

$$\theta = \frac{1.22\lambda}{D} = \frac{(1.22)(5.50 \times 10^{-7}\,\text{m})}{(5.1\,\text{m})} = 1.3 \times 10^{-7}\,\text{rad},$$

or 0.75×10^{-5} deg. (Note that this is equivalent to resolving two points less than 1 cm apart from a distance of 100 km!)

(*b*) For radio waves with $\lambda = 0.04\,\text{m}$, the resolution is

$$\theta = \frac{(1.22)(0.04\,\text{m})}{(300\,\text{m})} = 1.6 \times 10^{-4}\,\text{rad}.$$

The resolution is less because the wavelength is so much larger, but the larger objective is a plus.

† Earth-bound telescopes with large-diameter objectives are usually limited not by diffraction but by other effects such as turbulence in the atmosphere. The resolution of a high-quality microscope, on the other hand, normally *is* limited by diffraction; microscope objectives are complex compound lenses containing many elements of small diameter (since f is small), thus reducing aberrations.

NOTE In both cases, we determined the limit set by diffraction. The resolution for a visible-light Earth-bound telescope is not this good because of aberrations and, more importantly, turbulence in the atmosphere. In fact, large-diameter objectives are not justified by increased resolution, but by their greater light-gathering ability—they allow more light in, so fainter objects can be seen. Radiotelescopes are not hindered by atmospheric turbulence, and the resolution found in (*b*) is a good estimate.

Diffraction sets an ultimate limit on the detail that can be seen on any object. In Eq. 25–8 for resolving power, the focal length of a lens cannot practically be made less than (approximately) the radius of the lens, and even that is very difficult (see the lensmaker's equation, Eq. 23–10). In this best case, Eq. 25–8 gives, with $f \approx D/2$,

$$\mathrm{RP} \approx \frac{\lambda}{2}. \qquad \textbf{(25–9)}$$

Thus we can say, to within a factor of 2 or so, that

it is not possible to resolve detail of objects smaller than the wavelength of the radiation being used.

Resolution limited to λ

This is an important and useful rule of thumb.

Compound lenses in microscopes are now designed so well that the actual limit on resolution is often set by diffraction—that is, by the wavelength of the light used. To obtain greater detail, one must use radiation of shorter wavelength. The use of UV radiation can increase the resolution by a factor of perhaps 2. Far more important, however, was the discovery in the early twentieth century that electrons have wave properties (Chapter 27) and that their wavelengths can be very small. The wave nature of electrons is utilized in the electron microscope (Section 27–9), which can magnify 100 to 1000 times more than a visible-light microscope because of the much shorter wavelengths. X-rays, too, have very short wavelengths and are often used to study objects in great detail (Section 25–11).

25–9 Resolution of the Human Eye and Useful Magnification

The resolution of the human eye is limited by several factors, all of roughly the same order of magnitude. The resolution is best at the fovea, where the cone spacing is smallest, about $3\,\mu\mathrm{m}$ ($= 3000\,\mathrm{nm}$). The diameter of the pupil varies from about 0.1 cm to about 0.8 cm. So for $\lambda = 550\,\mathrm{nm}$ (where the eye's sensitivity is greatest), the diffraction limit is about $\theta \approx 1.22\lambda/D \approx 8 \times 10^{-5}\,\mathrm{rad}$ to $6 \times 10^{-4}\,\mathrm{rad}$. The eye is about 2 cm long, giving a resolving power (Eq. 25–8) of $s \approx (2 \times 10^{-2}\,\mathrm{m})(8 \times 10^{-5}\,\mathrm{rad}) \approx 2\,\mu\mathrm{m}$ at best, to about $15\,\mu\mathrm{m}$ at worst (pupil small). Spherical and chromatic aberration also limit the resolution to about $10\,\mu\mathrm{m}$. The net result is that the eye can resolve objects whose angular separation is about

$$5 \times 10^{-4}\,\mathrm{rad}$$

Best eye resolution

at best. This corresponds to objects separated by 1 cm at a distance of about 20 m.

The typical near point of a human eye is about 25 cm. At this distance, the eye can just resolve objects that are $(25\,\mathrm{cm})(5 \times 10^{-4}\,\mathrm{rad}) \approx 10^{-4}\,\mathrm{m} = \frac{1}{10}\,\mathrm{mm}$ apart. Since the best light microscopes can resolve objects no smaller than about 200 nm at best (Eq. 25–9 for violet light, $\lambda = 400\,\mathrm{nm}$), the useful magnification [= (resolution by naked eye)/(resolution by microscope)] is limited to about

$$\frac{10^{-4}\,\mathrm{m}}{200 \times 10^{-9}\,\mathrm{m}} \approx 500\times.$$

Maximum useful microscope magnification

In practice, magnifications of about 1000× are often used to minimize eyestrain. Any greater magnification would simply make visible the diffraction pattern produced by the microscope objective.

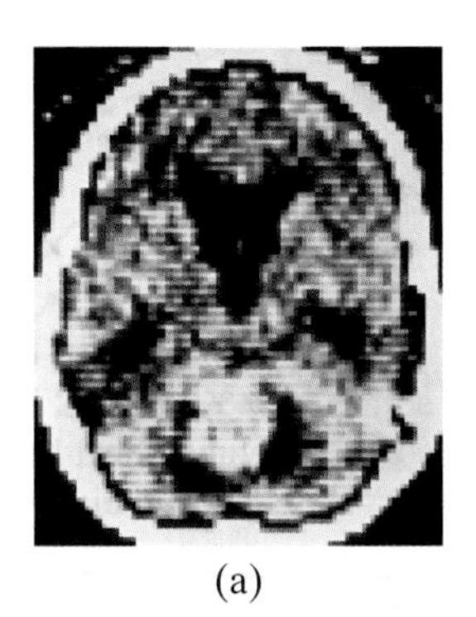

(a)

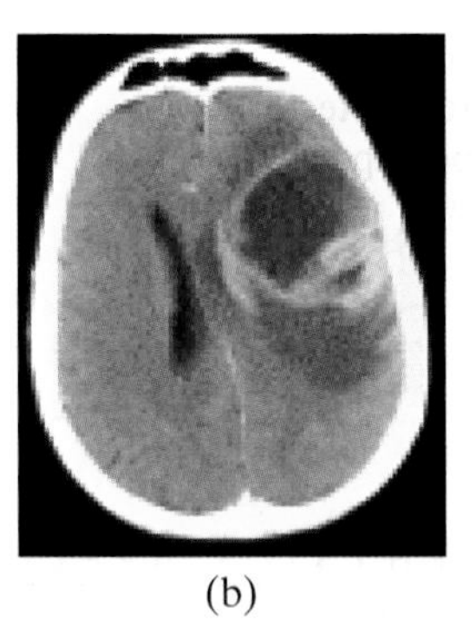

(b)

FIGURE 25–44 Two CT images, with different resolutions, each showing a cross section of a brain. Photo (a) is of low resolution; photo (b), of higher resolution, shows a brain tumor (dark area on the right).

* Image Formation

But how is the image formed? We can think of the slice to be imaged as being divided into many tiny picture elements (or pixels), which could be squares. (See, for example, Fig. 24–49.) For CT, the width of each pixel is chosen according to the width of the detectors and/or the width of the X-ray beams, and this determines the resolution of the image, which might be 1 mm. An X-ray detector measures the intensity of the transmitted beam. Subtracting this value from the intensity of the beam at the source, yields the total absorption (called a "projection") along that beam line. Complicated mathematical techniques are used to analyze all the absorption projections for the huge number of beam scans measured (see the next subsection), obtaining the absorption at each pixel and assigning each a "grayness value" according to how much radiation was absorbed. The image is made up of tiny spots (pixels) of varying shades of gray. Often the amount of absorption is color-coded. The colors in the resulting "false-color" image have nothing to do, however, with the actual color of the object.

Figure 25–44 illustrates what actual CT images look like. It is generally agreed that CT scanning has revolutionized some areas of medicine by providing much less invasive, and/or more accurate, diagnosis.

Computed tomography can also be applied to ultrasound imaging (Section 12–9) and to emissions from radioisotopes and nuclear magnetic resonance (Sections 31–8 and 31–9).

* Tomographic Image Reconstruction

How can the "grayness" of each pixel be determined even though all we can measure is the total absorption along each beam line in the slice? It can be done only by using the many beam scans made at a great many different angles. Suppose the image is to be an array of 100×100 elements for a total of 10^4 pixels. If we have 100 detectors and measure the absorption projections at 100 different angles, then we get 10^4 pieces of information. From this information, an image can be reconstructed, but not precisely. If more angles are measured, the reconstruction of the image can be done more accurately.

Image reconstruction

To suggest how mathematical reconstruction is done, we consider a very simple case using the "iterative" technique ("to iterate" is from the Latin "to repeat"). Suppose our sample slice is divided into the simple 2×2 pixels as shown in Fig. 25–45. The number in each pixel represents the amount of absorption by the material in that area (say, in tenths of a percent): that is, 4 represents twice as much absorption as 2. But we cannot directly measure these values—they are the unknowns we want to solve for. All we can measure are the projections—the total absorption along each beam line—and these are shown in the diagram as the sum of the absorptions for the pixels along each line at four different angles. These projections (given at the tip of each arrow) are what we can measure, and we now want to work back from them to see how close we can get to the true absorption value for each pixel. We start our analysis with each pixel being assigned a zero value, Fig. 25–46a. In the iterative technique, we use the projections to estimate the absorption value in each square, and repeat for each angle. The angle 1 projections are 7 and 13. We divide each of these equally between their two squares: each square in the left column gets $3\frac{1}{2}$ (half of 7), and each square in the right column gets $6\frac{1}{2}$ (half of 13); see Fig. 25–46b. Next we use

FIGURE 25–45 A simple 2×2 image showing true absorption values and measured projections.

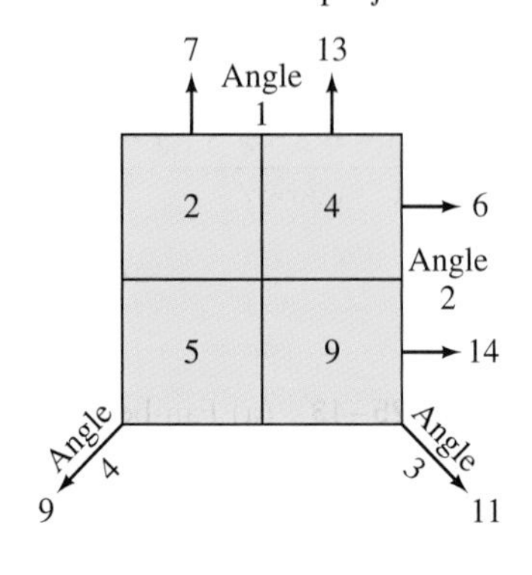

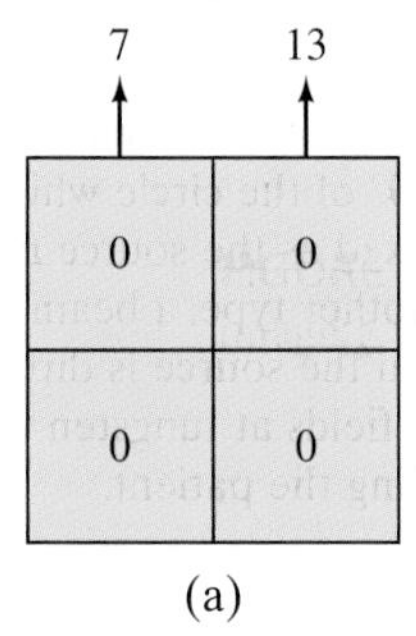

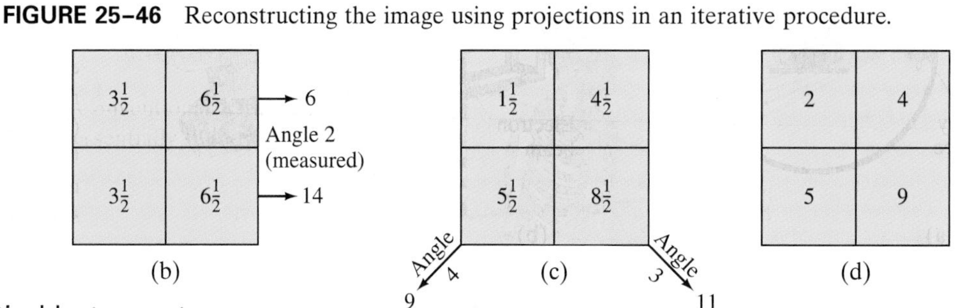

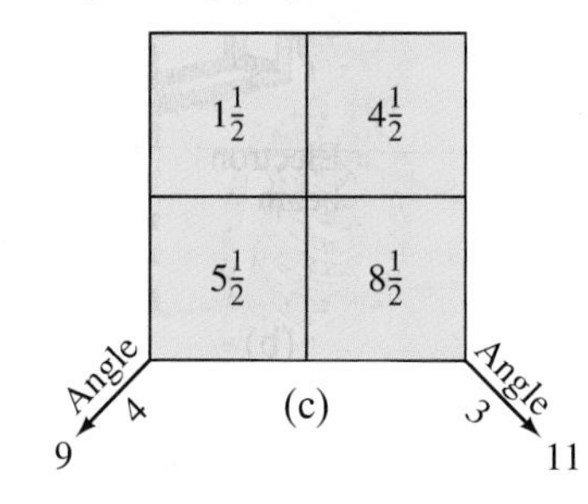

FIGURE 25–46 Reconstructing the image using projections in an iterative procedure.

the projections at angle 2. We calculate the difference between the measured projections at angle 2 (6 and 14) and the projections based on the previous estimate (top row: $3\frac{1}{2} + 6\frac{1}{2} = 10$; same for bottom row). Then we distribute this difference equally to the squares in that row. For the top row, we have

$$3\tfrac{1}{2} + \frac{6 - 10}{2} = 1\tfrac{1}{2} \quad \text{and} \quad 6\tfrac{1}{2} + \frac{6 - 10}{2} = 4\tfrac{1}{2};$$

and for the bottom row,

$$3\tfrac{1}{2} + \frac{14 - 10}{2} = 5\tfrac{1}{2} \quad \text{and} \quad 6\tfrac{1}{2} + \frac{14 - 10}{2} = 8\tfrac{1}{2}.$$

These values are inserted as shown in Fig. 25–46c. Next, the projection at angle 3 gives

$$\text{(upper left)} \quad 1\tfrac{1}{2} + \frac{11 - 10}{2} = 2 \quad \text{and} \quad \text{(lower right)} \quad 8\tfrac{1}{2} + \frac{11 - 10}{2} = 9;$$

and that for angle 4 gives

$$\text{(lower left)} \quad 5\tfrac{1}{2} + \frac{9 - 10}{2} = 5 \quad \text{and} \quad \text{(upper right)} \quad 4\tfrac{1}{2} + \frac{9 - 10}{2} = 4.$$

The result, shown in Fig. 25–46d, corresponds exactly to the true values. (In real situations, the true values are not known, which is why these computer techniques are required.) To obtain these numbers exactly, we used six pieces of information (two each at angles 1 and 2, one each at angles 3 and 4). For the much larger number of pixels used for actual images, exact values are generally not attained. Many iterations may be needed, and the calculation is considered sufficiently precise when the difference between calculated and measured projections is sufficiently small. The above example illustrates the "convergence" of the process: the first iteration (b to c in Fig. 25–46) changed the values by 2, the last iteration (c to d) by only $\frac{1}{2}$.

Summary

A **camera** lens forms an image on film, or on a charge-coupled device in a digital camera, by allowing light in through a shutter. The lens is focused by moving it relative to the film, and its ***f*-stop** (or lens opening) must be adjusted for the brightness of the scene and the chosen shutter speed. The f-stop is defined as the ratio of the focal length to the diameter of the lens opening.

The human **eye** also adjusts for the available light—by opening and closing the iris. It focuses not by moving the lens, but by adjusting the shape of the lens to vary its focal length. The image is formed on the retina, which contains an array of receptors known as rods and cones.

Diverging eye-glass or contact lenses are used to correct the defect of a nearsighted eye, which cannot focus well on distant objects. Converging lenses are used to correct for defects in which the eye cannot focus on close objects.

A **simple magnifier** is a converging lens that forms a virtual image of an object placed at (or within) the focal point. The **angular magnification**, when viewed by a relaxed normal eye, is

$$M = \frac{N}{f}, \qquad \textbf{(25–2a)}$$

where f is the focal length of the lens and N is the near point of the eye (25 cm for a "normal" eye).

An **astronomical telescope** consists of an **objective lens** or mirror, and an **eyepiece** that magnifies the real image formed by the objective. The **magnification** is equal to the ratio of the objective and eyepiece focal lengths, and the image is inverted:

$$M = -\frac{f_o}{f_e}. \qquad \textbf{(25–3)}$$

[*A compound **microscope** also uses objective and eyepiece lenses, and the final image is inverted. The total magnification is the product of the magnifications of the two lenses and is approximately

$$M \approx \frac{Nl}{f_e f_o}, \qquad \textbf{(25–6b)}$$

where l is the distance between the lenses, N is the near point of the eye, and f_o and f_e are the focal lengths of objective and eyepiece, respectively.]

Microscopes, telescopes, and other optical instruments are limited in the formation of sharp images by **lens aberrations**. These include **spherical aberration**, in which rays passing through the edge of a lens are not focused at the same point as those that pass near the center; and **chromatic aberration**, in which different colors are focused at different points. Compound lenses, consisting of several elements, can largely correct for aberrations.

The wave nature of light also limits the sharpness, or **resolution**, of images. Because of diffraction, it is *not possible to discern details smaller than the wavelength* of the radiation being used. This limits the useful magnification of a light microscope to about 500×.

[***X-rays** are a form of electromagnetic radiation of very short wavelength. They are produced when high-speed electrons, accelerated by high voltage in an evacuated tube, strike a glass or metal target.]

[***Computed tomography** (CT or CAT scans) uses many narrow X-ray beams through a section of the body to construct an image of that section.]

A reference frame that moves with constant velocity with respect to an inertial frame is itself also an inertial frame, since Newton's laws hold in it as well. When we say that we observe or make measurements from a certain reference frame, it means that we are at rest in that reference frame.

Relativity principle: the laws of physics are the same in all inertial reference frames

Both Galileo and Newton were aware of what we now call the **relativity principle** applied to mechanics: that *the basic laws of physics are the same in all inertial reference frames*. You may have recognized its validity in everyday life. For example, objects move in the same way in a smoothly moving (constant-velocity) train or airplane as they do on Earth. (This assumes no vibrations or rocking which would make the reference frame noninertial.) When you walk, drink a cup of soup, play pool, or drop a pencil on the floor while traveling in a train, airplane, or ship moving at constant velocity, the bodies move just as they do when you are at rest on Earth. Suppose you are in a car traveling rapidly at constant velocity. If you release a coin from above your head inside the car, how will it fall? It falls straight downward with respect to the car, and hits the floor directly below the point of release, Fig. 26–2a. (If you drop the coin out the car's window, this won't happen because the moving air drags the coin backward relative to the car.) This is just how objects fall on the Earth—straight down—and thus our experiment in the moving car is in accord with the relativity principle.

FIGURE 26–2 A coin is dropped by a person in a moving car. The upper views show the moment of the coin's release, the lower views are a short time later. (a) In the reference frame of the car, the coin falls straight down (and the tree moves to the left). (b) In a reference frame fixed on the Earth, the coin follows a curved (parabolic) path.

CAUTION

Laws are the same, but paths may be different in different reference frames

Note in this example, however, that to an observer on the Earth, the coin follows a curved path, Fig. 26–2b. The actual path followed by the coin is different as viewed from different frames of reference. This does not violate the relativity principle because this principle states that the *laws* of physics are the same in all inertial frames. The same law of gravity, and the same laws of motion, apply in both reference frames. And the acceleration of the coin is the same in both reference frames. The difference in Figs. 26–2a and b is that in the Earth's frame of reference, the coin has an initial velocity (equal to that of the car). The laws of physics therefore predict it will follow a parabolic path like any projectile (Chapter 3). In the car's reference frame, there is no initial velocity, and the laws of physics predict that the coin will fall straight down. The laws are the same in both reference frames, although the specific paths are different.

Galilean–Newtonian relativity involves certain unprovable assumptions that make sense from everyday experience. It is assumed that the lengths of objects are the same in one reference frame as in another, and that time passes at the same rate in different reference frames. In classical mechanics, then, space and time intervals are considered to be **absolute**: their measurement does not change from one reference frame to another. The mass of an object, as well as all forces, are assumed to be unchanged by a change in inertial reference frame.

The position of an object, however, is different when specified in different reference frames, and so is velocity. For example, a person may walk inside a bus toward the front with a speed of 2 m/s. But if the bus moves 10 m/s with respect to the Earth, the person is then moving with a speed of 12 m/s with respect to the Earth. The acceleration of a body, however, is the same in any inertial reference frame according to classical mechanics. This is because the change in velocity, and the time interval, will be the same. For example, the person in the bus may accelerate from 0 to 2 m/s in 1.0 seconds, so $a = 2\ \mathrm{m/s^2}$ in the reference frame of the bus. With respect to the Earth, the acceleration is

Position and velocity are different in different reference frames, but length is the same (classical)

$$\frac{(12\ \mathrm{m/s} - 10\ \mathrm{m/s})}{1.0\ \mathrm{s}} = 2\ \mathrm{m/s^2},$$

which is the same.

Since neither F, m, nor a changes from one inertial frame to another, then Newton's second law, $F = ma$, does not change. Thus Newton's second law satisfies the relativity principle. It is easily shown that the other laws of mechanics also satisfy the relativity principle.

That the laws of mechanics are the same in all inertial reference frames implies that no one inertial frame is special in any sense. We express this important conclusion by saying that **all inertial reference frames are equivalent** for the description of mechanical phenomena. No one inertial reference frame is any better than another. A reference frame fixed to a car or an aircraft traveling at constant velocity is as good as one fixed on the Earth. When you travel smoothly at constant velocity in a car or airplane, it is just as valid to say you are at rest and the Earth is moving as it is to say the reverse. There is no experiment you can do to tell which frame is "really" at rest and which is moving. Thus, there is no way to single out one particular reference frame as being at absolute rest.

All inertial reference frames are equally valid

A complication arose, however, in the last half of the nineteenth century. Maxwell's comprehensive and successful theory of electromagnetism (Chapter 22) predicted that light is an electromagnetic wave. Maxwell's equations gave the velocity of light c as $3.00 \times 10^8\ \mathrm{m/s}$; and this is just what is measured, within experimental error. The question then arose: in what reference frame does light have precisely the value predicted by Maxwell's theory? For it was assumed that light would have a different speed in different frames of reference. For example, if observers were traveling on a rocket ship at a speed of $1.0 \times 10^8\ \mathrm{m/s}$ away from a source of light, we might expect them to measure the speed of the light reaching them to be

$$(3.0 \times 10^8\ \mathrm{m/s}) - (1.0 \times 10^8\ \mathrm{m/s}) = 2.0 \times 10^8\ \mathrm{m/s}.$$

But Maxwell's equations have no provision for relative velocity. They predicted the speed of light to be $c = 3.0 \times 10^8\ \mathrm{m/s}$. This seemed to imply there must be some special reference frame where c would have this value.

We discussed in Chapters 11 and 12 that waves can travel on water and along ropes or strings, and sound waves travel in air and other materials. Nineteenth-century physicists viewed the material world in terms of the laws of mechanics, so it was natural for them to assume that light too must travel in some *medium*. They called this transparent medium the **ether** and assumed it permeated all space.† It was therefore assumed that the velocity of light given by Maxwell's equations must be with respect to the ether.

The "ether"

At first it appeared that Maxwell's equations did *not* satisfy the relativity principle. They were simplest in the frame where $c = 3.00 \times 10^8 \text{ m/s}$; that is, in a reference frame at rest in the ether. In any other reference frame, extra terms would have to be added to take into account the relative velocity. Thus, although most of the laws of physics obeyed the relativity principle, the laws of electricity and magnetism apparently did not. (Einstein's second postulate—see next Section—resolved this problem: Maxwell's equations do satisfy relativity.)

Scientists soon set out to determine the speed of the Earth relative to this absolute frame, whatever it might be. A number of clever experiments were designed. The most direct were performed by A. A. Michelson and E. W. Morley in the 1880s. They measured the difference in the speed of light in different directions using Michelson's interferometer (Section 24–9). They expected to find a difference depending on the orientation of their apparatus with respect to the ether. For just as a boat has different speeds relative to the land when it moves upstream, downstream, or across the stream, so too light would be expected to have different speeds depending on the velocity of the ether past the Earth.

The Michelson–Morley experiment

Strange as it may seem, they detected no difference at all. This was a great puzzle. A number of explanations were put forth over a period of years, but they led to contradictions or were otherwise not generally accepted. This **null result** was one of the great puzzles at the end of the nineteenth century.

The null result

Then in 1905, Albert Einstein proposed a radical new theory that reconciled these many problems in a simple way. But at the same time, as we shall see, it completely changed our ideas of space and time.

26–2 Postulates of the Special Theory of Relativity

The problems that existed at the start of the twentieth century with regard to electromagnetic theory and Newtonian mechanics were beautifully resolved by Einstein's introduction of the theory of relativity in 1905. Unaware of the Michelson–Morley null result, Einstein was motivated by certain questions regarding electromagnetic theory and light waves. For example, he asked himself: "What would I see if I rode a light beam?" The answer was that instead of a traveling electromagnetic wave, he would see alternating electric and magnetic fields at rest whose magnitude changed in space, but did not change in time. Such fields, he realized, had never been detected and indeed were not consistent with Maxwell's electromagnetic theory. He argued, therefore, that it was unreasonable to think that the speed of light relative to any observer could be reduced to zero, or in fact reduced at all. This idea became the second postulate of his theory of relativity.

†The medium for light waves could not be air, since light travels from the Sun to Earth through nearly empty space. Therefore, another medium was postulated, the ether. The ether was not only transparent but, because of difficulty in detecting it, was assumed to have zero density.

In his famous 1905 paper, Einstein proposed doing away completely with the idea of the ether and the accompanying assumption of an absolute reference frame at rest. This proposal was embodied in two postulates. The first postulate was an extension of the Galilean–Newtonian relativity principle to include not only the laws of mechanics but also those of the rest of physics, including electricity and magnetism:

***First postulate*[†] *(the relativity principle)*: The laws of physics have the same form in all inertial reference frames.**

The first postulate of special relativity

The second postulate is consistent with the first:

***Second postulate (constancy of the speed of light)*: Light propagates through empty space with a definite speed *c* independent of the speed of the source or observer.**

The second postulate of special relativity

These two postulates form the foundation of Einstein's **special theory of relativity**. It is called "special" to distinguish it from his later "general theory of relativity," which deals with noninertial (accelerating) reference frames (Chapter 33). The special theory, which is what we discuss here, deals only with inertial frames.

The second postulate may seem hard to accept, for it seems to violate common sense. First of all, we have to think of light traveling through empty space. Giving up the ether is not too hard, however, since it had never been detected. But the second postulate also tells us that the speed of light in vacuum is always the same, $3.00 \times 10^8\,\mathrm{m/s}$, no matter what the speed of the observer or the source. Thus, a person traveling toward or away from a source of light will measure the same speed for that light as someone at rest with respect to the source. This conflicts with our everyday experience: we would expect to have to add in the velocity of the observer. On the other hand, perhaps we can't expect our everyday experience to be helpful when dealing with the high velocity of light. Furthermore, the null result of the Michelson–Morley experiment is fully consistent with the second postulate.[‡]

Einstein's proposal has a certain beauty. By doing away with the idea of an absolute reference frame, it was possible to reconcile classical mechanics with Maxwell's electromagnetic theory. The speed of light predicted by Maxwell's equations *is* the speed of light in vacuum in *any* reference frame.

Einstein's theory required us to give up commonsense notions of space and time, and in the following Sections we will examine some strange but interesting consequences of special relativity. Our arguments for the most part will be simple ones. We will use a technique that Einstein himself did: we will imagine very simple experimental situations in which little mathematics is needed. In this way, we can see many of the consequences of relativity theory without getting involved in detailed calculations. Einstein called these "thought" experiments.

"Thought experiment"

26–3 Simultaneity

An important consequence of the theory of relativity is that we can no longer regard time as an absolute quantity. No one doubts that time flows onward and never turns back. But the time interval between two events, and even whether or not two events are simultaneous, depends on the observer's reference frame. By an "event," which we use a lot here, we mean something that happens at a particular place and at a particular time.

Event defined

[†]The first postulate can also be stated as: *There is no experiment you can do in an inertial reference frame to tell if you are at rest or moving uniformly at constant velocity.*

[‡]The Michelson–Morley experiment can also be considered as evidence for the first postulate, since it was intended to measure the motion of the Earth relative to an absolute reference frame. Its failure to do so implies the absence of any such preferred frame.

Two events are said to occur simultaneously if they occur at exactly the same time. But how do we know if two events occur precisely at the same time? If they occur at the same point in space—such as two apples falling on your head at the same time—it is easy. But if the two events occur at widely separated places, it is more difficult to know whether the events are simultaneous since we have to take into account the time it takes for the light from them to reach us. Because light travels at finite speed, a person who sees two events must calculate back to find out when they actually occurred. For example, if two events are *observed* to occur at the same time, but one actually took place farther from the observer than the other, then the more distant one must have occurred earlier, and the two events were not simultaneous.

We now imagine a simple thought experiment. Assume an observer, called O, is located exactly halfway between points A and B where two events occur, Fig. 26–3. Suppose the two events are lightning that strikes the points A and B, as shown. For brief events like lightning, only short pulses of light will travel outward from A and B and reach O. Observer O "sees" the events when the pulses of light reach point O. If the two pulses reach O at the same time, then the two events had to be simultaneous. This is because the two light pulses travel at the same speed (postulate 2), and since the distance OA equals OB, the time for the light to travel from A to O and B to O must be the same. Observer O can then definitely state that the two events occurred simultaneously. On the other hand, if O sees the light from one event before that from the other, then the former event occurred first.

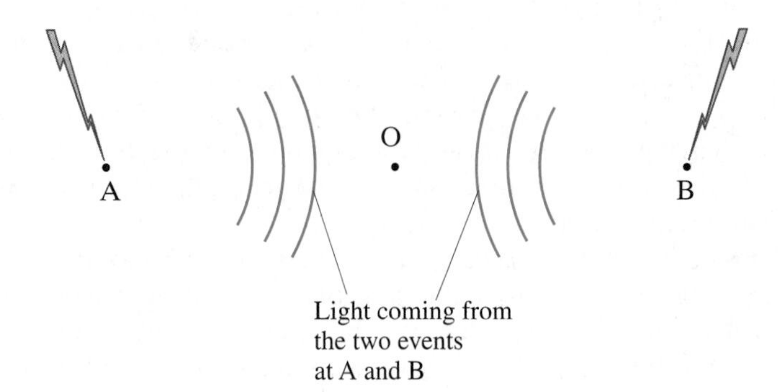

FIGURE 26–3 A moment after lightning strikes at points A and B, the pulses of light are traveling toward the observer O, but O "sees" the lightning only when the light reaches O.

The question we really want to examine is this: if two events are simultaneous to an observer in one reference frame, are they also simultaneous to another observer moving with respect to the first? Let us call the observers O_1 and O_2 and assume they are fixed in reference frames 1 and 2 that move with speed v relative to one another. These two reference frames can be thought of as trains (Fig. 26–4). O_2 says that O_1 is moving to the right with speed v, as in Fig. 26–4a; and O_1 says O_2 is moving to the left with speed v, as in Fig. 26–4b. Both viewpoints are legitimate according to the relativity principle. [There is no third point of view which will tell us which one is "really" moving.]

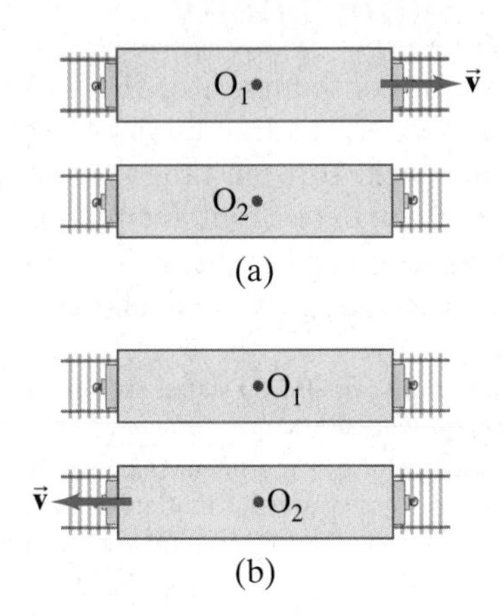

FIGURE 26–4 Observers O_1 and O_2, on two different trains (two different reference frames), are moving with relative speed v. O_2 says that O_1 is moving to the right (a); O_1 says that O_2 is moving to the left (b). Both viewpoints are legitimate—it all depends on your reference frame.

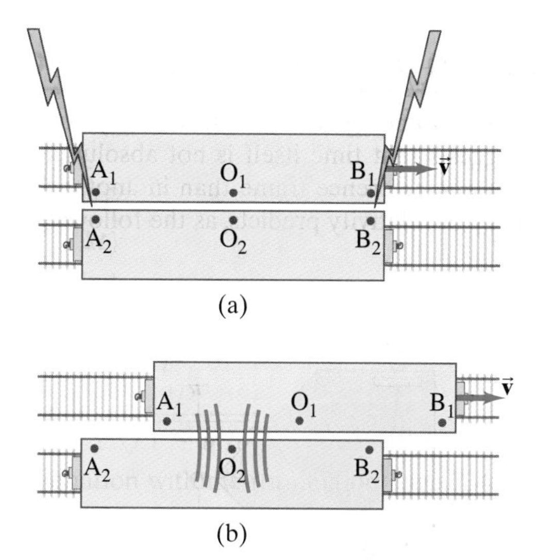

FIGURE 26–5 Thought experiment on simultaneity. To observer O_2, the reference frame of O_1 is moving to the right. In (a), one lightning bolt strikes the two reference frames at A_1 and A_2, and a second lightning bolt strikes at B_1 and B_2. (b) A moment later, the light from the two events reaches O_2 at the same time, so according to observer O_2, the two bolts of lightning strike simultaneously. But in O_1's reference frame, the light from B_1 has already reached O_1, whereas the light from A_1 has not yet reached O_1. So in O_1's reference frame, the event at B_1 must have preceded the event at A_1. Simultaneity in time is not absolute.

Now suppose that observers O_1 and O_2 observe and measure two lightning strikes. The lightning bolts mark both trains where they strike: at A_1 and B_1 on O_1's train, and at A_2 and B_2 on O_2's train, Fig. 26–5a. For simplicity, we assume that O_1 is exactly halfway between A_1 and B_1, and that O_2 is halfway between A_2 and B_2. Let us first put ourselves in O_2's reference frame, so we observe O_1 moving to the right with speed v. Let us also assume that the two events occur *simultaneously* in O_2's frame, and just at the instant when O_1 and O_2 are opposite each other, Fig. 26–5a. A short time later, Fig. 26–5b, the light from A_2 and B_2 reaches O_2 at the same time (we assumed this). Since O_2 knows (or measures) the distances O_2A_2 and O_2B_2 as equal, O_2 knows the two events are simultaneous in the O_2 reference frame.

But what does observer O_1 observe and measure? From our (O_2) reference frame, we can predict what O_1 will observe. We see that O_1 moves to the right during the time the light is traveling to O_1 from A_1 and B_1. As shown in Fig. 26–5b, we can see from our O_2 reference frame that the light from B_1 has already passed O_1, whereas the light from A_1 has not yet reached O_1. That is, O_1 observes the light coming from B_1 before observing the light coming from A_1. Given (1) that light travels at the same speed c in any direction and in any reference frame, and (2) that the distance O_1A_1 equals O_1B_1, then observer O_1 can only conclude that the event at B_1 occurred before the event at A_1. The two events are not simultaneous for O_1, even though they are for O_2.

Simultaneity is relative

We thus find that two events which take place at different locations and are simultaneous to one observer, are actually not simultaneous to a second observer who moves relative to the first.

It may be tempting to ask: "Which observer is right, O_1 or O_2?" The answer, according to relativity, is that they are *both* right. There is no "best" reference frame we can choose to determine which observer is right. Both frames are equally good. We can only conclude that *simultaneity is not an absolute concept*, but is relative. We are not aware of it in everyday life, however, because the effect is noticeable only when the relative speed of the two reference frames is very large (near c), or the distances involved are very large.

EXERCISE A Examine the experiment of Fig. 26–5 from O_1's reference frame. In this case, O_1 will be at rest and will see event B_1 occur before A_1. Will O_1 recognize that O_2, who is moving with speed v to the left, will see the two events as simultaneous? (*Hint*: draw a diagram equivalent to Fig. 26–5.)

Twin Paradox

Not long after Einstein proposed the special theory of relativity, an apparent paradox was pointed out. According to this **twin paradox**, suppose one of a pair of 20-year-old twins takes off in a spaceship traveling at very high speed to a distant star and back again, while the other twin remains on Earth. According to the Earth twin, the astronaut twin will age less. Whereas 20 years might pass for the Earth twin, perhaps only 1 year (depending on the spacecraft's speed) would pass for the traveler. Thus, when the traveler returns, the earthbound twin could expect to be 40 years old whereas the traveling twin would be only 21.

Twin paradox

This is the viewpoint of the twin on the Earth. But what about the traveling twin? If all inertial reference frames are equally good, won't the traveling twin make all the claims the Earth twin does, only in reverse? Can't the astronaut twin claim that since the Earth is moving away at high speed, time passes more slowly on Earth and the twin on Earth will age less? This is the opposite of what the Earth twin predicts. They cannot both be right, for after all the spacecraft returns to Earth and a direct comparison of ages and clocks can be made.

There is, however, no contradiction here. The consequences of the special theory of relativity—in this case, time dilation—can be applied only by observers in an inertial reference frame. The Earth is such a frame (or nearly so), whereas the spacecraft is not. The spacecraft accelerates at the start and end of its trip and when it turns around at the far point of its journey. During the acceleration, the twin on the spacecraft is not in an inertial frame. In between, the astronaut twin may be in an inertial frame (and is justified in saying the Earth twin's clocks run slow), but it is not always the same frame. So she cannot predict their relative ages when she returns to Earth. The Earth twin stays in the same inertial frame, and we can thus trust her predictions based on special relativity. Thus, there is no paradox. The prediction of the Earth twin that the traveling twin ages less is the proper one.

* Additional Example—Using γ

EXAMPLE 26–3 **γ for various speeds.** Determine the value of γ for a speed v equal to (*a*) 0, (*b*) $0.010c$, (*c*) $0.10c$, (*d*) $0.50c$, (*e*) $0.90c$, (*f*) $0.990c$.

APPROACH We simply plug into Eq. 26–2.

SOLUTION (*a*) For $v = 0, \gamma = 1/1 = 1$ exactly.
(*b*) For $v = 0.010c = 3.0 \times 10^6\,\text{m/s}$ (a pretty high speed):

$$\gamma = \frac{1}{\sqrt{1 - \left(\frac{0.010c}{c}\right)^2}}$$

$$= \frac{1}{\sqrt{1 - (0.010)^2}} = \frac{1}{\sqrt{0.99990}} = 1.00005.$$

Unless v is given to more significant figures, $\gamma = 1.0$ here. We see that γ is never less than 1.0 and will only exceed 1.0 significantly at higher speeds.
(*c*) For a speed 10 times higher, $v = 0.10c$, we get

$$\gamma = \frac{1}{\sqrt{1 - (0.10)^2}} = \frac{1}{\sqrt{0.99}} = 1.005.$$

(*d*) At a speed half the speed of light

$$\gamma = \frac{1}{\sqrt{1-(0.50)^2}} = 1.15.$$

(*e*) At $v = 0.90c$ we get $\gamma = 2.3$.
(*f*) At $v = 0.990c$ we get $\gamma = 7.1$.

Table 26–1 is a handy summary of these results.

TABLE 26–1 Values of γ

v	γ
0	1.000
0.01c	1.000
0.10c	1.005
0.50c	1.15
0.90c	2.3
0.99c	7.1

* Global Positioning System (GPS)

PHYSICS APPLIED
Global positioning system (GPS)

Airplanes, cars, boats, and hikers use **global positioning system (GPS)** receivers to tell them quite accurately where they are, at a given moment. The 24 global positioning system satellites send out precise time signals using atomic clocks. Your receiver compares the times received from at least four satellites, all of whose times are carefully synchronized to within 1 part in 10^{13}. By comparing the time differences with the known satellite positions and the fixed speed of light, the receiver can determine how far it is from each satellite and thus where it is on the Earth. It can do this to a typical accuracy of 15 m, if it has been constructed to make corrections such as the one below due to special relativity.

CONCEPTUAL EXAMPLE 26–4 **A relativity correction to GPS.** GPS satellites move at about $4\,\mathrm{km/s} = 4000\,\mathrm{m/s}$. Show that a good GPS receiver needs to correct for time dilation if it is to produce results consistent with atomic clocks accurate to 1 part in 10^{13}.

RESPONSE Let us calculate the magnitude of the time dilation effect by inserting $v = 4000\,\mathrm{m/s}$ into Eq. 26–1a:

$$\Delta t = \frac{1}{\sqrt{1-\frac{v^2}{c^2}}}\,\Delta t_0$$

$$= \frac{1}{\sqrt{1-\left(\frac{4\times10^3\,\mathrm{m/s}}{3\times10^8\,\mathrm{m/s}}\right)^2}}\,\Delta t_0$$

$$= \frac{1}{\sqrt{1-1.8\times10^{-10}}}\,\Delta t_0.$$

We use the binomial expansion: $(1\pm x)^n \approx 1 \pm nx$ for $x \ll 1$ (see Appendix A) which here is $(1-x)^{\frac{1}{2}} \approx 1+\frac{1}{2}x$. That is

$$\Delta t = \left(1+\tfrac{1}{2}(1.8\times10^{-10})\right)\Delta t_0 = (1+9\times10^{-11})\,\Delta t_0.$$

The time "error" divided by the time interval is

$$\frac{(\Delta t-\Delta t_0)}{\Delta t_0} = 1+9\times10^{-11}-1 = 9\times10^{-11} \approx 1\times10^{-10}.$$

Time dilation, if not accounted for, would introduce an error of about 1 part in 10^{10}, which is 1000 times greater than the precision of the atomic clocks. Not correcting for time dilation means a receiver could give much poorer position accuracy.

NOTE GPS devices must make other corrections as well, including effects associated with General Relativity.

EXAMPLE 26–10 **Mass change in a chemical reaction.** When two moles of hydrogen and one mole of oxygen react to form two moles of water, the energy released is 484 kJ. How much does the mass decrease in this reaction?

APPROACH We use Einstein's great concept of the interchangeability of mass and energy $(E = mc^2)$.

SOLUTION Using Eq. 26–9 we have for the change in mass Δm:

$$\Delta m = \frac{\Delta E}{c^2} = \frac{(-484 \times 10^3\,\mathrm{J})}{(3.00 \times 10^8\,\mathrm{m/s})^2} = -5.38 \times 10^{-12}\,\mathrm{kg}.$$

The initial mass of the system is $0.002\,\mathrm{kg} + 0.016\,\mathrm{kg} = 0.018\,\mathrm{kg}$. Thus the change in mass is relatively very tiny and can normally be neglected. [Conservation of mass is usually a reasonable principle to apply to chemical reactions.]

Units:
eV/c for p
eV/c^2 for m

In the tiny world of atoms and nuclei, it is common to quote energies in eV (electron volts) or multiples such as MeV $(10^6\,\mathrm{eV})$. Momentum (see Eq. 26–4) can be quoted in units of eV/c (or MeV/c). And mass can be quoted (from $E = mc^2$) in units of eV/c^2 (or MeV/c^2). Note the use of c to keep the units correct. The rest masses of the electron and the proton are readily shown to be $0.511\,\mathrm{MeV}/c^2$ and $938\,\mathrm{MeV}/c^2$, respectively. See also the Table inside the front cover.

EXAMPLE 26–11 **A 1-TeV proton.** The Tevatron accelerator at Fermilab in Illinois can accelerate protons to a kinetic energy of 1.0 TeV $(10^{12}\,\mathrm{eV})$. What is the speed of such a proton?

APPROACH We solve the kinetic energy formula, Eq. 26–6a, for v.

SOLUTION The rest energy of a proton is $E_0 = 938\,\mathrm{MeV}$ or $9.38 \times 10^8\,\mathrm{eV}$. Compared to the KE of $10^{12}\,\mathrm{eV}$, the rest energy can be neglected, so we simplify Eq. 26–6a to

$$\mathrm{KE} \approx \frac{m_0c^2}{\sqrt{1 - v^2/c^2}}.$$

We solve this for v in the following steps:

$$\sqrt{1 - \frac{v^2}{c^2}} = \frac{m_0c^2}{\mathrm{KE}};$$

$$1 - \frac{v^2}{c^2} = \left(\frac{m_0c^2}{\mathrm{KE}}\right)^2;$$

$$\frac{v^2}{c^2} = 1 - \left(\frac{m_0c^2}{\mathrm{KE}}\right)^2 = 1 - \left(\frac{9.38 \times 10^8\,\mathrm{eV}}{1.0 \times 10^{12}\,\mathrm{eV}}\right)^2;$$

$$v = \sqrt{1 - (9.38 \times 10^{-4})^2}\,c = 0.99999956c.$$

So the proton is traveling at a speed very nearly equal to c.

At low speeds, $v \ll c$, the relativistic formula for KE reduces to the classical one, as we show by using the binomial expansion, $(1 \pm x)^n = 1 \pm nx + n(n-1)x^2/2! + \cdots$. With $n = -\frac{1}{2}$, we expand the square root in Eq. 26–6a

$$\mathrm{KE} = m_0c^2\left(\frac{1}{\sqrt{1 - v^2/c^2}} - 1\right)$$

so that

$$\mathrm{KE} \approx m_0c^2\left(1 + \frac{1}{2}\frac{v^2}{c^2} + \cdots - 1\right)$$
$$\approx \tfrac{1}{2}m_0v^2.$$

The dots in the first expression represent very small terms in the expansion which we neglect since we assumed that $v \ll c$. Thus at low speeds, the relativistic form for kinetic energy reduces to the classical form, $\text{KE} = \frac{1}{2}mv^2$. This makes relativity a viable theory in that it can predict accurate results at low speed as well as at high. Indeed, the other equations of special relativity also reduce to their classical equivalents at ordinary speeds: length contraction, time dilation, and modifications to momentum as well as kinetic energy, all disappear for $v \ll c$ since $\sqrt{1 - v^2/c^2} \approx 1$.

A useful relation between the total energy E of a particle and its momentum p can also be derived. The momentum of a particle of rest mass m_0 and speed v is given by Eq. 26–4:

$$p = \frac{m_0 v}{\sqrt{1 - v^2/c^2}} = \gamma m_0 v.$$

Relativistic momentum

The total energy is (Eq. 26–7b)

$$E = \gamma m_0 c^2 = \frac{m_0 c^2}{\sqrt{1 - v^2/c^2}}.$$

We square this equation (and add a term "$v^2 - v^2$" which is zero, but will help us):

$$\begin{aligned} E^2 &= \frac{m_0^2 c^2 (v^2 - v^2 + c^2)}{1 - v^2/c^2} \\ &= p^2 c^2 + \frac{m_0^2 c^4 (1 - v^2/c^2)}{1 - v^2/c^2} \end{aligned}$$

or

$$E^2 = p^2 c^2 + m_0^2 c^4. \qquad \textbf{(26–10)}$$

Energy related to momentum

Thus, the total energy can be written in terms of the momentum p, or in terms of the kinetic energy (Eq. 26–7a), where we have assumed there is no potential energy.

* When Do We Use Relativistic Formulas?

From a practical point of view, we do not have much opportunity in our daily lives to use the mathematics of relativity. For example, the γ factor, $\gamma = 1/\sqrt{1 - v^2/c^2}$, which appears in many relativistic formulas, has a value of 1.005 when $v = 0.10c$. Thus, for speeds even as high as $0.10c = 3.0 \times 10^7\,\text{m/s}$, the factor $\sqrt{1 - v^2/c^2}$ in relativistic formulas gives a numerical correction of less than 1%. For speeds less than 0.10c, or unless mass and energy are interchanged, we don't usually need to use the more complicated relativistic formulas, and can use the simpler classical formulas.

If you are given a particle's rest mass m_0 and its kinetic energy KE, you can do a quick calculation to determine if you need to use relativistic formulas or if classical ones are good enough. You simply compute the ratio $\text{KE}/m_0 c^2$ because (Eq. 26–6b)

$$\frac{\text{KE}}{m_0 c^2} = \gamma - 1 = \frac{1}{\sqrt{1 - v^2/c^2}} - 1.$$

If this ratio comes out to be less than, say, 0.01, then $\gamma \leq 1.01$ and relativistic equations will correct the classical ones by about 1%. If your expected precision is no better than 1%, classical formulas are good enough. But if your precision is 1 part in 1000 (0.1%) then you would want to use relativistic formulas. If your expected precision is only 10%, you need relativity if $(\text{KE}/m_0 c^2) \gtrsim 0.1$.

EXERCISE E For 1% accuracy, does an electron with $\text{KE} = 100\,\text{eV}$ need to be treated relativistically? [*Hint*: the rest mass of an electron is 0.511 MeV.]

Planck's Quantum Hypothesis

In the year 1900, Max Planck (1858–1947) proposed a theory that was able to reproduce the graphs of Fig. 27–4. His theory, still accepted today, made a new and radical assumption: that the energy of the oscillations of atoms within molecules cannot have just any value; instead each has energy which is a multiple of a minimum value related to the frequency of oscillation by

$$E = hf.$$

Here h is a new constant, now called **Planck's constant**, whose value was estimated by Planck by fitting his formula for the blackbody radiation curve to experiment. The value accepted today is

$$h = 6.626 \times 10^{-34}\,\mathrm{J \cdot s}.$$

Planck's assumption suggests that the energy of any molecular vibration could be only some whole number multiple of hf:

Planck's quantum hypothesis

$$E = nhf, \qquad n = 1, 2, 3, \cdots, \tag{27–3}$$

where n is called a **quantum number** ("quantum" means "discrete amount" as opposed to "continuous"). This idea is often called **Planck's quantum hypothesis**, although little attention was brought to this point at the time. In fact, it appears that Planck considered it more as a mathematical device to get the "right answer" rather than as a discovery comparable to those of Newton. Planck himself continued to seek a classical explanation for the introduction of h. The recognition that this was an important and radical innovation did not come until later, after about 1905 when others, particularly Einstein, entered the field.

The quantum hypothesis, Eq. 27–3, states that the energy of an oscillator can be $E = hf$, or $2hf$, or $3hf$, and so on, but there cannot be vibrations with energies between these values. That is, energy would not be a continuous quantity as had been believed for centuries; rather it is **quantized**—it exists only in discrete amounts. The smallest amount of energy possible (hf) is called the **quantum of energy**. Recall from Chapter 11 that the energy of an oscillation is proportional to the amplitude squared. Thus another way of expressing the quantum hypothesis is that not just any amplitude of vibration is possible. The possible values for the amplitude are related to the frequency f.

A simple analogy may help. A stringed instrument such as a violin or guitar can be played over a continuous range of frequencies by moving your finger along the string. A flute or piano, on the other hand, is "quantized" in the sense that only certain frequencies (notes) can be played. Or compare a ramp, on which a box can be placed at any height, to a flight of stairs on which the box can have only certain discrete amounts of potential energy, as shown in Fig. 27–5.

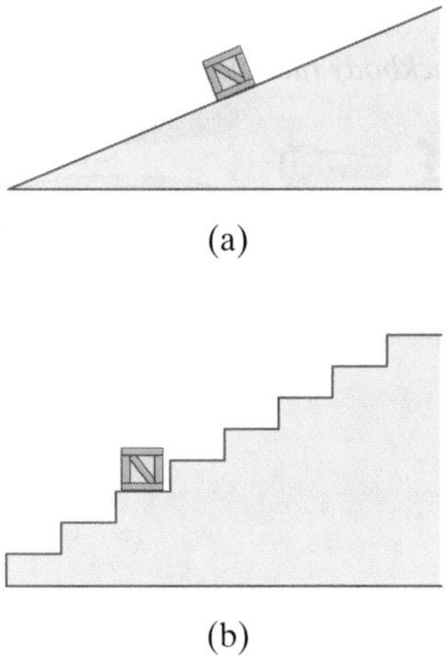

(a)

(b)

FIGURE 27–5 Ramp versus stair analogy. (a) On a ramp, a box can have continuous values of potential energy. (b) But on stairs, the box can have only discrete (quantized) values of energy.

27–3 Photon Theory of Light and the Photoelectric Effect

In 1905, the same year that he introduced the special theory of relativity, Einstein made a bold extension of the quantum idea by proposing a new theory of light. Planck's work had suggested that the vibrational energy of molecules in a radiating object is quantized with energy $E = nhf$, where n is an integer and f is the frequency of molecular vibration. Einstein argued that when light is emitted by a molecular oscillator, the molecule's vibrational energy of nhf must decrease by an amount hf (or by $2hf$, etc.) to another integer times hf, such as $(n-1)hf$. Then to conserve energy, the light ought to be emitted in packets, or *quanta*, each with an energy

Photon energy

$$E = hf, \tag{27–4}$$

where f is here the frequency of the emitted light.

Again h is Planck's constant. Since all light ultimately comes from a radiating source, this suggests that perhaps *light is transmitted as tiny particles*, or **photons**, as they are now called, in addition to the waves predicted by Maxwell's electromagnetic theory. The photon theory of light was also a radical departure from classical ideas. Einstein proposed a test of the quantum theory of light: quantitative measurements on the photoelectric effect.

Photons

When light shines on a metal surface, electrons are found to be emitted from the surface. This effect is called the **photoelectric effect** and it occurs in many materials, but is most easily observed with metals. It can be observed using the apparatus shown in Fig. 27–6. A metal plate P and a smaller electrode C are placed inside an evacuated glass tube, called a **photocell**. The two electrodes are connected to an ammeter and a source of emf, as shown. When the photocell is in the dark, the ammeter reads zero. But when light of sufficiently high frequency illuminates the plate, the ammeter indicates a current flowing in the circuit. We explain completion of the circuit by imagining that electrons, ejected by the impinging radiation, flow across the tube from the plate to the "collector" C as indicated in Fig. 27–6.

Photoelectric effect

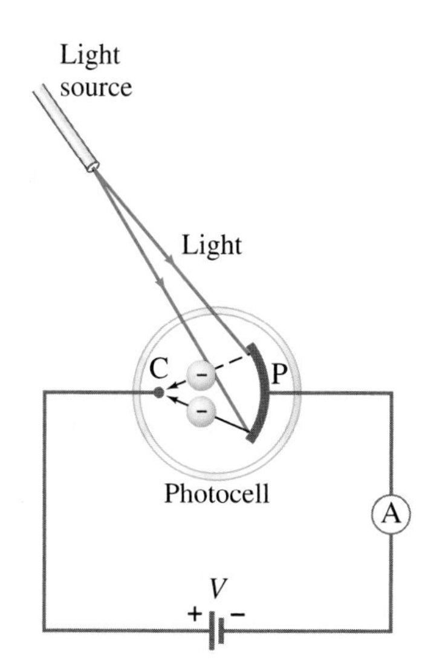

FIGURE 27–6 The photoelectric effect.

That electrons should be emitted when light shines on a metal is consistent with the electromagnetic (EM) wave theory of light: the electric field of an EM wave could exert a force on electrons in the metal and eject some of them. Einstein pointed out, however, that the wave theory and the photon theory of light give very different predictions on the details of the photoelectric effect. For example, one thing that can be measured with the apparatus of Fig. 27–6 is the maximum kinetic energy (KE_{max}) of the emitted electrons. This can be done by using a variable voltage source and reversing the terminals so that electrode C is negative and P is positive. The electrons emitted from P will be repelled by the negative electrode, but if this reverse voltage is small enough, the fastest electrons will still reach C and there will be a current in the circuit. If the reversed voltage is increased, a point is reached where the current reaches zero—no electrons have sufficient kinetic energy to reach C. This is called the *stopping potential*, or *stopping voltage*, V_0, and from its measurement, KE_{max} can be determined using conservation of energy (loss of KE = gain in PE):

$$\text{KE}_{\text{max}} = eV_0.$$

Now let us examine the details of the photoelectric effect from the point of view of the wave theory versus Einstein's particle theory.

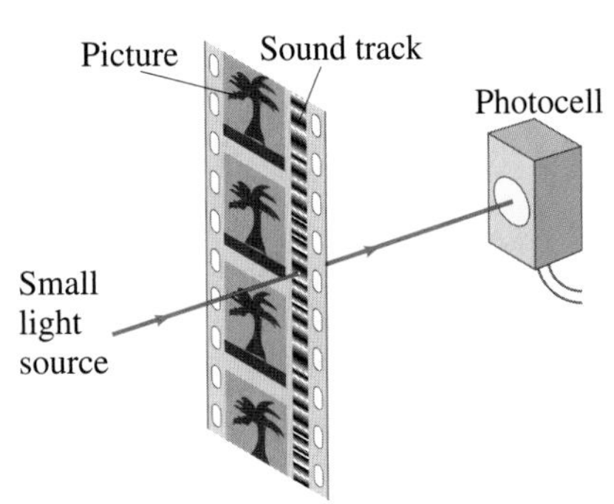

FIGURE 27–8 Optical sound track on movie film. In the projector, light from a small source (different from that for the picture) passes through the sound track on the moving film. The light and dark areas on the sound track vary the intensity of the transmitted light which reaches the photocell, whose current output is then a replica of the original sound. This output is amplified and sent to the loudspeakers. High-quality projectors can show movies containing several parallel sound tracks to go to different speakers around the theater.

Applications of the Photoelectric Effect

The photoelectric effect, besides playing an important historical role in confirming the photon theory of light, also has many practical applications. Burglar alarms and automatic door openers often make use of the photocell circuit of Fig. 27–6. When a person interrupts the beam of light, the sudden drop in current in the circuit activates a switch—often a solenoid—which operates a bell or opens the door. UV or IR light is sometimes used in burglar alarms because of its invisibility. Many smoke detectors use the photoelectric effect to detect tiny amounts of smoke that interrupt the flow of light and so alter the electric current. Photographic light meters use this circuit as well. Photocells are used in many other devices, such as absorption spectrophotometers, to measure light intensity. One type of film sound track is a variably shaded narrow section at the side of the film. Light passing through the film is thus "modulated," and the output electrical signal of the photocell detector follows the frequencies on the sound track. See Fig. 27–8. For many applications today, the vacuum-tube photocell of Fig. 27–6 has been replaced by a semiconductor device known as a **photodiode** (Section 29–8). In these semiconductors, the absorption of a photon liberates a bound electron, which changes the conductivity of the material, so the current through a photodiode is altered.

27–4 Energy, Mass, and Momentum of a Photon

We have just seen (Eq. 27–4) that the total energy of a single photon is given by $E = hf$. Because a photon always travels at the speed of light, it is truly a relativistic particle. Thus we must use relativistic formulas for dealing with its mass, energy, and momentum. The momentum of any particle of rest mass m_0 is given by $p = m_0 v/\sqrt{1 - v^2/c^2}$. Since $v = c$ for a photon, the denominator is zero. To avoid having an infinite momentum, we conclude that the photon's rest mass must be zero: $m_0 = 0$. This makes sense too because a photon can never be at rest (it goes at the speed of light). A photon's kinetic energy is its total energy:

$$\text{KE} = E = hf. \qquad \text{[photon]}$$

The momentum of a photon can be obtained from the relativistic formula (Eq. 26–10) $E^2 = p^2c^2 + m_0^2c^4$ where we set $m_0 = 0$ so $E^2 = p^2c^2$ or

$$p = \frac{E}{c}. \qquad \text{[photon]}$$

Since $E = hf$ for a photon, its momentum is related to its wavelength by

$$p = \frac{E}{c} = \frac{hf}{c} = \frac{h}{\lambda}. \qquad \textbf{(27–6)}$$

EXAMPLE 27–5 ESTIMATE Photons from a lightbulb. Estimate how many visible light photons a 100-W lightbulb emits per second. Assume the bulb has a typical efficiency of about 3% (that is, 97% of the energy goes to heat).

APPROACH Let's assume an average wavelength in the middle of the visible spectrum, $\lambda \approx 500\,\text{nm}$. The energy of each photon is $E = hf = hc/\lambda$. Only 3% of the 100-W power is emitted as light, or $3\,\text{W} = 3\,\text{J/s}$. The number of photons emitted per second equals the light output of 3 J per second divided by the energy of each photon.

SOLUTION The energy emitted in one second (= 3 J) is $E = Nhf$ where N is the number of photons emitted per second and $f = c/\lambda$. Hence

$$N = \frac{E}{hf} = \frac{E\lambda}{hc} = \frac{(3\,\text{J})(500 \times 10^{-9}\,\text{m})}{(6.63 \times 10^{-34}\,\text{J}\cdot\text{s})(3.0 \times 10^{8}\,\text{m/s})} \approx 8 \times 10^{18}$$

per second, or almost 10^{19} photons emitted per second, an enormous number.

EXAMPLE 27–6 **Photon momentum and force.** Suppose the 10^{19} photons emitted per second from the 100-W lightbulb in Example 27–5 were all focused onto a piece of black paper and absorbed. (*a*) Calculate the momentum of one photon and (*b*) estimate the force all these photons could exert on the paper.

APPROACH Each photon's momentum is obtained from Eq. 27–6, $p = h/\lambda$. Next, each absorbed photon's momentum changes from $p = h/\lambda$ to zero. We use Newton's second law, $F = \Delta p/\Delta t$, to get the force.

SOLUTION (*a*) Each photon has a momentum

$$p = \frac{h}{\lambda} = \frac{6.63 \times 10^{-34}\,\mathrm{J\cdot s}}{500 \times 10^{-9}\,\mathrm{m}} = 1.3 \times 10^{-27}\,\mathrm{kg\cdot m/s}.$$

(*b*) Using Newton's second law for $N = 10^{19}$ photons (Example 27–5) whose momentum changes from h/λ to 0, we obtain

$$F = \frac{\Delta p}{\Delta t} = \frac{Nh/\lambda - 0}{1\,\mathrm{s}} = N\frac{h}{\lambda} \approx \left(10^{19}\,\mathrm{s^{-1}}\right)\left(10^{-27}\,\mathrm{kg\cdot m/s}\right) \approx 10^{-8}\,\mathrm{N}.$$

This is a pretty tiny force, but we can see that a very strong light source could exert a measurable force, and near the Sun or a star the force due to photons in electromagnetic radiation could be considerable.

EXAMPLE 27–7 **Photosynthesis.** In *photosynthesis*, pigments such as chlorophyll in plants capture the energy of sunlight to change CO_2 to useful carbohydrate. About nine photons are needed to transform one molecule of CO_2 to carbohydrate and O_2. Assuming light of wavelength $\lambda = 670\,\mathrm{nm}$ (chlorophyll absorbs most strongly in the range 650 nm to 700 nm), how efficient is the photosynthetic process? The reverse chemical reaction releases an energy of 4.9 eV/molecule of CO_2.

PHYSICS APPLIED
Photosynthesis

APPROACH The efficiency is the minimum energy required (4.9 eV) divided by the actual energy absorbed, nine times the energy (hf) of one photon.

SOLUTION The energy of nine photons, each of energy $hf = hc/\lambda$ is $(9)\left(6.63 \times 10^{-34}\,\mathrm{J\cdot s}\right)\left(3.0 \times 10^{8}\,\mathrm{m/s}\right)/\left(6.7 \times 10^{-7}\,\mathrm{m}\right) = 2.7 \times 10^{-18}\,\mathrm{J}$ or 17 eV. Thus the process is $(4.9\,\mathrm{eV}/17\,\mathrm{eV}) = 29\%$ efficient.

* 27–5 Compton Effect

Besides the photoelectric effect, a number of other experiments were carried out in the early twentieth century which also supported the photon theory. One of these was the **Compton effect** (1923) named after its discoverer, A. H. Compton (1892–1962). Compton scattered short-wavelength light (actually X-rays) from various materials. He found that the scattered light had a slightly longer wavelength than did the incident light, and therefore a slightly lower frequency indicating a loss of energy. He explained this result on the basis of the photon theory as incident photons colliding with electrons of the material, Fig. 27–9. Using Eq. 27–6 for momentum of a photon, Compton applied the laws of conservation of momentum and energy to the collision of Fig. 27–9 and derived the following equation for the wavelength of the scattered photons:

$$\lambda' = \lambda + \frac{h}{m_0 c}(1 - \cos\phi), \qquad \textbf{(27–7)}$$

where m_0 is the rest mass of the electron. (The quantity $h/m_0 c$, which has the dimensions of length, is called the **Compton wavelength** of the electron.) We see that the predicted wavelength of scattered photons depends on the angle ϕ at which they are detected. Compton's measurements of 1923 were consistent with this formula. The wave theory of light predicts no such shift: an incoming EM wave of frequency f should set electrons into oscillation at frequency f; and such oscillating electrons would reemit EM waves of this same frequency f (Section 22–2), which would not change with angle (ϕ). Hence the Compton effect adds to the firm experimental foundation for the photon theory of light.

FIGURE 27–9 The Compton effect. A single photon of wavelength λ strikes an electron in some material, knocking it out of its atom. The scattered photon has less energy (since some is given to the electron) and hence has a longer wavelength λ'. Experiments found scattered X-rays of just the wavelengths predicted by conservation of energy and momentum using the photon model.

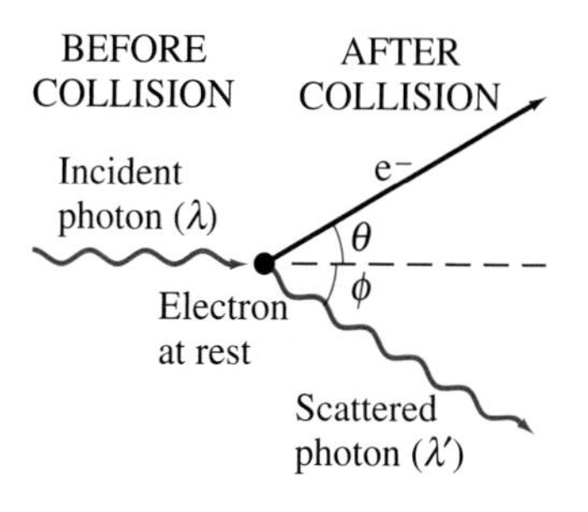

FIGURE 27–11 Niels Bohr (right), walking with Enrico Fermi along the Appian Way outside Rome. This photo shows one important way physics is done.

CAUTION

Not correct to say light is a wave and/or a particle. Light can **act** *like a wave or like a particle*

To clarify the situation, the great Danish physicist Niels Bohr (1885–1962, Fig. 27–11) proposed his famous **principle of complementarity**. It states that to understand an experiment, sometimes we find an explanation using wave theory and sometimes using particle theory. Yet we must be aware of both the wave and particle aspects of light if we are to have a full understanding of light. Therefore these two aspects of light complement one another.

It is not easy to "visualize" this duality. We cannot readily picture a combination of wave and particle. Instead, we must recognize that the two aspects of light are different "faces" that light shows to experimenters.

Part of the difficulty stems from how we think. Visual pictures (or models) in our minds are based on what we see in the everyday world. We apply the concepts of waves and particles to light because in the macroscopic world we see that energy is transferred from place to place by these two methods. We cannot see directly whether light is a wave or particle—so we do indirect experiments. To explain the experiments, we apply the models of waves or of particles to the nature of light. But these are abstractions of the human mind. When we try to conceive of what light really "is," we insist on a visual picture. Yet there is no reason why light should conform to these models (or visual images) taken from the macroscopic world. The "true" nature of light—if that means anything—is not possible to visualize. The best we can do is recognize that our knowledge is limited to the indirect experiments, and that in terms of everyday language and images, light reveals both wave and particle properties.

It is worth noting that Einstein's equation $E = hf$ itself links the particle and wave properties of a light beam. In this equation, E refers to the energy of a particle; and on the other side of the equation, we have the frequency f of the corresponding wave.

27–8 Wave Nature of Matter

In 1923, Louis de Broglie (1892–1987) extended the idea of the wave–particle duality. He much appreciated the symmetry in nature, and argued that if light sometimes behaves like a wave and sometimes like a particle, then perhaps those things in nature thought to be particles—such as electrons and other material objects—might also have wave properties. De Broglie proposed that the wavelength of a material particle would be related to its momentum in the same way as for a photon, Eq. 27–6, $p = h/\lambda$. That is, for a particle having linear momentum $p = mv$, the wavelength λ is given by

de Broglie wavelength

$$\lambda = \frac{h}{p}, \tag{27–8}$$

and is valid classically ($p = m_0 v$ for $v \ll c$) and relativistically ($p = \gamma m_0 v = m_0 v/\sqrt{1 - v^2/c^2}$). This is sometimes called the **de Broglie wavelength** of a particle.

EXAMPLE 27–10 **Wavelength of a ball.** Calculate the de Broglie wavelength of a 0.20-kg ball moving with a speed of 15 m/s.

APPROACH We simply use Eq. 27–8.

SOLUTION $\lambda = \frac{h}{p} = \frac{h}{mv} = \frac{(6.6 \times 10^{-34}\,\text{J}\cdot\text{s})}{(0.20\,\text{kg})(15\,\text{m/s})} = 2.2 \times 10^{-34}\,\text{m}.$

The wavelength of Example 27–10 is an unimaginably small wavelength. Even if the speed were extremely small, say 10^{-4} m/s, the wavelength would be about 10^{-29} m. Indeed, the wavelength of any ordinary object is much too small to be measured and detected. The problem is that the properties of waves, such as interference and diffraction, are significant only when the size of objects or slits is not much larger than the wavelength. And there are no known objects or slits to diffract waves only 10^{-30} m long, so the wave properties of ordinary objects go undetected.

But tiny elementary particles, such as electrons, are another matter. Since the mass m appears in the denominator of Eq. 27–8, a very small mass should have a much larger wavelength.

EXAMPLE 27–11 Wavelength of an electron. Determine the wavelength of an electron that has been accelerated through a potential difference of 100 V.

APPROACH If the kinetic energy is much less than the rest energy, we can use classical $\text{KE} = \frac{1}{2}mv^2$ (see end of Section 26–9). For an electron, $m_0c^2 = 0.511\text{ MeV}$. We then apply conservation of energy: the KE acquired by the electron equals its loss in PE. After solving for v, we use Eq. 27–8 to find the de Broglie wavelength.

SOLUTION The gain in kinetic energy will equal the loss in potential energy $(\Delta\text{PE} = eV - 0)$: $\text{KE} = eV$, so $\text{KE} = 100\text{ eV}$. The ratio $\text{KE}/m_0c^2 = 100\text{ eV}/(0.511 \times 10^6\text{ eV}) \approx 10^{-4}$, so relativity is not needed. Thus

$$\frac{1}{2}mv^2 = eV$$

and

$$v = \sqrt{\frac{2\,eV}{m}} = \sqrt{\frac{(2)(1.6 \times 10^{-19}\text{ C})(100\text{ V})}{(9.1 \times 10^{-31}\text{ kg})}} = 5.9 \times 10^6\text{ m/s}.$$

Then

$$\lambda = \frac{h}{mv} = \frac{(6.63 \times 10^{-34}\text{ J}\cdot\text{s})}{(9.1 \times 10^{-31}\text{ kg})(5.9 \times 10^6\text{ m/s})} = 1.2 \times 10^{-10}\text{ m},$$

or 0.12 nm.

EXERCISE E As a particle travels faster, does its de Broglie wavelength decrease, increase, or remain the same?

From Example 27–11, we see that electrons can have wavelengths on the order of 10^{-10} m, and even smaller. Although small, this wavelength can be detected: the spacing of atoms in a crystal is on the order of 10^{-10} m and the orderly array of atoms in a crystal could be used as a type of diffraction grating, as was done earlier for X-rays (see Section 25–11). C. J. Davisson and L. H. Germer performed the crucial experiment; they scattered electrons from the surface of a metal crystal and, in early 1927, observed that the electrons were scattered into a pattern of regular peaks. When they interpreted these peaks as a diffraction pattern, the wavelength of the diffracted electron wave was found to be just that predicted by de Broglie, Eq. 27–8. In the same year, G. P. Thomson (son of J. J. Thomson) used a different experimental arrangement and also detected diffraction of electrons. (See Fig. 27–12. Compare it to X-ray diffraction, Section 25–11.) Later experiments showed that protons, neutrons, and other particles also have wave properties.

FIGURE 27–12 Diffraction pattern of electrons scattered from aluminum foil, as recorded on film.

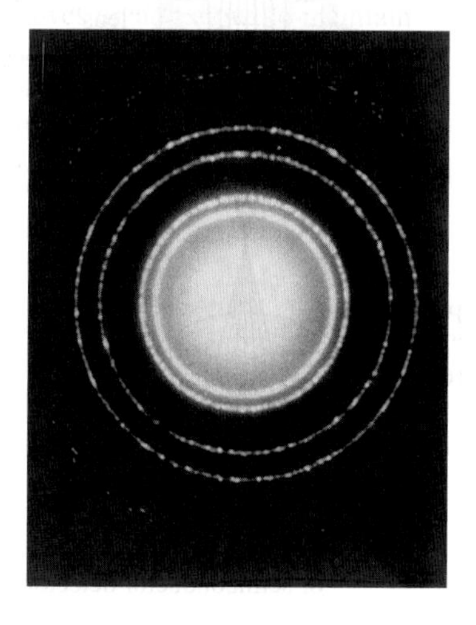

Thus the wave–particle duality applies to material objects as well as to light. The principle of complementarity applies to matter as well. That is, we must be aware of both the particle and wave aspects in order to have an understanding of matter, including electrons. But again we must recognize that a visual picture of a "wave–particle" is not possible.

Wave–particle duality and complementarity apply to matter as well as light

EXAMPLE 27–14 **Absorption wavelength.** Use Figure 27–27 to determine the maximum wavelength that hydrogen in its ground state can absorb. What would be the next smaller wavelength that would work?

APPROACH Maximum wavelength corresponds to minimum energy, and this would be the jump from the ground state up to the first excited state (Fig. 27–27). The next smaller wavelength occurs for the jump from the ground state to the second excited state. In each case, the energy difference can be used to find the wavelength.

SOLUTION The energy needed to jump from the ground state to the first excited state is $13.6\,\text{eV} - 3.4\,\text{eV} = 10.2\,\text{eV}$; the required wavelength, as we saw in Example 27–12, is 122 nm. The energy to jump from the ground state to the second excited state is $13.6\,\text{eV} - 1.5\,\text{eV} = 12.1\,\text{eV}$, which corresponds to a wavelength

$$\lambda = \frac{c}{f} = \frac{hc}{hf} = \frac{hc}{E_3 - E_1}$$

$$= \frac{(6.63 \times 10^{-34}\,\text{J}\cdot\text{s})(3.00 \times 10^8\,\text{m/s})}{(12.1\,\text{eV})(1.60 \times 10^{-19}\,\text{J/eV})} = 103\,\text{nm}.$$

Additional Examples

EXAMPLE 27–15 **Ionization energy.** (*a*) Use the Bohr model to determine the ionization energy of the He^+ ion, which has a single electron. (*b*) Also calculate the minimum wavelength a photon must have to cause ionization.

APPROACH We want to determine the minimum energy required to lift the electron from its ground state and to barely reach the free state at $E = 0$. The ground state energy of He^+ is given by Eq. 27–15b with $n = 1$ and $Z = 2$.

SOLUTION (*a*) Since all the symbols in Eq. 27–15b are the same as for the calculation for hydrogen, except that Z is 2 instead of 1, we see that E_1 will be $Z^2 = 2^2 = 4$ times the E_1 for hydrogen. That is,

$$E_1 = 4(-13.6\,\text{eV}) = -54.4\,\text{eV}.$$

Thus, to ionize the He^+ ion should require 54.4 eV, and this value agrees with experiment.

(*b*) The minimum wavelength photon that can cause ionization will have energy $hf = 54.4\,\text{eV}$ and wavelength

$$\lambda = \frac{c}{f} = \frac{hc}{hf} = \frac{(6.63 \times 10^{-34}\,\text{J}\cdot\text{s})(3.00 \times 10^8\,\text{m/s})}{(54.4\,\text{eV})(1.60 \times 10^{-19}\,\text{J/eV})} = 22.8\,\text{nm}.$$

NOTE If the atom absorbed a photon of greater energy (wavelength shorter than 22.8 nm), the atom could still be ionized and the freed electron would have kinetic energy of its own.

In this last Example, we saw that E_1 for the He^+ ion is four times more negative than that for hydrogen. Indeed, the energy-level diagram for He^+ looks just like that for hydrogen, Fig. 27–27, except that the numerical values for each energy level are four times larger. Note, however, that we are talking here about the He^+ *ion*. Normal (neutral) helium has two electrons and its energy level diagram is entirely different.

CONCEPTUAL EXAMPLE 27–16 **Hydrogen at 20°C.** Estimate the average kinetic energy of whole hydrogen atoms (not just the electrons) at room temperature, and use the result to explain why nearly all H atoms are in the ground state at room temperature, and hence emit no light.

RESPONSE According to kinetic theory (Chapter 13), the average KE of atoms or molecules in a gas is given by Eq. 13–8:

$$\overline{\text{KE}} = \tfrac{3}{2}kT,$$

where $k = 1.38 \times 10^{-23}\,\text{J/K}$ is Boltzmann's constant, and T is the kelvin (absolute) temperature. Room temperature is about $T = 300\,\text{K}$, so

$$\overline{\text{KE}} = \tfrac{3}{2}(1.38 \times 10^{-23}\,\text{J/K})(300\,\text{K}) = 6.2 \times 10^{-21}\,\text{J},$$

or, in electron volts:

$$\overline{\text{KE}} = \frac{6.2 \times 10^{-21}\,\text{J}}{1.6 \times 10^{-19}\,\text{J/eV}} = 0.04\,\text{eV}.$$

The average KE of an atom as a whole is thus very small compared to the energy between the ground state and the next higher energy state ($13.6\,\text{eV} - 3.4\,\text{eV} = 10.2\,\text{eV}$). Any atoms in excited states quickly fall to the ground state and emit light. Once in the ground state, collisions with other atoms can transfer energy of only 0.04 eV on the average. A small fraction of atoms can have much more energy (see Section 13–11 on the distribution of molecular speeds), but even a KE that is 10 times the average is not nearly enough to excite atoms above the ground state. Thus, at room temperature, nearly all atoms are in the ground state. Atoms can be excited to upper states by very high temperatures, or by passing a current of high energy electrons through the gas, as in a discharge tube (Fig. 27–20).

Correspondence Principle

We should note that Bohr made some radical assumptions that were at variance with classical ideas. He assumed that electrons in fixed orbits do not radiate light even though they are accelerating (moving in a circle), and he assumed that angular momentum is quantized. Furthermore, he was not able to say how an electron moved when it made a transition from one energy level to another. On the other hand, there is no real reason to expect that in the tiny world of the atom electrons would behave as ordinary-sized objects do. Nonetheless, he felt that where quantum theory overlaps with the macroscopic world, it should predict classical results. This is the **correspondence principle**, already mentioned in regard to relativity (Section 26–11). This principle does work for Bohr's theory of the hydrogen atom. The orbit sizes and energies are quite different for $n = 1$ and $n = 2$, say. But orbits with $n = 100{,}000{,}000$ and 100,000,001 would be very close in size and energy (see Fig. 27–27). Indeed, jumps between such large orbits, which would approach macroscopic sizes, would be imperceptible. Such orbits would thus appear to be continuously spaced, which is what we expect in the everyday world.

Correspondence principle

Finally, it must be emphasized that the well-defined orbits of the Bohr model do not actually exist. The Bohr model is only a model, not reality. The idea of electron orbits was rejected a few years later, and today electrons are thought of (Chapter 28) as forming "probability clouds."

Questions

1. What can be said about the relative temperatures of whitish-yellow, reddish, and bluish stars? Explain.
2. If energy is radiated by all objects, why can we not see them in the dark? (See also Section 14–8.)
3. Does a lightbulb at a temperature of 2500 K produce as white a light as the Sun at 6000 K? Explain.
4. Darkrooms for developing black-and-white film were sometimes lit by a red bulb. Why red? Would such a bulb work in a darkroom for developing color photographs?
5. If the threshold wavelength in the photoelectric effect increases when the emitting metal is changed to a different metal, what can you say about the work functions of the two metals?
6. Explain why the existence of a cutoff frequency in the photoelectric effect more strongly favors a particle theory rather than a wave theory of light.
7. UV light causes sunburn, whereas visible light does not. Suggest a reason.
8. If an X-ray photon is scattered by an electron, does its wavelength change? If so, does it increase or decrease?
9. In both the photoelectric effect and in the Compton effect, a photon collides with an electron causing the electron to fly off. What then, is the difference between the two processes?
10. Consider a point source of light. How would the intensity of light vary with distance from the source according to (*a*) wave theory, (*b*) particle (photon) theory? Would this help to distinguish the two theories?
11. Explain how the photoelectric circuit of Fig. 27–6 could be used in (*a*) a burglar alarm, (*b*) a smoke detector, (*c*) a photographic light meter.
12. Why do we say that light has wave properties? Why do we say that light has particle properties?
13. Why do we say that electrons have wave properties? Why do we say that electrons have particle properties?
14. What is the difference between a photon and an electron? Be specific: make a list.
15. If an electron and a proton travel at the same speed, which has the shorter wavelength? Explain.
16. In Rutherford's planetary model of the atom, what keeps the electrons from flying off into space?
17. How can you tell if there is oxygen near the surface of the Sun?
18. When a wide spectrum of light passes through hydrogen gas at room temperature, absorption lines are observed that correspond only to the Lyman series. Why don't we observe the other series?
19. Explain how the closely spaced energy levels for hydrogen near the top of Fig. 27–27 correspond to the closely spaced spectral lines at the top of Fig. 27–22.
20. Is it possible for the de Broglie wavelength of a "particle" to be greater than the dimensions of the particle? To be smaller? Is there any direct connection?
21. In a helium atom, which contains two electrons, do you think that on average the electrons are closer to the nucleus or farther away than in a hydrogen atom? Why?
22. How can the spectrum of hydrogen contain so many lines when hydrogen contains only one electron?
23. The Lyman series is brighter than the Balmer series, because this series of transitions ends up in the most common state for hydrogen, the ground state. Why then was the Balmer series discovered first?
24. Use conservation of momentum to explain why photons emitted by hydrogen atoms have slightly less energy than that predicted by Eq. 27–10.
25. The work functions for sodium and cesium are 2.28 eV and 2.14 eV, respectively. For incident photons of a given frequency, which metal will give a higher maximum kinetic energy for the electrons?
26. (*a*) Does a beam of infrared photons always have less energy than a beam of ultraviolet photons? Explain. (*b*) Does a single infrared photon always have less energy than a single ultraviolet photon?
27. Light of 450-nm wavelength strikes a metal surface, and a stream of electrons emerges from the metal. If light of the same intensity but of wavelength 400 nm strikes the surface, are more electrons emitted? Does the energy of the emitted electrons change? Explain.
28. Suppose we obtain an emission spectrum for hydrogen at very high temperature (when some of the atoms are in excited states), and an absorption spectrum at room temperature, when all atoms are in the ground state. Will the two spectra contain identical lines?

Problems

27–1 Discovery of the Electron

1. (I) What is the value of e/m for a particle that moves in a circle of radius 7.0 mm in a 0.86-T magnetic field if a perpendicular 320-V/m electric field will make the path straight?
2. (II) (*a*) What is the velocity of a beam of electrons that go undeflected when passing through crossed (perpendicular) electric and magnetic fields of magnitude 1.88×10^4 V/m and 2.90×10^{-3} T, respectively? (*b*) What is the radius of the electron orbit if the electric field is turned off?
3. (II) An oil drop whose mass is determined to be 2.8×10^{-15} kg is held at rest between two large plates separated by 1.0 cm when the potential difference between them is 340 V. How many excess electrons does this drop have?

27–2 Planck's Quantum Hypothesis

4. (I) How hot is a metal being welded if it radiates most strongly at 440 nm?
5. (I) Estimate the peak wavelength for radiation from (*a*) ice at 0°C, (*b*) a floodlamp at 3500 K, (*c*) helium at 4 K, (*d*) for the universe at $T = 2.725$ K, assuming blackbody emission. In what region of the EM spectrum is each?
6. (I) (*a*) What is the temperature if the peak of a blackbody spectrum is at 18.0 nm? (*b*) What is the wavelength at the peak of a blackbody spectrum if the body is at a temperature of 2000 K?

7. (I) An HCl molecule vibrates with a natural frequency of 8.1×10^{13} Hz. What is the difference in energy (in joules and electron volts) between possible values of the oscillation energy?
8. (II) The steps of a flight of stairs are 20.0 cm high (vertically). If a 68.0-kg person stands with both feet on the same step, what is the gravitational potential energy of this person, relative to the ground, on (*a*) the first step, (*b*) the second step, (*c*) the third step, (*d*) the nth step? (*e*) What is the change in energy as the person descends from step 6 to step 2?
9. (II) Estimate the peak wavelength of light issuing from the pupil of the human eye (which approximates a blackbody) assuming normal body temperature.

27–3 and 27–4 Photons and the Photoelectric Effect

10. (I) What is the energy of photons (joules) emitted by an 88.5-MHz FM radio station?
11. (I) What is the energy range (in joules and eV) of photons in the visible spectrum, of wavelength 400 nm to 750 nm?
12. (I) A typical gamma ray emitted from a nucleus during radioactive decay may have an energy of 300 keV. What is its wavelength? Would we expect significant diffraction of this type of light when it passes through an everyday opening, like a door?
13. (I) About 0.1 eV is required to break a "hydrogen bond" in a protein molecule. Calculate the minimum frequency and maximum wavelength of a photon that can accomplish this.
14. (I) Calculate the momentum of a photon of yellow light of wavelength 6.00×10^{-7} m.
15. (I) What is the momentum of a $\lambda = 0.010$ nm X-ray photon?
16. (II) The human eye can respond to as little as 10^{-18} J of light energy. For a wavelength at the peak of visual sensitivity, 550 nm, how many photons lead to an observable flash?
17. (II) What minimum frequency of light is needed to eject electrons from a metal whose work function is 4.3×10^{-19} J?
18. (II) What is the longest wavelength of light that will emit electrons from a metal whose work function is 3.10 eV?
19. (II) The work functions for sodium, cesium, copper, and iron are 2.3, 2.1, 4.7, and 4.5 eV, respectively. Which of these metals will not emit electrons when visible light shines on it?
20. (II) In a photoelectric-effect experiment it is observed that no current flows unless the wavelength is less than 570 nm. (*a*) What is the work function of this material? (*b*) What is the stopping voltage required if light of wavelength 400 nm is used?
21. (II) What is the maximum kinetic energy of electrons ejected from barium $(W_0 = 2.48\text{ eV})$ when illuminated by white light, $\lambda = 400$ to 750 nm?
22. (II) Barium has a work function of 2.48 eV. What is the maximum kinetic energy of electrons if the metal is illuminated by UV light of wavelength 365 nm? What is their speed?
23. (II) When UV light of wavelength 285 nm falls on a metal surface, the maximum kinetic energy of emitted electrons is 1.40 eV. What is the work function of the metal?
24. (II) The threshold wavelength for emission of electrons from a given surface is 350 nm. What will be the maximum kinetic energy of ejected electrons when the wavelength is changed to (*a*) 280 nm, (*b*) 360 nm?
25. (II) A certain type of film is sensitive only to light whose wavelength is less than 660 nm. What is the energy (eV and kcal/mol) needed for the chemical reaction to occur which causes the film to change?
26. (II) When 230-nm light falls on a metal, the current through a photoelectric circuit (Fig. 27–6) is brought to zero at a stopping voltage of 1.64 V. What is the work function of the metal?
27. (II) In a photoelectric experiment using a clean sodium surface, the maximum energy of the emitted photons was measured for a number of different incident frequencies, with the following results.

Frequency (10^{14} Hz)	Energy (eV)
11.8	2.60
10.6	2.11
9.9	1.81
9.1	1.47
8.2	1.10
6.9	0.57

Plot the graph of these results and find: (*a*) Planck's constant; (*b*) the cutoff frequency of sodium; (*c*) the work function.

28. (II). Show that the energy E (in electron volts) of a photon whose wavelength is λ (nm) is given by

$$E = \frac{1.240 \times 10^3 \text{ eV}\cdot\text{nm}}{\lambda\,(\text{nm})}.$$

* 27–4 Compton Effect

* 29. (II) The quantity h/m_0c, which has the dimensions of length, is called the *Compton wavelength*. Determine the Compton wavelength for (*a*) an electron, (*b*) a proton. (*c*) Show that if a photon has wavelength equal to the Compton wavelength of a particle, the photon's energy is equal to the rest energy of the particle.

* 30. (II) X-rays of wavelength $\lambda = 0.120$ nm are scattered from carbon. What is the Compton wavelength shift for photons detected at angles (relative to the incident beam) of (*a*) 45°, (*b*) 90°, (*c*) 180°?

* 31. (III) In the Compton effect, a 0.100-nm photon strikes a free electron in a head-on collision and knocks it into the forward direction. The rebounding photon recoils directly backward. Use conservation of (relativistic) energy and momentum to determine (*a*) the kinetic energy of the electron, and (*b*) the wavelength of the recoiling photon. (*Note*: use Eq. 27–6, but not Eq. 27–7.)

27–6 Pair Production

32. (I) How much total kinetic energy will an electron–positron pair have if produced by a 3.84-MeV photon?
33. (II) What is the longest wavelength photon that could produce a proton–antiproton pair? (Each has a mass of 1.67×10^{-27} kg.)
34. (II) What is the minimum photon energy needed to produce a $\mu^+ - \mu^-$ pair? The mass of each μ (muon) is 207 times the mass of the electron. What is the wavelength of such a photon?
35. (II) An electron and a positron, each moving at 1.0×10^5 m/s, collide head on, disappear, and produce two photons, each with the same energy and momentum moving in opposite directions. What is the energy and momentum of each photon?
36. (II) A gamma-ray photon produces an electron–positron pair, each with a kinetic energy of 245 keV. What was the energy and wavelength of the photon?

Thus you won't know its *future* position. The same would be true, but to a much lesser extent, if you observe the Ping-pong ball using light. In order to "see" the ball, at least one photon must scatter from it, and the reflected photon must enter your eye or some other detector. When a photon strikes an ordinary-sized object, it does not appreciably alter the motion or position of the object. But when a photon strikes a very tiny object like an electron, it can transfer momentum to the object and thus greatly change the object's motion and position in an unpredictable way. The mere act of measuring the position of an object at one time makes our knowledge of its future position imprecise.

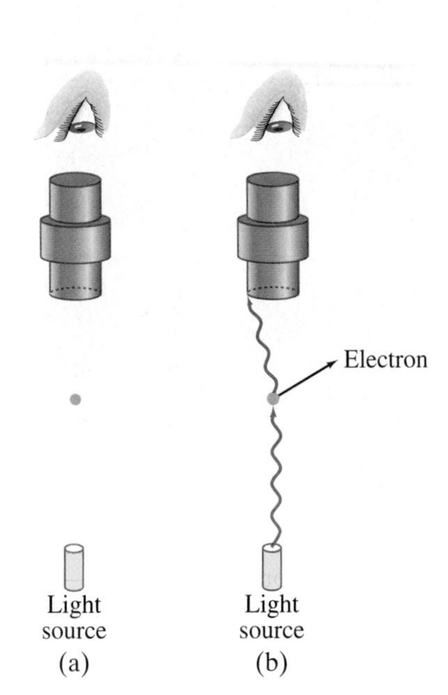

FIGURE 28–5 Thought experiment for observing an electron with a powerful light microscope. At least one photon must scatter from the electron (transferring some momentum to it) and enter the microscope.

Now let us see where the wave–particle duality comes in. Imagine a thought experiment in which we are trying to measure the position of an object, say an electron, with photons, Fig. 28–5. (The arguments would be similar if we were using, instead, an electron microscope.) As we saw in Chapter 25, objects can be seen to an accuracy at best of about the wavelength of the radiation used. If we want an accurate position measurement, we must use a short wavelength. But a short wavelength corresponds to high frequency and large momentum ($p = h/\lambda$); and the more momentum the photons have, the more momentum they can give the object when they strike it. If we use photons of longer wavelength, and correspondingly smaller momentum, the object's motion when struck by the photons will not be affected as much. But the longer wavelength means lower resolution, so the object's position will be less accurately known. Thus the act of observing produces an uncertainty in both the *position* and the *momentum* of the electron. This is the essence of the *uncertainty principle* first enunciated by Heisenberg in 1927.

Quantitatively, we can make an approximate calculation of the magnitude of this effect. If we use light of wavelength λ, the position can be measured at best to an accuracy of about λ. That is, the uncertainty in the position measurement, Δx, is approximately

$$\Delta x \approx \lambda.$$

Suppose that the object can be detected by a single photon. The photon has a momentum $p_x = h/\lambda$. When the photon strikes our object, it will give some or all of this momentum to the object, Fig. 28–5. Therefore, the final x momentum of our object will be uncertain in the amount

$$\Delta p_x \approx \frac{h}{\lambda}$$

since we can't tell beforehand how much momentum will be transferred. The product of these uncertainties is

$$(\Delta x)(\Delta p_x) \approx h.$$

The uncertainties could be worse than this, depending on the apparatus and the number of photons needed for detection. A more careful mathematical calculation shows the product of the uncertainties as, at best, about

UNCERTAINTY PRINCIPLE (Δx and Δp)

$$(\Delta x)(\Delta p_x) \gtrsim \frac{h}{2\pi}. \qquad \textbf{(28–1)}$$

This is a mathematical statement of the **Heisenberg uncertainty principle**, or, as it is sometimes called, the **indeterminancy principle**. It tells us that we cannot measure both the position *and* momentum of an object precisely at the same time. The more accurately we try to measure the position, so that Δx is small, the greater will be the uncertainty in momentum, Δp_x. If we try to measure the momentum very precisely, then the uncertainty in the position becomes large.

The uncertainty principle does not forbid individual precise measurements, however. For example, in principle we could measure the position of an object exactly. But then its momentum would be completely unknown. Thus, although we might know the position of the object exactly at one instant, we could have no idea at all where it would be a moment later. The uncertainties expressed here are inherent in nature, and reflect the best precision theoretically attainable even with the best instruments.

Uncertainties not due to instrument deficiency, but inherent in nature (wave–particle)

Another useful form of the uncertainty principle relates energy and time, and we examine this as follows. The object to be detected has an uncertainty in position $\Delta x \approx \lambda$. The photon that detects it travels with speed c, and it takes a time $\Delta t \approx \Delta x/c \approx \lambda/c$ to pass through the distance of uncertainty. Hence, the measured time when our object is at a given position is uncertain by about

$$\Delta t \approx \frac{\lambda}{c}.$$

Since the photon can transfer some or all of its energy ($= hf = hc/\lambda$) to our object, the uncertainty in energy of our object as a result is

$$\Delta E \approx \frac{hc}{\lambda}.$$

The product of these two uncertainties is

$$(\Delta E)(\Delta t) \approx h.$$

A more careful calculation gives

$$(\Delta E)(\Delta t) \gtrsim \frac{h}{2\pi}. \qquad \textbf{(28–2)}$$

UNCERTAINTY PRINCIPLE (ΔE and Δt)

This form of the uncertainty principle tells us that the energy of an object can be uncertain (or can be interpreted as briefly nonconserved) by an amount ΔE for a time $\Delta t \approx h/(2\pi\, \Delta E)$.

The quantity $(h/2\pi)$ appears so often in quantum mechanics that for convenience it is given the symbol $\hbar$ ("h-bar"). That is,

$$\hbar = \frac{h}{2\pi} = \frac{6.626 \times 10^{-34}\,\mathrm{J\cdot s}}{2\pi} = 1.055 \times 10^{-34}\,\mathrm{J\cdot s}.$$

By using this notation, Eqs. 28–1 and 28–2 for the uncertainty principle can be written

$$(\Delta x)(\Delta p_x) \gtrsim \hbar \qquad \text{and} \qquad (\Delta E)(\Delta t) \gtrsim \hbar.$$

We have been discussing the position and velocity of an electron as if it were a particle. But it isn't simply a particle. Indeed, we have the uncertainty principle because an electron—and matter in general—has wave as well as particle properties. What the uncertainty principle really tells us is that if we insist on thinking of the electron as a particle, then there are certain limitations on this simplified view—namely, that the position and velocity cannot both be known precisely at the same time; and even that the electron does not *have* a precise position and momentum at the same time (because it is not simply a particle). Similarly, the energy can be uncertain in the amount ΔE for a time $\Delta t \approx \hbar/\Delta E$.

Because Planck's constant, h, is so small, the uncertainties expressed in the uncertainty principle are usually negligible on the macroscopic level. But at the level of atomic sizes, the uncertainties are significant. Because we consider ordinary objects to be made up of atoms containing nuclei and electrons, the uncertainty principle is relevant to our understanding of all of nature. The uncertainty principle expresses, perhaps most clearly, the probabilistic nature of quantum mechanics. It thus is often used as a basis for philosophic discussion.

Helium, Z = 2

n	l	m_l	m_s
1	0	0	$\frac{1}{2}$
1	0	0	$-\frac{1}{2}$

Lithium, Z = 3

n	l	m_l	m_s
1	0	0	$\frac{1}{2}$
1	0	0	$-\frac{1}{2}$
2	0	0	$\frac{1}{2}$

Sodium, Z = 11

n	l	m_l	m_s
1	0	0	$\frac{1}{2}$
1	0	0	$-\frac{1}{2}$
2	0	0	$\frac{1}{2}$
2	0	0	$-\frac{1}{2}$
2	1	1	$\frac{1}{2}$
2	1	1	$-\frac{1}{2}$
2	1	0	$\frac{1}{2}$
2	1	0	$-\frac{1}{2}$
2	1	−1	$\frac{1}{2}$
2	1	−1	$-\frac{1}{2}$
3	0	0	$\frac{1}{2}$

Let us now look at the structure of some of the simpler atoms when they are in the ground state. After hydrogen, the next simplest atom is *helium* with two electrons. Both electrons can have $n = 1$, since one can have spin up $(m_s = +\frac{1}{2})$ and the other spin down $(m_s = -\frac{1}{2})$, thus satisfying the exclusion principle. Since $n = 1$, then l and m_l must be zero (Table 28–1). Thus the two electrons have the quantum numbers indicated in the Table in the margin.

Lithium has three electrons, two of which can have $n = 1$. But the third cannot have $n = 1$ without violating the exclusion principle. Hence the third electron must have $n = 2$. It happens that the $n = 2$, $l = 0$ level has a lower energy than $n = 2$, $l = 1$, so the electrons in the ground state have the quantum numbers indicated in the Table in the margin. The quantum numbers of the third electron could also be, say, $(n, l, m_l, m_s) = (3, 1, -1, \frac{1}{2})$. But the atom in this case would be in an excited state since it would have greater energy. It would not be long before it jumped to the ground state with the emission of a photon. At room temperature, unless extra energy is supplied (as in a discharge tube), the vast majority of atoms are in the ground state.

We can continue in this way to describe the quantum numbers of each electron in the ground state of larger and larger atoms. The quantum numbers for sodium, with its eleven electrons, are shown in the Table in the margin.

Figure 28–10 shows a simple energy level diagram where occupied states are shown as up or down arrows $(m_s = +\frac{1}{2}$ or $-\frac{1}{2})$, and possible empty states are shown as a small circle.

The ground-state configuration for all atoms is given in the **periodic table**, which is displayed inside the back cover of this book, and discussed in the next Section.

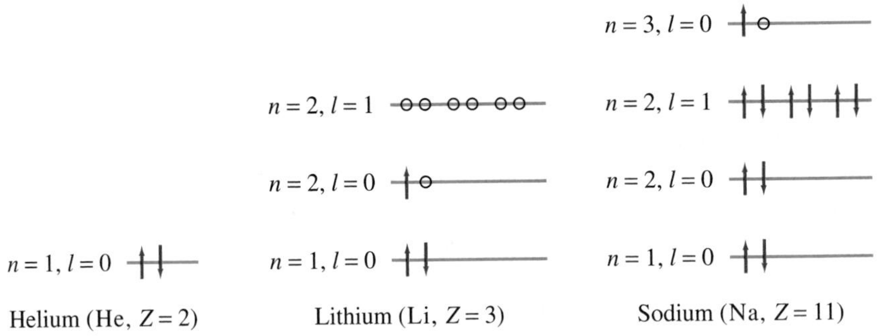

FIGURE 28–10 Energy level diagram showing occupied states (arrows) and unoccupied states (○) for He, Li, and Na. Note that we have shown the $n = 2$, $l = 1$ level of Li even though it is empty.

28–8 The Periodic Table of Elements

More than a century ago, Dmitri Mendeleev (1834–1907) arranged the (then) known elements into what we now call the **periodic table** of the elements. The atoms were arranged according to increasing mass, but also so that elements with similar chemical properties would fall in the same column. Today's version is shown inside the back cover of this book. Each square contains the atomic number Z, the symbol for the element, and the atomic mass (in atomic mass units). Finally, the lower left corner shows the configuration of the ground state of the atom. This requires some explanation. Electrons with the same value of n are referred to as being in the same **shell**. Electrons with $n = 1$ are in one shell (the K shell), those with $n = 2$ are in a second shell (the L shell), those with $n = 3$ are in the third (M) shell, and so on. Electrons with the same values of n and l are referred to as being in the same **subshell**. Letters are often used to specify the value of l as shown in Table 28–2. That is, $l = 0$ is the *s* subshell; $l = 1$ is the *p* subshell; $l = 2$ is the *d* subshell; beginning with $l = 3$, the letters follow the alphabet, *f*, *g*, *h*, *i*, and so on. (The first letters *s*, *p*, *d*, and *f* were originally abbreviations of "sharp," "principal," "diffuse," and "fundamental," experimental terms referring to the spectra.)

TABLE 28–2 Value of *l*

Value of *l*	Letter Symbol	Maximum Number of Electrons in Subshell
0	*s*	2
1	*p*	6
2	*d*	10
3	*f*	14
4	*g*	18
5	*h*	22
⋮	⋮	⋮

The Pauli exclusion principle limits the number of electrons possible in each shell and subshell. For any value of l, there are $2l + 1$ possible m_l values (m_l can be any integer from 1 to l, from -1 to $-l$, or zero), and two possible m_s values. There can be, therefore, at most $2(2l + 1)$ electrons in any l subshell. For example, for $l = 2$, five m_l values are possible $(2, 1, 0, -1, -2)$, and for each of these, m_s can be $+\frac{1}{2}$ or $-\frac{1}{2}$ for a total of $2(5) = 10$ states. Table 28–2 lists the maximum number of electrons that can occupy each subshell.

Since the energy levels depend almost entirely on the values of n and l, it is customary to specify the electron configuration simply by giving the n value and the appropriate letter for l, with the number of electrons in each subshell given as a superscript. The ground-state configuration of sodium, for example, is written as $1s^2 2s^2 2p^6 3s^1$. This is simplified in the periodic table by specifying the configuration only of the outermost electrons and any other nonfilled subshells (see Table 28–3 here, and the periodic table inside the back cover).

TABLE 28–3 Electron Configuration of Some Elements

Z (Number of Electrons)	Element[†]	Ground State Configuration (outer electrons)
1	H	$1s^1$
2	He	$1s^2$
3	Li	$2s^1$
4	Be	$2s^2$
5	B	$2s^2 2p^1$
6	C	$2s^2 2p^2$
7	N	$2s^2 2p^3$
8	O	$2s^2 2p^4$
9	F	$2s^2 2p^5$
10	Ne	$2s^2 2p^6$
11	Na	$3s^1$
12	Mg	$3s^2$
13	Al	$3s^2 3p^1$
14	Si	$3s^2 3p^2$
15	P	$3s^2 3p^3$
16	S	$3s^2 3p^4$
17	Cl	$3s^2 3p^5$
18	Ar	$3s^2 3p^6$
19	K	$4s^1$
20	Ca	$4s^2$
21	Sc	$3d^1 4s^2$
22	Ti	$3d^2 4s^2$
23	V	$3d^3 4s^2$
24	Cr	$3d^5 4s^1$
25	Mn	$3d^5 4s^2$
26	Fe	$3d^6 4s^2$

[†] Names of elements can be found in Appendix B.

CONCEPTUAL EXAMPLE 28–5 **Electron configurations.** Which of the following electron configurations are possible, and which are not: (*a*) $1s^2 2s^2 2p^6 3s^3$; (*b*) $1s^2 2s^2 2p^6 3s^2 3p^5 4s^2$; (*c*) $1s^2 2s^2 2p^6 2d^1$?

RESPONSE (*a*) This is not allowed, because too many electrons (3) are shown in the s subshell of the M ($n = 3$) shell. The s subshell has $m_l = 0$, with two slots only, for "spin up" and "spin down" electrons.
(*b*) This is allowed, but it is an excited state. One of the electrons from the $3p$ subshell has jumped up to the $4s$ subshell. Since there are 19 electrons, the element is potassium.
(*c*) This is not allowed, because there is no d ($l = 2$) subshell in the $n = 2$ shell (Table 28–1). The outermost electron will have to be (at least) in the $n = 3$ shell.

EXERCISE D Write the complete ground-state configuration for gallium, with its 31 electrons.

The grouping of atoms in the periodic table is according to increasing atomic number, Z. There is also a strong regularity according to chemical properties. Although this is treated in chemistry textbooks, we discuss it here briefly because it is a result of quantum mechanics. See the periodic table on the inside back cover.

All the noble gases (in column VIII of the periodic table) have completely filled shells or subshells. That is, their outermost subshell is completely full, and the electron distribution is spherically symmetric. With such full spherical symmetry, other electrons are not attracted nor are electrons readily lost (ionization energy is high). This is why the noble gases are nonreactive (more on this when we discuss molecules and bonding in Chapter 29). Column VII contains the **halogens**, which lack one electron from a filled shell. Because of the shapes of the orbits (see Section 29–1), an additional electron can be accepted from another atom, and hence these elements are quite reactive. They have a valence of -1, meaning that when an extra electron is acquired, the resulting ion has a net charge of $-1e$. Column I of the periodic table contains the **alkali metals**, all of which have a single outer s electron. This electron spends most of its time outside the inner closed shells and subshells which shield it from most of the nuclear charge. Indeed, it is relatively far from the nucleus and is attracted to it by a net charge of only about $+1e$, because of the shielding effect of the other electrons. Hence this outer electron is easily removed and can spend much of its time around another atom, forming a molecule. This is why the alkali metals are highly reactive and have a valence of $+1$. The other columns of the periodic table can be treated similarly.

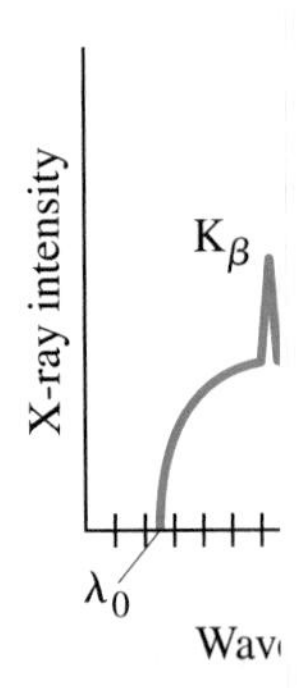

FIGURE 28–1 Spectrum of X molybdenum t tube operated

FIGURE 28–1 photon produc decelerated by target atom.

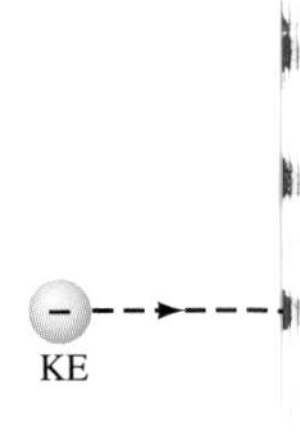

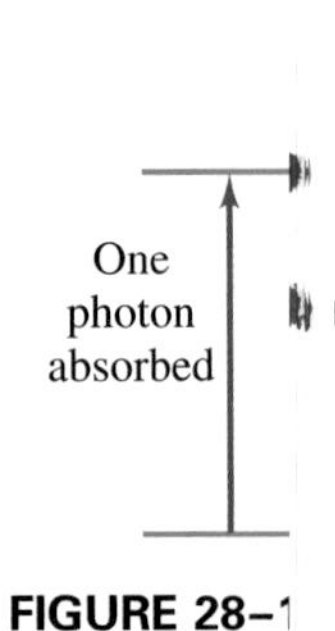

FIGURE 28–1

PHYSI

Fl

Ionic bond

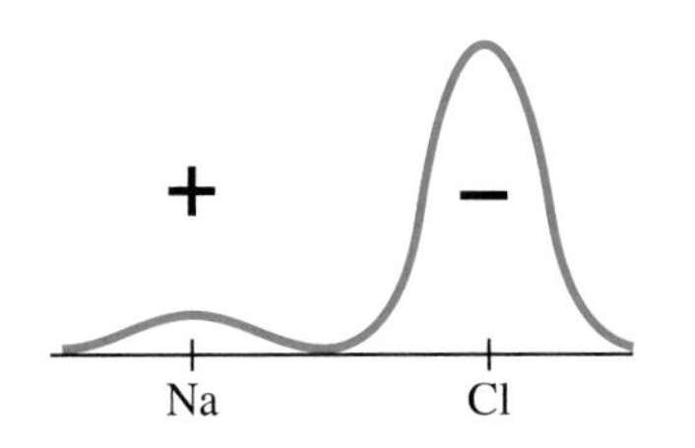

FIGURE 29–3 Probability distribution for the last electron of Na in NaCl.

FIGURE 29–4 In a neutral sodium atom, the 10 inner electrons shield the nucleus, so the single outer electron is attracted by a net charge of $+1e$.

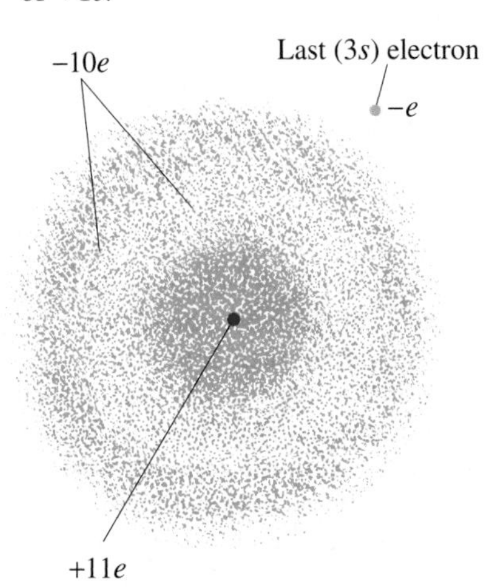

Ionic Bonds

An **ionic bond** is, in a sense, a special case of the covalent bond. Instead of the electrons being shared equally, they are shared unequally. For example, in sodium chloride (NaCl), the outer electron of the sodium spends nearly all its time around the chlorine (Fig. 29–3). The chlorine atom acquires a net negative charge as a result of the extra electron, whereas the sodium atom is left with a net positive charge. The electrostatic attraction between these two charged atoms holds them together. The resulting bond is called an *ionic bond* because it is created by the attraction between the two ions (Na^+ and Cl^-). But to understand the ionic bond, we must understand why the extra electron from the sodium spends so much of its time around the chlorine. After all, the chlorine is neutral; why should it attract another electron?

The answer lies in the probability distributions of the two neutral atoms. Sodium contains 11 electrons, 10 of which are in spherically symmetric closed shells (Fig. 29–4). The last electron spends most of its time beyond these closed shells. Because the closed shells have a total charge of $-10e$ and the nucleus has charge $+11e$, the outermost electron in sodium "feels" a net attraction due to $+1e$. It is not held very strongly. On the other hand, 12 of chlorine's 17 electrons form closed shells, or subshells (corresponding to $1s^2 2s^2 2p^6 3s^2$). These 12 form a spherically symmetric shield around the nucleus. The other five electrons are in $3p$ states whose probability distributions are not spherically symmetric and have a form similar to those for the $2p$ states in hydrogen shown in Fig. 28–9b and c. Four of these $3p$ electrons can have "doughnut-shaped" distributions symmetric about the z axis, as shown in Fig. 29–5. The fifth can have a "barbell-shaped" distribution (as for $m_l = 0$ in Fig. 28–9b), which in Fig. 29–5 is shown only in dashed outline because it is half empty. That is, the exclusion principle allows one more electron to be in this state (it will have spin opposite to that of the electron already there). If an extra electron—say from a Na atom—happens to be in the vicinity, it can be in this state, say at point x in Fig. 29–5. It could experience an attraction due to as much as $+5e$ because the $+17e$ of the nucleus is partly shielded at this point by the 12 inner electrons. Thus, the outer electron of a sodium atom will be more strongly attracted by the $+5e$ of the chlorine atom than by the $+1e$ of its own atom. This, combined with the strong attraction between the two ions when the extra electron stays with the Cl^-, produces the charge distribution of Fig. 29–3, and hence the ionic bond.

FIGURE 29–5 Neutral chlorine atom. The $+17e$ of the nucleus is shielded by the 12 electrons in the inner shells and subshells. Four of the five $3p$ electrons are shown in doughnut-shaped clouds, and the fifth is in the (dashed-line) cloud concentrated about the z axis (vertical). An extra electron at x will be attracted by a net charge that can be as much as $+5e$.

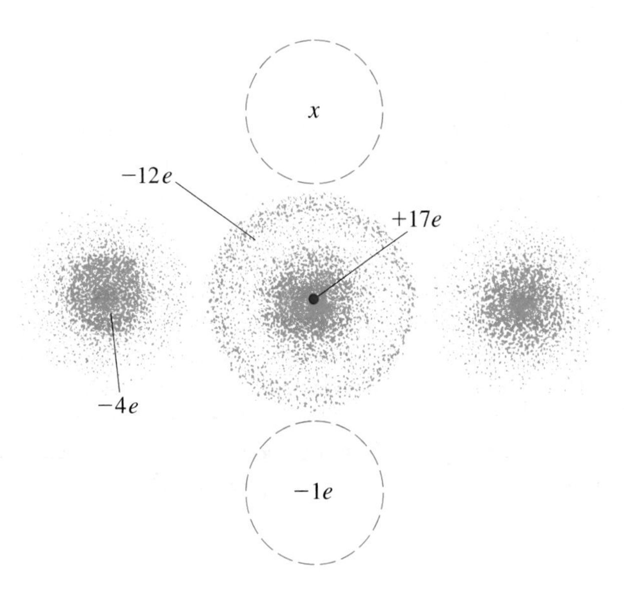

Partial Ionic Character of Covalent Bonds

A pure covalent bond in which the electrons are shared equally occurs mainly in symmetrical molecules such as H_2, O_2, and Cl_2. When the atoms involved are different from each other, it is usual to find that the shared electrons are more likely to be in the vicinity of one atom than the other. The extreme case is an ionic bond; in intermediate cases the covalent bond is said to have a *partial ionic character*. The molecules themselves are **polar**—that is, one part (or parts) of the molecule has a net positive charge and other parts a net negative charge. An example is the water molecule, H_2O (Fig. 29–6). The shared electrons are more likely to be found around the oxygen atom than around the two hydrogens.

Polar molecules

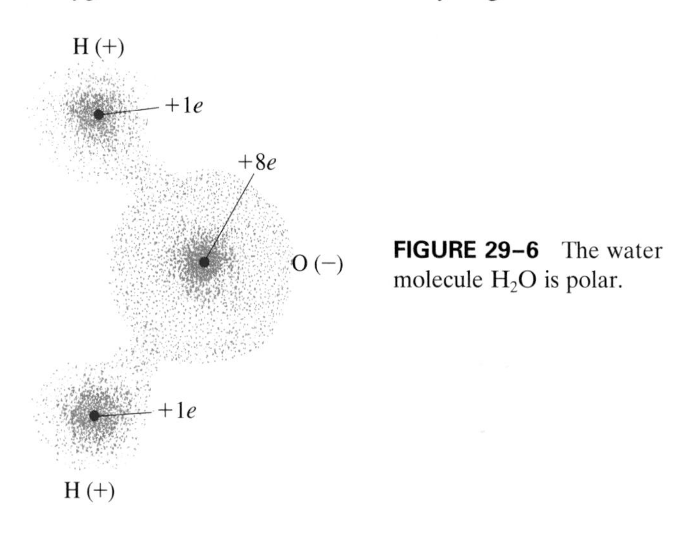

FIGURE 29–6 The water molecule H_2O is polar.

The reason is similar to that discussed above in connection with ionic bonds. Oxygen has eight electrons ($1s^2 2s^2 2p^4$), of which four form a spherically symmetric core and the other four could have, for example, a doughnut-shaped distribution. The barbell-shaped distribution on the z axis (like that shown dashed in Fig. 29–5) could be empty, so electrons from hydrogen atoms can be attracted by a net charge of $+4e$. They are also attracted by the H nuclei, so they partly orbit the H atoms as well as the O atom. The net effect is that there is a net positive charge on each H atom (less than $+1e$), because the electrons spend only part of their time there. And, there is a net negative charge on the O atom.

* 29–2 Potential-Energy Diagrams for Molecules

It is useful to analyze the interaction between two objects—say, between two atoms or molecules—with the use of a potential-energy diagram, a plot of the potential energy versus the separation distance.

For the simple case of two point charges, q_1 and q_2, the PE is given by (we combine Eqs. 17–2 and 17–5)

$$\text{PE} = \frac{1}{4\pi\epsilon_0}\frac{q_1 q_2}{r},$$

where r is the distance between the charges, and the constant $(1/4\pi\epsilon_0)$ is equal to $9.0 \times 10^9\,\text{N}\cdot\text{m}^2/\text{C}^2$. If the two charges have the same sign, the PE is positive for all values of r, and a graph of PE versus r in this case is shown in Fig. 29–7a. The force is repulsive (the charges have the *same* sign) and the curve rises as r decreases; this makes sense since work is done to bring the charges together, thereby increasing their potential energy. If, on the other hand, the two charges are of the *opposite* sign, the PE is negative because the product $q_1 q_2$ is negative. The force is attractive in this case, and the graph of PE versus r looks like Fig. 29–7b. The PE becomes more *negative* as r decreases.

FIGURE 29–7 Potential energy as a function of separation for two point charges of (a) like sign and (b) opposite sign.

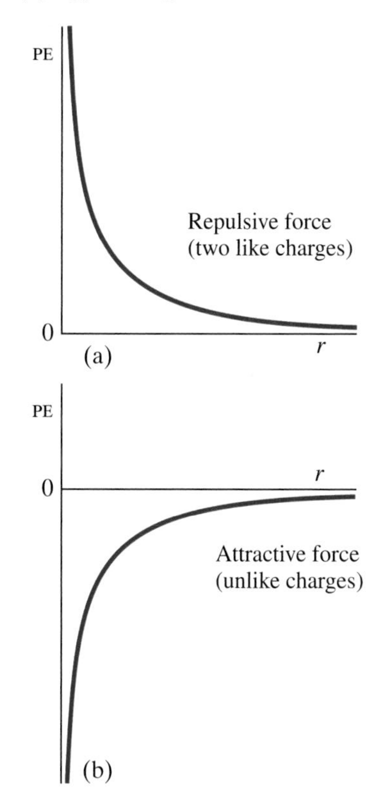

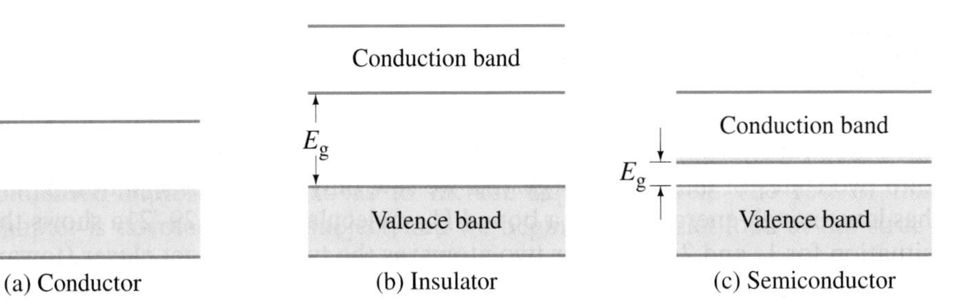

FIGURE 29–23 Energy bands for (a) a conductor, (b) an insulator, which has a large energy gap E_g, and (c) a semiconductor, which has a small energy gap E_g. Shading represents occupied states. Pale shading in (c) represents electrons that can pass from the top of the valence band to the bottom of the conduction band due to thermal agitation at room temperature (exaggerated).

Semiconductors (pure)

Figure 29–23 compares the relevant energy bands (a) for conductors, (b) for insulators, and also (c) for the important class of materials known as **semiconductors**. The bands for a pure (or **intrinsic**) semiconductor, such as silicon or germanium, are like those for an insulator, except that the unfilled conduction band is separated from the filled valence band by a much smaller energy gap, E_g, typically on the order of 1 eV. At room temperature, a few electrons can acquire enough thermal energy to reach the conduction band, and so a very small current may flow when a voltage is applied. At higher temperatures, more electrons have enough energy to jump the gap. Often this effect can more than offset the effects of more frequent collisions due to increased disorder at higher temperature, so the resistivity of semiconductors can *decrease* with increasing temperature (see Table 18–1). But this is not the whole story of semiconductor conduction. When a potential difference is applied to a semiconductor, the few electrons in the conduction band move toward the positive electrode. Electrons in the valence band try to do the same thing, and a few can because there are a small number of unoccupied states which were left empty by the electrons reaching the conduction band. Such unfilled electron states are called **holes**. Each electron in the valence band that fills a hole in this way as it moves toward the positive electrode leaves behind its own hole, so the holes migrate toward the negative electrode. As the electrons tend to accumulate at one side of the material, the holes tend to accumulate on the opposite side. We will look at this phenomenon in more detail in the next Section.

Holes (in a semiconductor)

EXAMPLE 29–4 Calculating the energy gap. It is found that the conductivity of a certain semiconductor increases when light of wavelength 345 nm or shorter strikes it, suggesting that electrons are being promoted from the valence band to the conduction band. What is the energy gap, E_g, for this semiconductor?

APPROACH The longest wavelength (lowest energy) photon to cause an increase in conductivity has $\lambda = 345\,\text{nm}$, and its energy $(= hf)$ equals the energy gap.

SOLUTION The gap energy equals the energy of a $\lambda = 345$-nm photon:

$$E_g = hf = \frac{hc}{\lambda} = \frac{(6.63 \times 10^{-34}\,\text{J}\cdot\text{s})(3.00 \times 10^{8}\,\text{m/s})}{(345 \times 10^{-9}\,\text{m})(1.60 \times 10^{-19}\,\text{J/eV})} = 3.6\,\text{eV}.$$

CONCEPTUAL EXAMPLE 29–5 Which is transparent? The energy gap for silicon is 1.14 eV at room temperature, whereas that of zinc sulfide (ZnS) is 3.6 eV. Which one of these is opaque to visible light, and which is transparent?

RESPONSE Visible light photons span energies from roughly 1.8 eV to 3.2 eV ($E = hf = hc/\lambda$ where $\lambda = 400\,\text{nm}$ to $700\,\text{nm}$ and $1\,\text{eV} = 1.6 \times 10^{-19}\,\text{J}$.) Light is absorbed by the electrons in a material. Silicon's energy gap is small enough to absorb these photons, thus bumping electrons well up into the conduction band, so silicon is opaque. On the other hand, zinc sulfide's energy gap is too large to absorb visible photons, so the light can pass through the material; it can be transparent.

* 29–7 Semiconductors and Doping

Nearly all electronic devices today use semiconductors. The most common are silicon (Si) and germanium (Ge). An atom of silicon or germanium has four outer electrons that act to hold the atoms in the regular lattice structure of the crystal, shown schematically in Fig. 29–24a. Germanium and silicon acquire properties useful for electronics only when a tiny amount of impurity is introduced into the crystal structure (perhaps 1 part in 10^6 or 10^7). This is called **doping** the semiconductor. Two kinds of doped semiconductor can be made, depending on the type of impurity used. If the impurity is an element whose atoms have five outer electrons, such as arsenic, we have the situation shown in Fig. 29–24b, with the arsenic atoms holding positions in the crystal lattice where normally silicon atoms would be. Only four of arsenic's electrons fit into the bonding structure. The fifth does not fit in and can move relatively freely, somewhat like the electrons in a conductor. Because of this small number of extra electrons, a doped semiconductor becomes slightly conducting. The density of conduction electrons in an intrinsic (undoped) semiconductor at room temperature is very low, usually less than 1 per 10^9 atoms. With an impurity concentration of 1 in 10^6 or 10^7 when doped, the conductivity will be much higher and it can be controlled with great precision. An arsenic-doped silicon crystal is called an ***n*-type semiconductor** because *negative* charges (electrons) carry the electric current.

Doped semiconductors

n-type

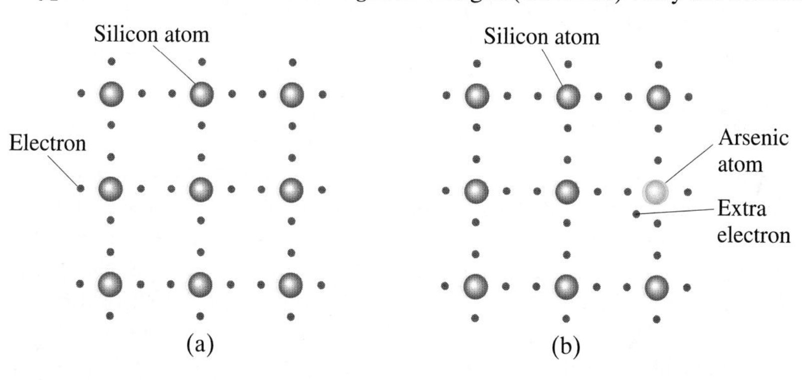

FIGURE 29–24 Two-dimensional representation of a silicon crystal. (a) Four (outer) electrons surround each silicon atom. (b) Silicon crystal doped with a small percentage of arsenic atoms: the extra electron doesn't fit into the crystal lattice and so is free to move about. This is an n-type semiconductor.

In a ***p*-type semiconductor**, a small percentage of semiconductor atoms are replaced by atoms with three outer electrons—such as gallium. As shown in Fig. 29–25a, there is a "hole" in the lattice structure near a gallium atom since it has only three outer electrons. Electrons from nearby silicon atoms can jump into this hole and fill it. But this leaves a hole where that electron had previously been, Fig. 29–25b. The vast majority of atoms are silicon, so holes are almost always next to a silicon atom. Since silicon atoms require four outer electrons to be neutral, this means that there is a net positive charge at the hole. Whenever an electron moves to fill a hole, the positive hole is then at the previous position of that electron. Another electron can then fill this hole, and the hole thus moves to a new location; and so on. This type of semiconductor is called *p-type* because it is the positive holes that seem to carry the electric current. Note, however, that both *p-type* and *n-type* semiconductors have *no net charge* on them.

p-type

CAUTION

p-type semiconductors act as though + charges move—but electrons actually do the moving

Holes are positive

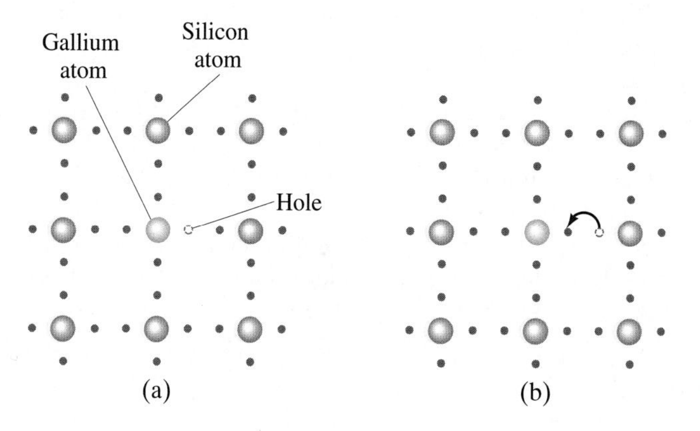

FIGURE 29–25 A p-type semiconductor, gallium-doped silicon. (a) Gallium has only three outer electrons, so there is an empty spot, or *hole* in the structure. (b) Electrons from silicon atoms can jump into the hole and fill it. As a result, the hole moves to a new location (to the right in this figure), to where the electron used to be.

According to the band theory (Section 29–6), in a doped semiconductor the impurity provides additional energy states between the bands as shown in Fig. 29–26. In an n-type semiconductor, the impurity energy level lies just below the conduction band, Fig. 29–26a. Electrons in this energy level need only about 0.05 eV in Si (even less in Ge) to reach the conduction band; this is on the order of the thermal energy, $\frac{3}{2}kT$ (= 0.04 eV at 300 K), so transitions occur readily at room temperature. This energy level can thus supply electrons to the conduction band, so it is called a **donor** level. In p-type semiconductors, the impurity energy level is just above the valence band (Fig. 29–26b). It is called an **acceptor** level because electrons from the valence band can easily jump into it. Positive holes are left behind in the valence band, and as other electrons move into these holes, the holes move about as discussed earlier.

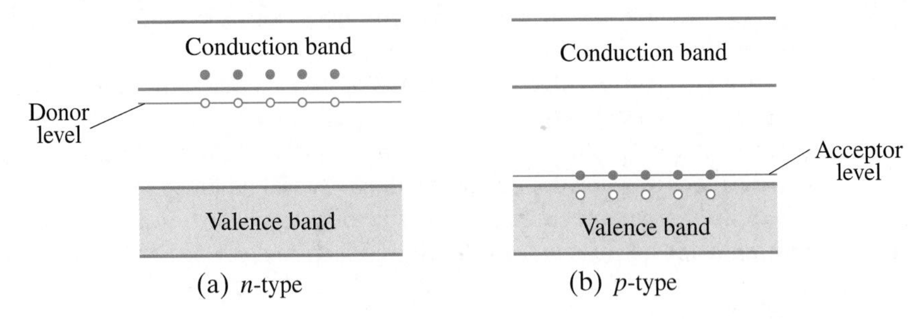

FIGURE 29–26 Impurity energy levels in doped semiconductors.

* 29–8 Semiconductor Diodes

Semiconductor diodes and transistors are essential components of modern electronic devices. The miniaturization achieved today allows many thousands of diodes, transistors, resistors, and so on, to be placed on a single *chip* less than a millimeter on a side. We now discuss, briefly and qualitatively, the operation of diodes and transistors.

pn *junction diode*

When an n-type semiconductor is joined to a p-type, a ***pn* junction diode** is formed. Separately, the two semiconductors are electrically neutral. When joined, a few electrons near the junction diffuse from the n-type into the p-type semiconductor, where they fill a few of the holes. The n-type is left with a positive charge, and the p-type acquires a net negative charge. Thus a potential difference is established, with the n side positive relative to the p side, and this prevents further diffusion of electrons.

If a battery is connected to a diode with the positive terminal to the p side and the negative terminal to the n side as in Fig. 29–27a, the externally applied voltage opposes the internal potential difference and the diode is said to be **forward biased**. If the voltage is great enough (about 0.3 V for Ge, 0.6 V for Si at room temperature), a current will flow. The positive holes in the p-type semiconductor are repelled by the positive terminal of the battery, and the electrons in the n-type are repelled by the negative terminal of the battery. The holes and electrons meet at the junction, and the electrons cross over and fill the holes. A current is flowing. Meanwhile, the positive terminal of the battery is continually pulling electrons off the p end, forming new holes, and electrons are being supplied by the negative terminal at the n end. Consequently, a large current flows through the diode.

When the diode is **reverse biased**, as in Fig. 29–27b, the holes in the p end are attracted to the battery's negative terminal and the electrons in the n end are attracted to the positive terminal. The current carriers do not meet near the junction and, ideally, no current flows.

A graph of current versus voltage for a typical diode is shown in Fig. 29–28. As can be seen, a real diode does allow a small amount of reverse current to flow.[†] For most practical purposes, this is negligible.

FIGURE 29–27 Schematic diagram showing how a semiconductor diode operates. Current flows when the voltage is connected in forward bias, as in (a), but not when connected in reverse bias, as in (b).

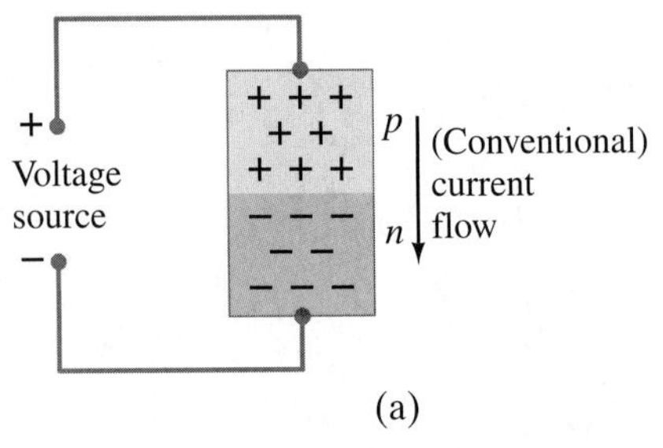

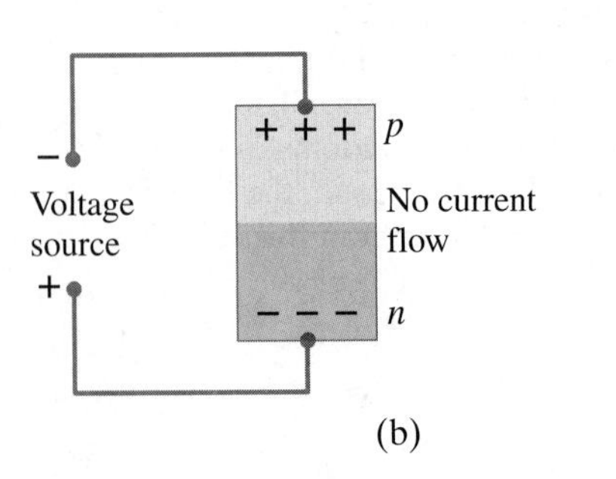

[†] At room temperature it is a few μA in Ge, and a few pA in Si. The reverse current increases rapidly with temperature, however, and may render a diode ineffective above 200°C.

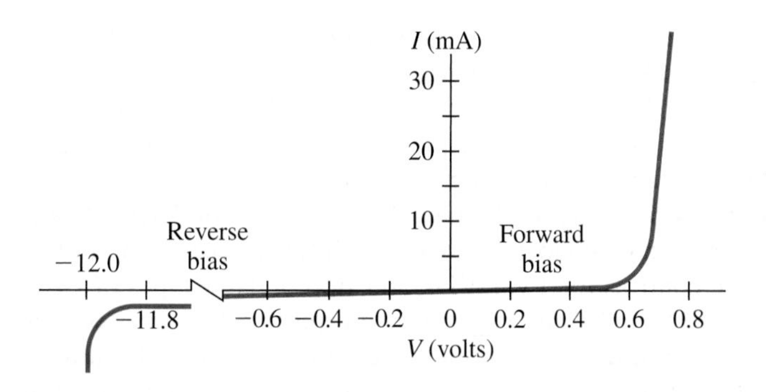

FIGURE 29–28 Current through a diode as a function of applied voltage.

EXAMPLE 29–6 **A diode.** The diode whose current–voltage characteristics are shown in Fig. 29–28 is connected in series with a 4.0-V battery and a resistor. If a current of 15 mA is to pass through the diode, what resistance must the resistor have?

APPROACH We use Fig. 29–28, where we see that the voltage drop across the diode is about 0.7 V when the current is 15 mA. Then we use simple circuit analysis and Ohm's law (Chapters 18 and 19).

SOLUTION The voltage drop across the resistor is $4.0\,\mathrm{V} - 0.7\,\mathrm{V} = 3.3\,\mathrm{V}$, so $R = V/I = (3.3\,\mathrm{V})/(1.5 \times 10^{-2}\,\mathrm{A}) = 220\,\Omega$.

The symbol for a diode is

[diode]

where the arrow represents the direction conventional (+) current flows readily.

Since a *pn* junction diode allows current to flow only in one direction (as long as the voltage is not too high), it can serve as a **rectifier**—to change ac into dc. A simple rectifier circuit is shown in Fig. 29–29a. The ac source applies a voltage across the diode alternately positive and negative. Only during half of each cycle will a current pass through the diode; only then is there a current through the resistor R. Hence, a graph of the voltage V_{ab} across R as a function of time looks like the output voltage shown in Fig. 29–29b. This **half-wave rectification** is not exactly dc, but it is unidirectional. More useful is a **full-wave rectifier** circuit, which uses two diodes (or sometimes four) as shown in Fig. 29–30a. At any given instant, either one diode or the other will conduct current to the right. Therefore, the output across the load resistor R will be as shown in Fig. 29–30b. Actually this is the voltage if the capacitor C were not in the circuit. The capacitor tends to store charge and, if the time constant RC is sufficiently long, helps to smooth out the current as shown in Fig. 29–30c. (The variation in output shown in Fig. 29–30c is called *ripple voltage*.)

Rectifier circuits are important because most line voltage in buildings is ac, and most electronic devices require a dc voltage for their operation. Hence, diodes are found in nearly all electronic devices including radio and TV sets, calculators, and computers.

FIGURE 29–29 (a) A simple (half-wave) rectifier circuit using a semiconductor diode. (b) AC source input voltage, and output voltage across R, as functions of time.

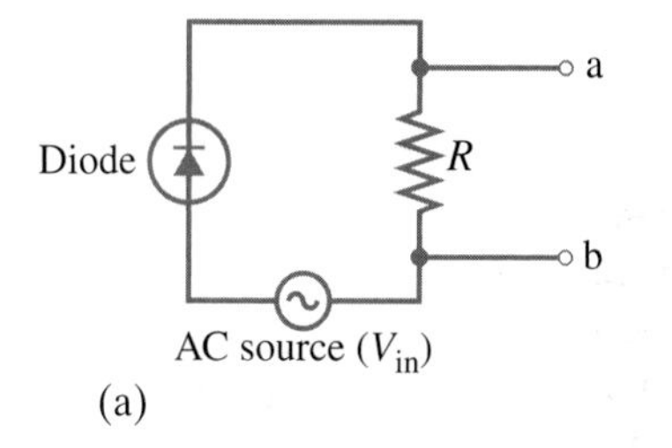

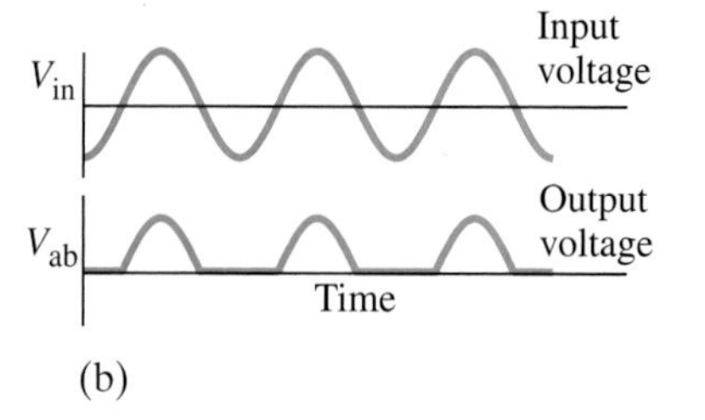

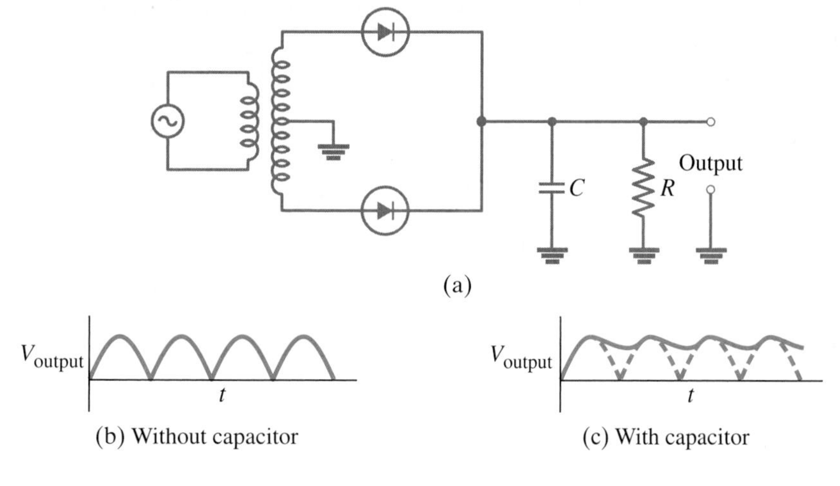

FIGURE 29–30 (a) Full-wave-rectifier circuit (including a transformer so the magnitude of the voltage can be changed). (b) Output voltage in the absence of capacitor C. (c) Output voltage with the capacitor in the circuit.

Additional Example

EXAMPLE 30–7 **KE of the α in $^{232}_{92}U$ decay.** For the $^{232}_{92}U$ decay of Example 30–6, how much of the 5.4-MeV disintegration energy will be carried off by the α particle?

APPROACH In any reaction, momentum must be conserved as well as energy.

SOLUTION Before disintegation, the nucleus can be assumed to be at rest, so the total momentum was zero. After disintegration, the total vector momentum must still be zero so the magnitude of the α particle's momentum must equal the magnitude of the daughter's momentum (Fig. 30–6):

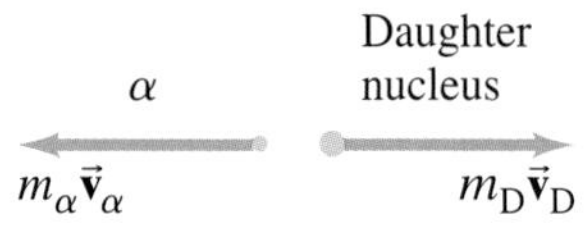

$$m_\alpha v_\alpha = m_D v_D.$$

Thus $v_\alpha = m_D v_D/m_\alpha$ and the α's kinetic energy is

$$\text{KE}_\alpha = \tfrac{1}{2}m_\alpha v_\alpha^2 = \tfrac{1}{2}m_\alpha\left(\frac{m_D v_D}{m_\alpha}\right)^2 = \tfrac{1}{2}m_D v_D^2\left(\frac{m_D}{m_\alpha}\right) = \left(\frac{m_D}{m_\alpha}\right)\text{KE}_D$$

$$= \left(\frac{228.028731\text{ u}}{4.002603\text{ u}}\right)\text{KE}_D = 57\text{KE}_D.$$

The total disintegration energy is $Q = \text{KE}_\alpha + \text{KE}_D = 57\text{KE}_D + \text{KE}_D = 58\text{KE}_D$. Hence

$$\text{KE}_\alpha = \frac{57}{58}Q = 5.3\text{ MeV}.$$

The lighter α particle carries off (57/58) or 98% of the total KE.

Daughter nucleus, α, $m_\alpha\vec{\mathbf{v}}_\alpha$, $m_D\vec{\mathbf{v}}_D$

FIGURE 30–6 Momentum conservation in Example 30–7.

Why α particles?

Why, you may wonder, do nuclei emit this combination of four nucleons called an α particle? Why not just four separate nucleons, or even one? The answer is that the α particle is very strongly bound, so that its mass is significantly less than that of four separate nucleons. As we saw in Example 30–3, two protons and two neutrons separately have a total mass of about 4.032980 u (electrons included). The total mass of $^{228}_{90}Th$ plus four separate nucleons is 232.061711 u, which is greater than the mass of the parent (232.037146). Such a decay could not occur because it would violate the conservation of energy. Similarly, it is almost always true that the emission of a single nucleon is energetically not possible.

Smoke Detectors—An Application

PHYSICS APPLIED
Smoke detector

One widespread application of nuclear physics is present in nearly every home in the form of an ordinary **smoke detector**. The most common type of detector contains about 0.2 mg of the radioactive americium isotope, $^{241}_{95}Am$, in the form of AmO_2. The radiation continually ionizes the nitrogen and oxygen molecules in the air space between two oppositely charged plates. The resulting conductivity allows a small steady current. If smoke enters, the radiation is absorbed by the smoke particles rather than by the air molecules, thus reducing the current. The current drop is detected by the device's electronics and sets off the alarm. The radiation dose that escapes from an intact americium smoke detector is much less than the natural radioactive background, and so can be considered relatively harmless. There is no question that smoke detectors save lives and reduce property damage.

30–5 Beta Decay

β^- Decay

Transmutation of elements also occurs when a nucleus decays by β decay—that is, with the emission of an electron or β^- particle. The nucleus ${}^{14}_{6}\text{C}$, for example, emits an electron when it decays:

$${}^{14}_{6}\text{C} \rightarrow {}^{14}_{7}\text{N} + \text{e}^- + \text{a neutrino},$$

where e^- is the symbol for the electron. (The symbol ${}^{\;0}_{-1}\text{e}$ is sometimes used for the electron whose charge corresponds to $Z = -1$ and, since it is not a nucleon and has very small mass, then $A = 0$.) The particle known as the neutrino, whose charge $q = 0$ and whose rest mass is very small or zero, was not initially detected and was only later hypothesized to exist, as we shall discuss later in this Section. No nucleons are lost when an electron is emitted, and the total number of nucleons, A, is the same in the daughter nucleus as in the parent. But because an electron has been emitted from the nucleus itself, the charge on the daughter nucleus is $+1e$ greater than that on the parent. The parent nucleus in the decay written above had $Z = +6$, so from charge conservation the nucleus remaining behind must have a charge of $+7e$. So the daughter nucleus has $Z = 7$, which is nitrogen.

It must be carefully noted that the electron emitted in β decay is *not* an orbital electron. Instead, the electron is created *within the nucleus itself.* What happens is that one of the neutrons changes to a proton and in the process (to conserve charge) emits an electron. Indeed, free neutrons actually do decay in this fashion:

CAUTION

β-decay e^- comes from nucleus (not an orbital electron)

$$\text{n} \rightarrow \text{p} + \text{e}^- + \text{a neutrino}.$$

Because of their origin in the nucleus, the electrons emitted in β decay are often referred to as "β particles," rather than as electrons, to remind us of their origin. They are, nonetheless, indistinguishable from orbital electrons.

EXAMPLE 30–8 Energy release in ${}^{14}_{6}\text{C}$ decay. How much energy is released when ${}^{14}_{6}\text{C}$ decays to ${}^{14}_{7}\text{N}$ by β emission?

APPROACH We find the mass difference before and after decay, Δm. The energy released is $E = (\Delta m)c^2$. The masses given in Appendix B are those of the neutral atom, and we have to keep track of the electrons involved. Assume the parent nucleus has six orbiting electrons so it is neutral; its mass is 14.003242 u. The daughter in this decay, ${}^{14}_{7}\text{N}$, is not neutral since it has the same six orbital electrons circling it but the nucleus has a charge of $+7e$. However, the mass of this daughter with its six electrons, plus the mass of the emitted electron (which makes a total of seven electrons), is just the mass of a neutral nitrogen atom.

CAUTION

Be careful with atomic and electron masses in β decay

SOLUTION The total mass in the final state is

$$(\text{mass of } {}^{14}_{7}\text{N nucleus} + 6 \text{ electrons}) + (\text{mass of 1 electron}),$$

and this is equal to

$$\text{mass of neutral } {}^{14}_{7}\text{N (includes 7 electrons)},$$

which, from Appendix B is a mass of 14.003074 u. So the mass difference is $14.003242\,\text{u} - 14.003074\,\text{u} = 0.000168\,\text{u}$, which is equivalent to an energy change $\Delta m\, c^2 = (0.000168\,\text{u})(931.5\,\text{MeV/u}) = 0.156\,\text{MeV}$ or 156 keV.

NOTE The neutrino doesn't contribute to either the mass or charge balance since it has $q = 0$ and $m \approx 0$.

According to Example 30–8, we would expect the emitted electron to have a kinetic energy of 156 keV. (The daughter nucleus, because its mass is very much larger than that of the electron, recoils with very low velocity and hence gets very little of the kinetic energy.) Indeed, very careful measurements indicate that a few emitted β particles do have kinetic energy close to this calculated value. But the vast majority of emitted electrons have somewhat less energy. In fact, the energy of the emitted electron can be anywhere from zero up to the maximum value as calculated above. This range of electron kinetic energy was found for any β decay. It was as if the law of conservation of energy was being violated, and indeed Bohr actually considered this possibility. Careful experiments indicated that linear momentum and angular momentum also did not seem to be conserved. Physicists were troubled at the prospect of having to give up these laws, which had worked so well in all previous situations. In 1930, Wolfgang Pauli proposed an alternate solution: perhaps a new particle that was very difficult to detect was emitted during β decay in addition to the electron. This hypothesized particle could be carrying off the energy, momentum, and angular momentum required to maintain the conservation laws. This new particle was named the **neutrino**—meaning "little neutral one"—by the great Italian physicist Enrico Fermi (1901–1954; Fig. 30–7), who in 1934 worked out a detailed theory of β decay. (It was Fermi who, in this theory, postulated the existence of the fourth force in nature which we call the *weak nuclear force*.) The electron neutrino has zero charge, spin of $\frac{1}{2}\hbar$, and was long thought to have zero rest mass, although today it seems possible it does have a very tiny rest mass ($< 0.6\,\mathrm{eV}/c^2$). If its rest mass is zero, it is much like a photon in that it is neutral and travels at the speed of light. But the neutrino is far more difficult to detect. In 1956, complex experiments produced further evidence for the existence of the neutrino; but by then, most physicists had already accepted its existence.

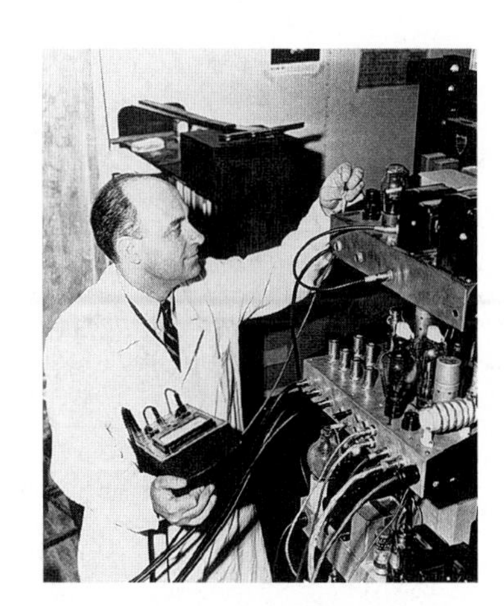

FIGURE 30–7 Enrico Fermi. Fermi contributed significantly to both theoretical and experimental physics, a feat almost unique in modern times.

The symbol for the neutrino is the Greek letter nu (ν). The correct way of writing the decay of $^{14}_{6}\mathrm{C}$ is then

β^- decay

$$^{14}_{6}\mathrm{C} \rightarrow {}^{14}_{7}\mathrm{N} + \mathrm{e}^- + \bar{\nu}.$$

The bar ($^-$) over the neutrino symbol is to indicate that it is an "antineutrino." (Why this is called an antineutrino rather than simply a neutrino need not concern us now; it is discussed in Chapter 32.)

β^+ Decay

Many isotopes decay by electron emission. They are always isotopes that have too many neutrons compared to the number of protons. That is, they are isotopes that lie above the stable isotopes plotted in Fig. 30–2. But what about unstable isotopes that have too few neutrons compared to their number of protons—those that fall below the stable isotopes of Fig. 30–2? These, it turns out, decay by emitting a **positron** instead of an electron. A positron (sometimes called an e^+ or β^+ particle) has the same mass as the electron, but it has a positive charge of $+1e$. Because it is so like an electron, except for its charge, the positron is called the **antiparticle**† to the electron. An example of a β^+ decay is that of $^{19}_{10}\mathrm{Ne}$:

Positron (β^+) decay

$$^{19}_{10}\mathrm{Ne} \rightarrow {}^{19}_{9}\mathrm{F} + \mathrm{e}^+ + \nu,$$

where e^+ (or $^{0}_{1}\mathrm{e}$) stands for a positron. Note that the ν emitted here is a neutrino, whereas that emitted in β^- decay is called an antineutrino. Thus an antielectron (= positron) is emitted with a neutrino, whereas an antineutrino is emitted with an electron; this gives a certain balance as discussed in Chapter 32.

†Discussed in Chapter 32. Briefly, an antiparticle has the same mass as its corresponding particle, but opposite charge.

We can write β^- and β^+ decay, in general, as follows:

$$^A_Z\text{N} \rightarrow {}^{A}_{Z+1}\text{N}' + e^- + \bar{\nu} \qquad [\beta^- \text{ decay}]$$

$$^A_Z\text{N} \rightarrow {}^{A}_{Z-1}\text{N}' + e^+ + \nu, \qquad [\beta^+ \text{ decay}]$$

where N is the parent nucleus and N′ is the daughter.

Electron Capture

Besides β^- and β^+ emission, there is a third related process. This is **electron capture** (abbreviated EC in Appendix B) and occurs when a nucleus absorbs one of its orbiting electrons. An example is ^7_4Be, which as a result becomes ^7_3Li. The process is written

Electron capture

$$^7_4\text{Be} + e^- \rightarrow {}^7_3\text{Li} + \nu,$$

or, in general,

$$^A_Z\text{N} + e^- \rightarrow {}^{A}_{Z-1}\text{N}' + \nu. \qquad [\text{electron capture}]$$

Usually it is an electron in the innermost (K) shell that is captured, in which case it is called "K-capture." The electron disappears in the process, and a proton in the nucleus becomes a neutron; a neutrino is emitted as a result. This process is inferred experimentally by detection of emitted X-rays (due to other electrons jumping down to fill the empty state) of just the proper energy.

K-capture

In β decay, it is the weak nuclear force that plays the crucial role. The neutrino is unique in that it interacts with matter only via the weak force, which is why it is so hard to detect.

30–6 Gamma Decay

Gamma rays are photons having very high energy. They have their origin in the decay of a nucleus, much like emission of photons by excited atoms. Like an atom, a nucleus itself can be in an excited state. When it jumps down to a lower energy state, or to the ground state, it emits a photon which we call a γ ray. The possible energy levels of a nucleus are much farther apart than those of an atom: on the order of keV or MeV, as compared to a few eV for electrons in an atom. Hence, the emitted photons have energies that can range from a few keV to several MeV. For a given decay, the γ ray always has the same energy. Since a γ ray carries no charge, there is no change in the element as a result of a γ decay.

How does a nucleus get into an excited state? It may occur because of a violent collision with another particle. More commonly, the nucleus remaining after a previous radioactive decay may be in an excited state. A typical example is shown in the energy-level diagram of Fig. 30–8. $^{12}_5\text{B}$ can decay by β decay directly to the ground state of $^{12}_6\text{C}$; or it can go by β decay to an excited state of $^{12}_6\text{C}$, which then decays by emission of a 4.4-MeV γ ray to the ground state.

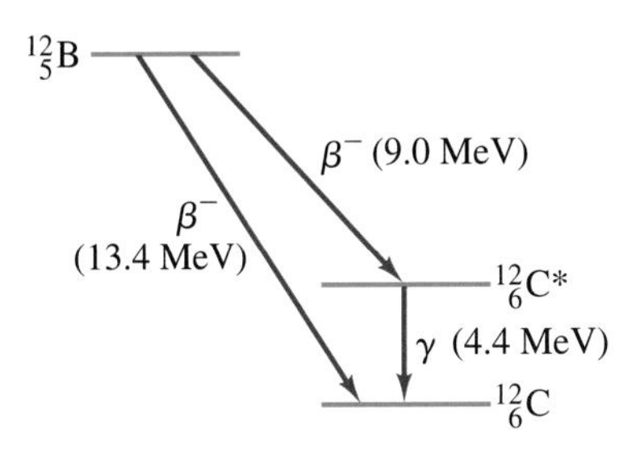

FIGURE 30–8 Energy-level diagram showing how $^{12}_5\text{B}$ can decay to the ground state of $^{12}_6\text{C}$ by β decay (total energy released = 13.4 MeV), or can instead β decay to an excited state of $^{12}_6\text{C}$ (indicated by *), which subsequently decays to its ground state by emitting a 4.4-MeV γ ray.

We can write γ decay as

$$^A_Z\text{N}^* \rightarrow {}^A_Z\text{N} + \gamma, \qquad [\gamma \text{ decay}]$$

where the asterisk means "excited state" of that nucleus.

* Isomers; Internal Conversion

In some cases, a nucleus may remain in an excited state for some time before it emits a γ ray. The nucleus is then said to be in a **metastable state** and is called an **isomer**.

An excited nucleus can sometimes return to the ground state by another process known as **internal conversion** with no γ ray emitted. In this process, the excited nucleus interacts with one of the orbital electrons and ejects this electron from the atom with the same kinetic energy (minus the binding energy of the electron) that an emitted γ ray would have had.

Internal conversion

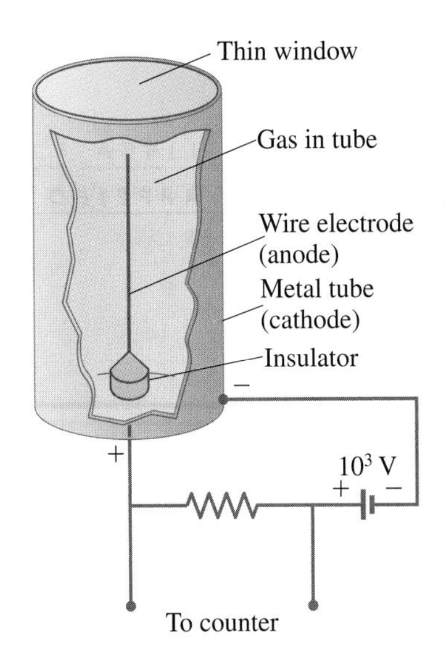

FIGURE 30–13 Diagram of a Geiger counter.

30–13 Detection of Radiation

Individual particles such as electrons, protons, α particles, neutrons, and γ rays are not detected directly by our senses. Consequently, a variety of instruments have been developed to detect them.

One of the most common is the **Geiger counter**. As shown in Fig. 30–13, it consists of a cylindrical metal tube filled with a certain type of gas. A long wire runs down the center and is kept at a high positive voltage $(\approx 10^3\,\text{V})$ with respect to the outer cylinder. The voltage is just slightly less than that required to ionize the gas atoms. When a charged particle enters through the thin "window" at one end of the tube, it ionizes a few atoms of the gas. The freed electrons are attracted toward the positive wire, and as they are accelerated they strike and ionize additional atoms. An "avalanche" of electrons is quickly produced, and when it reaches the wire anode, it produces a voltage pulse. The pulse, after being amplified, can be sent to an electronic counter, which counts how many particles have been detected. Or the pulses can be sent to a loudspeaker and each detection of a particle is heard as a "click." Only a fraction of the radiation emitted by a sample is detected by any detector.

Scintillators

A **scintillation counter** makes use of a solid, liquid, or gas known as a **scintillator** or **phosphor**. The atoms of a scintillator are easily excited when struck by an incoming particle and emit visible light when they return to their ground states. Typical scintillators are crystals of NaI and certain plastics. One face of a solid scintillator is cemented to a photomultiplier tube, and the whole is wrapped with opaque material to keep it light tight or is placed within a light-tight container. The **photomultiplier** (PM) **tube** converts the energy of the scintillator-emitted photon(s) into an electric signal. A PM tube is a vacuum tube containing several electrodes (typically 8 to 14), called *dynodes*, which are maintained at successively higher voltages as shown in Fig. 30–14. At its top surface is a photoelectric surface, called the *photocathode*, whose work function (Section 27–3) is low enough that an electron is easily released when struck by a photon from the scintillator. Such an electron is accelerated toward the first dynode. When it strikes the first dynode, the electron has acquired sufficient kinetic energy so that it can eject two to five more electrons. These, in turn, are accelerated to the second dynode, and a multiplication process begins. The number of electrons striking the last dynode may be 10^6 or more. Thus the passage of a particle through the scintillator results in an electric signal at the

PM tube

FIGURE 30–14 Scintillation counter with a photomultiplier tube.

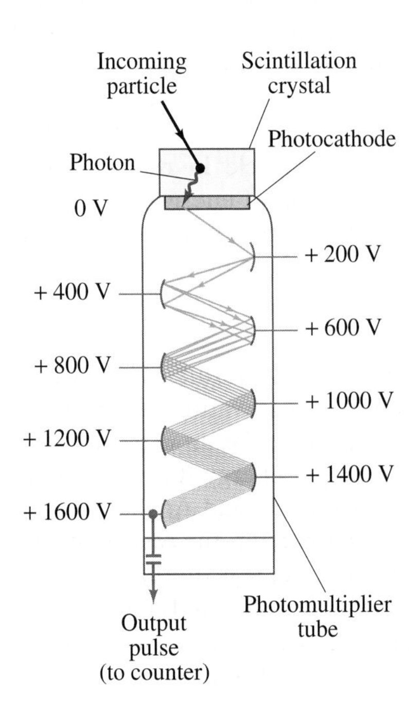

output of the PM tube that can be sent to an electronic counter just as for a Geiger tube. Solid scintillators are much more dense than the gas of a Geiger counter, and so are much more efficient detectors—especially for γ rays, which interact less with matter than do β rays. Scintillators that can measure the total energy deposited are much used today and are called **calorimeters**.

Calorimeter

In tracer work (Section 31–7), **liquid scintillators** are often used. Radioactive samples taken at different times or from different parts of an organism are placed directly in small bottles containing the liquid scintillator. This is particularly convenient for detection of β rays from ${}^{3}_{1}\mathrm{H}$ and ${}^{14}_{6}\mathrm{C}$, which have very low energies and have difficulty passing through the outer covering of a crystal scintillator or Geiger tube. A PM tube is still used to produce the electric signal.

Semiconductor detector (pn junction)

A **semiconductor detector** consists of a reverse-biased *pn* junction diode (Section 29–8). A particle passing through the junction can excite electrons into the conduction band, leaving holes in the valence band. The freed charges produce a short electrical pulse that can be counted just as for Geiger and scintillation counters. Silicon wafer semiconductors have their surface etched into tiny pixels, thus providing detailed particle position information.

Hospital workers and others who work around radiation carry *film badges* which detect the accumulation of radiation. The film inside is periodically replaced and developed, the darkness being related to total exposure (see Section 31–5).

Emulsions

The devices discussed so far are used for counting the number of particles (or decays of a radioactive isotope). Other devices allow the track of charged particles to be *seen*. The simplest is the **photographic emulsion**, which can be small and portable, used now particularly for cosmic-ray studies from balloons. A charged particle passing through a layer of photographic emulsion ionizes the atoms along its path. These points undergo a chemical change, and when the emulsion is developed the particle's path is revealed. (One type of neutrino, τ, was indirectly discovered using an emulsion at Fermilab; see Chapter 32.)

Cloud chamber

In a **cloud chamber**, a gas is cooled to a temperature slightly below its usual condensation point ("supercooled"), and gas molecules condense on any ionized molecules present. Ions produced when a charged particle passes through serve as centers on which tiny droplets form (Fig. 30–15). Light scatters more from these droplets than from the gas background, so a photo of the cloud chamber at the right moment shows the track of the particle. An important instrument in the early days of nuclear physics, it is little used today.

Bubble chamber

The **bubble chamber**, invented in 1952 by D. A. Glaser (1926–), makes use of a superheated liquid kept close to its normal boiling point. The bubbles characteristic of boiling form around ions produced by the passage of a charged particle. A photograph of the interior of the chamber reveals paths of particles that recently passed through. Because the bubble chamber uses a liquid, often liquid hydrogen, it is a much more efficient device than a cloud chamber for observing the tracks of charged particles and their interactions with the nuclei of the liquid. A magnetic field is usually applied across the chamber and the momentum of the moving particles can be determined from the radius of curvature of their paths.

Path of particle

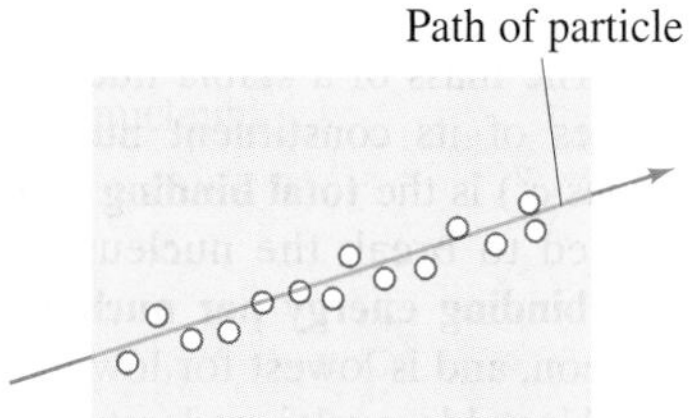

FIGURE 30–15 In a cloud or bubble chamber, droplets or bubbles are formed around ions produced by the passage of a charged particle.

Problems

30–1 Nuclear Properties

1. (I) A pi meson has a mass of $139\,\text{MeV}/c^2$. What is this in atomic mass units?
2. (I) What is the approximate radius of an alpha particle (^4_2He)?
3. (II) What is the rest mass of a bare α particle in MeV/c^2?
4. (II) (*a*) What is the approximate radius of a $^{64}_{29}\text{Cu}$ nucleus? (*b*) Approximately what is the value of A for a nucleus whose radius is 3.9×10^{-15} m?
5. (II) (*a*) Show that the density of nuclear matter is essentially the same for all nuclei. (*b*) What would be the radius of the Earth if it had its actual mass but had the density of nuclei? (*c*) What would be the radius of a $^{238}_{92}\text{U}$ nucleus if it had the density of the Earth?
6. (II) (*a*) What is the fraction of the hydrogen atom's mass that is in the nucleus? (*b*) What is the fraction of the hydrogen atom's volume that is occupied by the nucleus?
7. (II) Approximately how many nucleons are there in a 1.0-kg object? Does it matter what the object is made of? Why or why not?
8. (III) How much energy must an α particle have to just "touch" the surface of a $^{238}_{92}\text{U}$ nucleus?

30–2 Binding Energy

9. (I) Estimate the total binding energy for $^{40}_{20}\text{Ca}$, using Fig. 30–1.
10. (I) Use Fig. 30–1 to estimate the total binding energy of (*a*) $^{238}_{92}\text{U}$, and (*b*) $^{84}_{36}\text{Kr}$.
11. (II) Use Appendix B to calculate the binding energy of ^2_1H (deuterium).
12. (II) Calculate the binding energy per nucleon for a $^{14}_7\text{N}$ nucleus.
13. (II) Determine the binding energy of the last neutron in a $^{40}_{19}\text{K}$ nucleus.
14. (II) Calculate the total binding energy, and the binding energy per nucleon, for (*a*) ^6_3Li, (*b*) $^{208}_{82}\text{Pb}$. Use Appendix B.
15. (II) Compare the average binding energy of a nucleon in $^{23}_{11}\text{Na}$ to that in $^{24}_{11}\text{Na}$.
16. (III) How much energy is required to remove (*a*) a proton, (*b*) a neutron, from $^{16}_8\text{O}$? Explain the difference in your answers.
17. (III) (*a*) Show that the nucleus ^8_4Be (mass = 8.005305 u) is unstable and will decay into two α particles. (*b*) Is $^{12}_6\text{C}$ stable against decay into three α particles? Show why or why not.

30–3 to 30–7 Radioactive Decay

18. (I) How much energy is released when tritium, ^3_1H, decays by β^- emission?
19. (I) What is the maximum kinetic energy of an electron emitted in the β decay of a free neutron?
20. (I) Show that the decay $^{11}_6\text{C} \rightarrow {}^{10}_5\text{B} + \text{p}$ is not possible because energy would not be conserved.
21. (II) $^{22}_{11}\text{Na}$ is radioactive. (*a*) Is it a β^- or β^+ emitter? (*b*) Write down the decay reaction, and estimate the maximum kinetic energy of the emitted β.
22. (II) Give the result of a calculation that shows whether or not the following decays are possible:
    (*a*) $^{236}_{92}\text{U} \rightarrow {}^{235}_{92}\text{U} + \text{n}$;
    (*b*) $^{16}_8\text{O} \rightarrow {}^{15}_8\text{O} + \text{n}$;
    (*c*) $^{23}_{11}\text{Na} \rightarrow {}^{22}_{11}\text{Na} + \text{n}$.
23. (II) A $^{238}_{92}\text{U}$ nucleus emits an α particle with kinetic energy = 4.20 MeV. (*a*) What is the daughter nucleus, and (*b*) what is the approximate atomic mass (in u) of the daughter atom? Ignore recoil of the daughter nucleus.
24. (II) When $^{23}_{10}\text{Ne}$ (mass = 22.9945 u) decays to $^{23}_{11}\text{Na}$ (mass = 22.9898 u), what is the maximum kinetic energy of the emitted electron? What is its minimum energy? What is the energy of the neutrino in each case? Ignore recoil of the daughter nucleus.
25. (II) A nucleus of mass 238 u, initially at rest, emits an α particle with a KE of 5.0 MeV. What is the KE of the recoiling daughter nucleus?
26. (II) What is the maximum KE of the emitted β particle during the decay of $^{60}_{27}\text{Co}$?
27. (II) The nuclide $^{32}_{15}\text{P}$ decays by emitting an electron whose maximum kinetic energy can be 1.71 MeV. (*a*) What is the daughter nucleus? (*b*) Calculate the daughter's atomic mass (in u).
28. (II) The isotope $^{218}_{84}\text{Po}$ can decay by either α or β^- emission. What is the energy release in each case? The mass of $^{218}_{84}\text{Po}$ is 218.008965 u.
29. (II) How much energy is released in electron capture by beryllium: $^7_4\text{Be} + {}^{\;\,0}_{-1}\text{e} \rightarrow {}^7_3\text{Li} + \nu$?
30. (II) A photon with a wavelength of 1.00×10^{-13} m is ejected from an atom. Calculate its energy and explain why it is a γ ray from the nucleus or a photon from the atom.
31. (II) Determine the maximum kinetic energy of β^+ particles released when $^{11}_6\text{C}$ decays to $^{11}_5\text{B}$. What is the maximum energy the neutrino can have? What is its minimum energy?
32. (II) How much recoil energy does a $^{40}_{19}\text{K}$ nucleus get when it emits a 1.46-MeV gamma ray?
33. (III) What is the energy of the α particle emitted in the decay $^{210}_{84}\text{Po} \rightarrow {}^{206}_{82}\text{Pb} + \alpha$? Take into account the recoil of the daughter nucleus.
34. (III) The α particle emitted when $^{238}_{92}\text{U}$ decays has 4.20 MeV of kinetic energy. Calculate the recoil kinetic energy of the daughter nucleus and the Q-value of the decay.
35. (III) Show that when a nucleus decays by β^+ decay, the total energy released is equal to

$$(M_\text{P} - M_\text{D} - 2m_\text{e})c^2,$$

where M_P and M_D are the masses of the parent and daughter atoms (neutral), and m_e is the mass of an electron or positron.

30–8 to 30–11 Half-Life, Decay Rates, Decay Series, Dating

36. (I) A radioactive material produces 1280 decays per minute at one time, and 4.6 h later produces 320 decays per minute. What is its half-life?

37. (I) (*a*) What is the decay constant of $^{238}_{92}U$ whose half-life is 4.5×10^9 yr? (*b*) The decay constant of a given nucleus is $8.2 \times 10^{-5}\,s^{-1}$. What is its half-life?

38. (I) What is the activity of a sample of $^{14}_{6}C$ that contains 3.1×10^{20} nuclei?

39. (I) What fraction of a sample of $^{68}_{32}Ge$, whose half-life is about 9 months, will remain after 3.0 yr?

40. (I) What fraction of a sample is left after exactly 6 half-lives?

41. (II) How many nuclei of $^{238}_{92}U$ remain in a rock if the activity registers 640 decays per second?

42. (II) In a series of decays, the nuclide $^{235}_{92}U$ becomes $^{207}_{82}Pb$. How many α and β^- particles are emitted in this series?

43. (II) The iodine isotope $^{131}_{53}I$ is used in hospitals for diagnosis of thyroid function. If 682 μg are ingested by a patient, determine the activity (*a*) immediately, (*b*) 1.0 h later when the thyroid is being tested, and (*c*) 6 months later. Use Appendix B.

44. (II) $^{124}_{55}Cs$ has a half-life of 30.8 s. (*a*) If we have 8.8 μg initially, how many Cs nuclei are present? (*b*) How many are present 2.0 min later? (*c*) What is the activity at this time? (*d*) After how much time will the activity drop to less than about 1 per second?

45. (II) Calculate the mass of a sample of pure $^{40}_{19}K$ with an initial decay rate of $2.0 \times 10^5\,s^{-1}$. The half-life of $^{40}_{19}K$ is 1.28×10^9 yr.

46. (II) Calculate the activity of a pure 9.7-μg sample of $^{32}_{15}P$ $\left(T_{\frac{1}{2}} = 1.23 \times 10^6\,s\right)$.

47. (II) The activity of a sample of $^{35}_{16}S$ $\left(T_{\frac{1}{2}} = 7.55 \times 10^6\,s\right)$ is 2.65×10^5 decays per second. What is the mass of the sample?

48. (II) A sample of $^{233}_{92}U$ $\left(T_{\frac{1}{2}} = 1.59 \times 10^5\,yr\right)$ contains 7.50×10^{19} nuclei. (*a*) What is the decay constant? (*b*) Approximately how many disintegrations will occur per minute?

49. (II) The activity of a sample drops by a factor of 10 in 8.6 minutes. What is its half-life?

50. (II) A 285-g sample of pure carbon contains 1.3 parts in 10^{12} (atoms) of $^{14}_{6}C$. How many disintegrations occur per second?

51. (II) A sample of $^{40}_{19}K$ is decaying at a rate of 6.70×10^2 decays/s. What is the mass of the sample?

52. (II) The rubidium isotope $^{87}_{37}Rb$, a β emitter with a half-life of 4.75×10^{10} yr, is used to determine the age of rocks and fossils. Rocks containing fossils of ancient animals contain a ratio of $^{87}_{38}Sr$ to $^{87}_{37}Rb$ of 0.0160. Assuming that there was no $^{87}_{38}Sr$ present when the rocks were formed, estimate the age of these fossils. [*Hint*: use Eq. 30–3.]

53. (II) Use Fig. 30–11 and calculate the relative decay rates for α decay of $^{218}_{84}Po$ and $^{214}_{84}Po$.

54. (II) $^{7}_{4}Be$ decays with a half-life of about 53 d. It is produced in the upper atmosphere, and filters down onto the Earth's surface. If a plant leaf is detected to have 450 decays/s of $^{7}_{4}Be$, (*a*) how long do we have to wait for the decay rate to drop to 15 per second? (*b*) Estimate the initial mass of $^{7}_{4}Be$ on the leaf.

55. (II) Two of the naturally occurring radioactive decay sequences start with $^{232}_{90}Th$, and $^{235}_{92}U$. The first five decays of these two sequences are:

$$\alpha, \beta, \beta, \alpha, \alpha$$

and

$$\alpha, \beta, \alpha, \beta, \alpha.$$

Determine the resulting intermediate daughter nuclei in each case.

56. (II) An ancient wooden club is found that contains 290 g of carbon and has an activity of 8.0 decays per second. Determine its age assuming that in living trees the ratio of $^{14}C/^{12}C$ atoms is about 1.3×10^{-12}.

57. (III) At $t = 0$, a pure sample of radioactive nuclei contains N_0 nuclei whose decay constant is λ. Determine a formula for the number of daughter nuclei, N_D, as a function of time; assume the daughter is stable and that $N_D = 0$ at $t = 0$.

General Problems

58. Which radioactive isotope of lead is being produced in a reaction where the measured activity of a sample drops to 1.050% of its original activity in 4.00 h?

59. An old wooden tool is found to contain only 6.0% of $^{14}_{6}C$ that a sample of fresh wood would. How old is the tool?

60. A neutron star consists of neutrons at approximately nuclear density. Estimate, for a 10-km-diameter neutron star, (*a*) its mass number, (*b*) its mass (kg), and (*c*) the acceleration of gravity at its surface.

61. The $^{3}_{1}H$ isotope of hydrogen, which is called *tritium* (because it contains three nucleons), has a half-life of 12.33 yr. It can be used to measure the age of objects up to about 100 yr. It is produced in the upper atmosphere by cosmic rays and brought to Earth by rain. As an application, determine approximately the age of a bottle of wine whose $^{3}_{1}H$ radiation is about $\frac{1}{10}$ that present in new wine.

62. Some elementary particle theories (Section 32–11) suggest that the proton may be unstable, with a half-life $\geq 10^{32}$ yr. How long would you expect to wait for one proton in your body to decay (consider that your body is all water)?

Possible Fusion Reactors

Fusion reactor

The possibility of utilizing the energy released in fusion to make a power reactor is very attractive. The fusion reactions most likely to succeed in a reactor involve the isotopes of hydrogen, ${}^{2}_{1}\mathrm{H}$ (deuterium) and ${}^{3}_{1}\mathrm{H}$ (tritium), and are as follows, with the energy released given in parentheses:

Fusion reactions for possible reactor

$$ {}^{2}_{1}\mathrm{H} + {}^{2}_{1}\mathrm{H} \rightarrow {}^{3}_{1}\mathrm{H} + {}^{1}_{1}\mathrm{H} \qquad (4.03\,\mathrm{MeV}) \qquad \textbf{(31–8a)} $$

$$ {}^{2}_{1}\mathrm{H} + {}^{2}_{1}\mathrm{H} \rightarrow {}^{3}_{2}\mathrm{He} + \mathrm{n} \qquad (3.27\,\mathrm{MeV}) \qquad \textbf{(31–8b)} $$

$$ {}^{2}_{1}\mathrm{H} + {}^{3}_{1}\mathrm{H} \rightarrow {}^{4}_{2}\mathrm{He} + \mathrm{n}. \qquad (17.59\,\mathrm{MeV}) \qquad \textbf{(31–8c)} $$

Comparing these energy yields with that for the fission of ${}^{235}_{92}\mathrm{U}$, we can see that the energy released in fusion reactions can be greater for a given mass of fuel than in fission. Furthermore, as fuel, a fusion reactor could use deuterium, which is very plentiful in the water of the oceans (the natural abundance of ${}^{2}_{1}\mathrm{H}$ is 0.0115% on average, or about 1 g of deuterium per 80 L of water). The simple proton–proton reaction of Eq. 31–6a, which could use a much more plentiful source of fuel, ${}^{1}_{1}\mathrm{H}$, has such a small probability of occurring that it cannot be considered a possibility on Earth.

Although a useful fusion reactor has not yet been achieved, considerable progress has been made in overcoming the inherent difficulties. The problems are associated with the fact that all nuclei have a positive charge and repel each other. However, if they can be brought close enough together so that the short-range attractive nuclear force can come into play, the latter can pull the nuclei together and fusion will occur. For the nuclei to get close enough together, they must have large kinetic energy to overcome the electric repulsion. High kinetic energies are easily attainable with particle accelerators (Chapter 32), but the number of particles involved is too small. To produce realistic amounts of energy, we must deal with matter in bulk, for which high kinetic energy means higher temperatures. Indeed, very high temperatures are required for fusion to occur, and fusion devices are often referred to as **thermonuclear devices**. The Sun and other stars are very hot, many millions of degrees, so the nuclei are moving fast enough for fusion to take place, and the energy released keeps the temperature high so that further fusion reactions can occur. The Sun and the stars represent huge self-sustaining thermonuclear reactors that stay together because of their great gravitational mass; but on Earth, containment of the fast-moving nuclei at the high temperatures and densities required has proven difficult.

It was realized after World War II that the temperature produced within a fission (or "atomic") bomb was close to $10^{8}\,\mathrm{K}$. This suggested that a fission bomb could be used to ignite a fusion bomb (popularly known as a thermonuclear or hydrogen bomb) to release the vast energy of fusion. The uncontrollable release of fusion energy in an H-bomb (in 1952) was relatively easy to obtain. But to realize usable energy from fusion at a slow and controlled rate turned out to be a serious challenge.

EXAMPLE 31–9 ESTIMATE Temperature needed for d–t fusion. Estimate the temperature required for deuterium–tritium fusion (d–t) to occur.

APPROACH We assume the nuclei approach head-on, each with kinetic energy KE, and that the nuclear force comes into play when the distance between their centers equals the sum of their nuclear radii. The electrostatic potential energy (Section 17–5) of the two particles at this distance must equal the total kinetic energy of the two particles when far apart. The average kinetic energy is related to Kelvin temperature by Eq. 13–8.

SOLUTION The radii of the two nuclei ($Z_1 = 2$ and $Z_2 = 3$) are given by Eq. 30–1: $r_d \approx 1.5\,\text{fm}$, $r_t \approx 1.7\,\text{fm}$, so $r_d + r_t = 3.2 \times 10^{-15}\,\text{m}$. We equate the kinetic energy of the two initial particles to the potential energy when very close:

$$2\text{KE} \approx \frac{1}{4\pi\epsilon_0}\frac{e^2}{(r_d + r_t)}$$

$$\approx \left(9.0 \times 10^9\,\frac{\text{N}\cdot\text{m}^2}{\text{C}^2}\right)\frac{(1.6 \times 10^{-19}\,\text{C})^2}{(3.2 \times 10^{-15}\,\text{m})(1.6 \times 10^{-19}\,\text{J/eV})} \approx 0.45\,\text{MeV}.$$

Thus, $\text{KE} \approx 0.22\,\text{MeV}$, and if we ask that the average kinetic energy be this high, then from Eq. 13–8, $\frac{3}{2}kT = \overline{\text{KE}}$, we have

$$T = \frac{2\overline{\text{KE}}}{3k} = \frac{2(0.22\,\text{MeV})(1.6 \times 10^{-13}\,\text{J/MeV})}{3(1.38 \times 10^{-23}\,\text{J/K})} \approx 2 \times 10^9\,\text{K}.$$

NOTE More careful calculations show that the temperature required for fusion is actually about an order of magnitude less than this rough estimate, partly because it is not necessary that the *average* kinetic energy be 0.22 MeV—a small percentage with this much energy (particles in the high-energy tail of the Maxwell distribution, Fig. 13–18) would be sufficient. Reasonable estimates for a usable fusion reactor are in the range $T \gtrsim 2$ to $4 \times 10^8\,\text{K}$.

It is not only a high temperature that is required for a fusion reactor. There must also be a high density of nuclei to ensure a sufficiently high collision rate. A real difficulty with controlled fusion is to contain nuclei long enough and at a high enough density for sufficient reactions to occur that a usable amount of energy is obtained. At the temperatures needed for fusion, the atoms are ionized, and the resulting collection of nuclei and electrons is referred to as a **plasma**. Ordinary materials vaporize at a few thousand degrees at best, and hence cannot be used to contain a high-temperature plasma. Two major containment techniques are *magnetic confinement* and *inertial confinement.*

In **magnetic confinement**, magnetic fields are used to try to contain the hot plasma. One possibility is a torus-shaped design, Fig. 31–12, originally called a **tokamak**.

The second method for containing the fuel for fusion is **inertial confinement**: a small pellet of deuterium and tritium is struck simultaneously from several directions by very intense laser beams (Fig. 31–13). The intense influx of energy heats and ionizes the pellet into a plasma, compressing it and heating it to temperatures at which fusion occurs. The confinement time is on the order of 10^{-11} to $10^{-9}\,\text{s}$, during which time the ions do not move appreciably because of their own inertia, fusion takes place, and the pellet explodes.

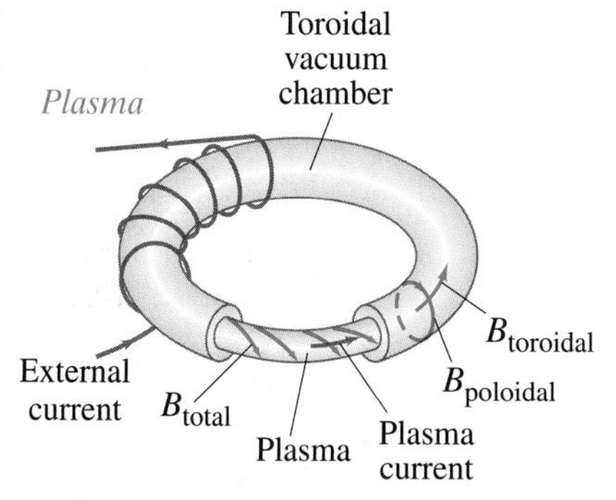

FIGURE 31–12 Tokamak configuration, showing the total $\vec{\mathbf{B}}$ field due to external current plus current in the plasma itself.

(a)

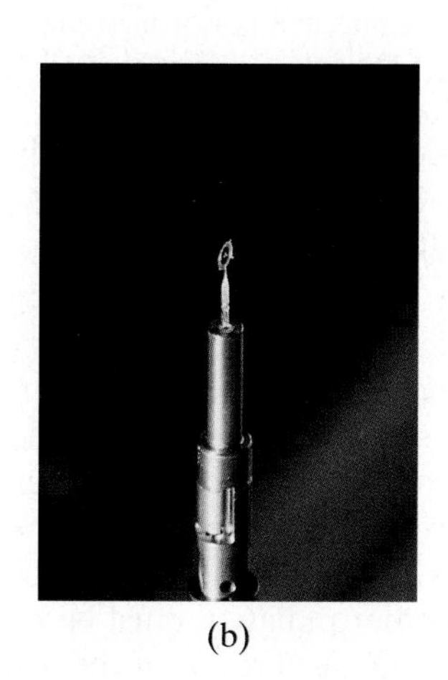

(b)

FIGURE 31–13 (a) Target chamber (5 m in diameter) of the NOVA laser at Lawrence Livermore Laboratory, into which 10 laser beams converge on a target. (b) A 1-mm-diameter DT (deuterium–tritium) target, on its support, at the center of the target chamber.

31–4 Passage of Radiation Through Matter; Radiation Damage

When we speak of *radiation*, we include α, β, γ, and X-rays, as well as protons, neutrons, and other particles such as pions (see Chapter 32). Because charged particles can ionize the atoms or molecules of any material they pass through, they are referred to as **ionizing radiation**. And because radiation produces ionization, it can cause considerable damage to materials, particularly to biological tissue.

Charged particles, such as α and β rays and protons, cause ionization because of electric forces. That is, when they pass through a material, they can attract or repel electrons strongly enough to remove them from the atoms of the material. Since the α and β rays emitted by radioactive substances have energies on the order of 1 MeV (10^4 to 10^7 eV), whereas ionization of atoms and molecules requires on the order of 10 eV, it is clear that a single α or β particle can cause thousands of ionizations.

Neutral particles also give rise to ionization when they pass through materials. For example, X-ray and γ-ray photons can ionize atoms by knocking out electrons by means of the photoelectric and Compton effects (Chapter 27). Furthermore, if a γ ray has sufficient energy (greater than 1.02 MeV), it can undergo pair production: an electron and a positron are produced (Section 27–6). The charged particles produced in all of these processes can themselves go on to produce further ionization. Neutrons, on the other hand, interact with matter mainly by collisions with nuclei, with which they interact strongly. Often the nucleus is broken apart by such a collision, altering the molecule of which it was a part. And the fragments produced can in turn cause ionization.

Radiation passing through matter can do considerable damage. Metals and other structural materials become brittle and their strength can be weakened if the radiation is very intense, as in nuclear reactor power plants and for space vehicles that must pass through areas of intense cosmic radiation.

Biological damage

The radiation damage produced in biological organisms is due primarily to ionization produced in cells. Several related processes can occur. Ions or radicals are produced that are highly reactive and take part in chemical reactions that interfere with the normal operation of the cell. All forms of radiation can ionize atoms by knocking out electrons. If these are bonding electrons, the molecule may break apart, or its structure may be altered so that it does not perform its normal function or may perform a harmful function. In the case of proteins, the loss of one molecule is not serious if there are other copies of it in the cell and additional copies can be made from the gene that codes for it. However, large doses of radiation may damage so many molecules that new copies cannot be made quickly enough, and the cell dies. Damage to the DNA is more serious, since a cell may have only one copy. Each alteration in the DNA can affect a gene and alter the molecule it codes for, so that needed proteins or other materials may not be made at all. Again the cell may die. The death of a single cell is not normally a problem, since the body can replace it with a new one. (There are exceptions, such as neurons, which are *not* replaceable, so their loss is serious.) But if many cells die, the organism may not be able to recover. On the other hand, a cell may survive but be defective. It may go on dividing and produce many more defective cells, to the detriment of the whole organism. Thus radiation can cause cancer—the rapid uncontrolled production of cells.

Radiation damage to biological organisms is often separated into categories. *Somatic damage* refers to any part of the body except the reproductive organs. Somatic damage affects that particular organism, causing cancer and, at high doses, radiation sickness (characterized by nausea, fatigue, loss of body hair, and other symptoms) or even death. *Genetic damage* refers to damage to reproductive cells, causing mutations, the majority of which are harmful and are transmitted to future generations. The possible damage done by the medical use of X-rays and other radiation must be balanced against the medical benefits and prolongation of life as a result of their diagnostic use.

31–5 Measurement of Radiation—Dosimetry

Although the passage of ionizing radiation through the human body can cause considerable damage, radiation can also be used to treat certain diseases, particularly cancer, often by using very narrow beams directed at a cancerous tumor in order to destroy it (Section 31–6). It is therefore important to be able to quantify the amount, or **dose**, of radiation. This is the subject of **dosimetry**.

The strength of a source can be specified at a given time by stating the **source activity**, or how many disintegrations occur per second. The traditional unit is the **curie** (Ci), defined as

Source activity

$$1\ \text{Ci} = 3.70 \times 10^{10}\ \text{disintegrations per second.}$$

The curie (unit)

(This figure comes from the original definition as the activity of exactly one gram of radium.) Although the curie is still in common use, the SI unit for source activity is the **becquerel** (Bq), defined as

$$1\ \text{Bq} = 1\ \text{disintegration/s.}$$

The becquerel (unit)

Commercial suppliers of **radionuclides** (radioactive nuclides) specify the activity at a given time. Since the activity decreases over time, more so for short-lived isotopes, it is important to take this into account.

The source activity $(\Delta N/\Delta t)$ is related to the number of radioactive nuclei present, N, and to the half-life, $T_{\frac{1}{2}}$, by (see Section 30–8):

$$\frac{\Delta N}{\Delta t} = \lambda N = \frac{0.693}{T_{\frac{1}{2}}} N.$$

EXAMPLE 31–10 Radioactivity taken up by cells. In a certain experiment, $0.016\ \mu\text{Ci}$ of ${}^{32}_{15}\text{P}$ is injected into a medium containing a culture of bacteria. After 1.0 h the cells are washed and a detector that is 70% efficient (counts 70% of emitted β rays) records 720 counts per minute from all the cells. What percentage of the original ${}^{32}_{15}\text{P}$ was taken up by the cells?

APPROACH The half-life of ${}^{32}_{15}\text{P}$ is about 14 days (Appendix B), so we can ignore any loss of activity over 1 hour. From the given activity, we find how many β rays are emitted. We can compare 70% of this to the $(720/\text{min})/(60\ \text{s/min}) = 12$ per second detected.

SOLUTION The total number of disintegrations per second originally was $(0.016 \times 10^{-6})(3.7 \times 10^{10}) = 590$. The counter could be expected to count 70% of this, or 410 per second. Since it counted $720/60 = 12$ per second, then $12/410 = 0.029$ or 2.9% was incorporated into the cells.

Another type of measurement is the exposure or **absorbed dose**—that is, the effect the radiation has on the absorbing material. The earliest unit of dosage was the **roentgen** (R), defined in terms of the amount of ionization produced by the radiation (1.6×10^{12} ion pairs per gram of dry air at standard conditions). Today, 1 R is defined as the amount of X or γ radiation that deposits 0.878×10^{-2} J of energy per kilogram of air. The roentgen was largely superseded by another unit of absorbed dose applicable to any type of radiation, the **rad**: *1 rad is that amount of radiation which deposits energy at a rate of 1.00×10^{-2} J/kg in any absorbing material.* (This is quite close to the roentgen for X- and γ rays.) The proper SI unit for absorbed dose is the **gray** (Gy):

Absorbed dose

The rad (unit)

$$1\ \text{Gy} = 1\ \text{J/kg} = 100\ \text{rad.} \qquad \textbf{(31–9)}$$

The gray (unit)

The absorbed dose depends not only on the strength of a given radiation beam (number of particles per second) and the energy per particle, but also on the type of material that is absorbing the radiation. Bone, for example, absorbs more of the radiation normally used than does flesh, so the same beam passing through a human body deposits a greater dose (in rads or grays) in bone than in flesh.

* 31–6 Radiation Therapy

Radiation can be very useful (not just a danger)

The applications of radioactivity and radiation to human beings and other organisms is a vast field that has filled many books. In the medical field there are two basic aspects: (1) **radiation therapy**—the treatment of disease (mainly cancer)—which we discuss in this Section; and (2) the *diagnosis* of disease, which we discuss in the following Sections of this Chapter.

PHYSICS APPLIED

Radiation therapy

Radiation can cause cancer. It can also be used to treat it. Rapidly growing cancer cells are especially susceptible to destruction by radiation. Nonetheless, large doses are needed to kill the cancer cells, and some of the surrounding normal cells are inevitably killed as well. It is for this reason that cancer patients receiving radiation therapy often suffer side effects characteristic of radiation sickness. To minimize the destruction of normal cells, a narrow beam of γ or X-rays is often used when a cancerous tumor is well localized. The beam is directed at the tumor, and the source (or body) is rotated so that the beam passes through various parts of the body to keep the dose at any one place as low as possible—except at the tumor and its immediate surroundings, where the beam passes at all times (Fig. 31–15). The radiation may be from a radioactive source such as $^{60}_{27}Co$, or it may be from an X-ray machine that produces photons in the range 200 keV to 5 MeV. Protons, neutrons, electrons, and pions, which are produced in particle accelerators (Section 32–1), are also being used in cancer therapy.

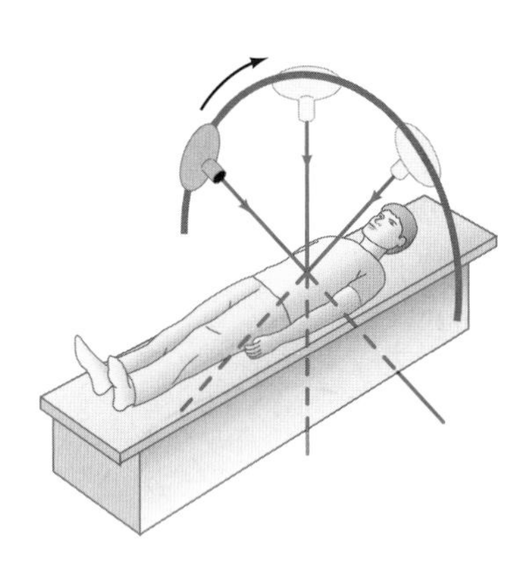

FIGURE 31–15 Radiation source rotates so that the beam always passes through the diseased tissue, but minimizing the dose in the rest of the body.

In some cases, a tiny radioactive source may be inserted directly inside a tumor, which will eventually kill the majority of the cells. A similar technique is used to treat cancer of the thyroid with the radioactive isotope $^{131}_{53}I$. The thyroid gland concentrates iodine present in the bloodstream, particularly in any area where abnormal growth is taking place. Its intense radioactivity can destroy the defective cells.

Although radiation can increase the lifespan of many patients, it is not always completely effective. It may not be possible to kill all the cancer cells, so a recurrence of the disease is possible. Many cases, especially when the cancerous cells are not well localized in one area, are difficult to treat without damaging healthy organs.

Another application of radiation is for sterilizing bandages, surgical equipment, and even packaged foods, since bacteria and viruses can be killed or deactivated by large doses of radiation.

* 31–7 Tracers and Imaging in Research and Medicine

PHYSICS APPLIED

Tracers in medicine and biology

Radioactive isotopes are commonly used in biological and medical research as **tracers**. A given compound is artificially synthesized using a radioactive isotope such as $^{14}_{6}C$ or $^{3}_{1}H$. Such "tagged" molecules can then be traced as they move through an organism or as they undergo chemical reactions. The presence of these tagged molecules (or parts of them, if they undergo chemical change) can be detected by a Geiger or scintillation counter, which detects emitted radiation (see Section 30–13). How food molecules are digested, and to what parts of the body they are diverted, can be traced in this way. Radioactive tracers have been used to determine how amino acids and other essential compounds are synthesized by organisms. The permeability of cell walls to various molecules and ions can he determined using radioactive isotopes: the tagged molecule or ion is injected into the extracellular fluid, and the radioactivity present inside and outside the cells is measured as a function of time.

In a technique known as **autoradiography**, the position of the radioactive isotopes is detected on film. For example, the distribution of carbohydrates produced in the leaves of plants from absorbed CO_2 can be observed by keeping the plant in an atmosphere where the carbon atom in the CO_2 is ${}^{14}_{6}C$. After a time, a leaf is placed firmly on a photographic plate and the emitted radiation darkens the film most strongly where the isotope is most strongly concentrated (Fig. 31–16a). Autoradiography using labeled nucleotides (components of DNA) has revealed much about the details of DNA replication (Fig. 31–16b).

For medical diagnosis, the radionuclide commonly used today is ${}^{99m}_{43}Tc$, a long-lived excited state of technetium-99 (the "m" in the symbol stands for "metastable" state). It is formed when ${}^{99}_{42}Mo$ decays. The great usefulness of ${}^{99m}_{43}Tc$ derives from its convenient half-life of 6 h (short, but not too short) and the fact that it can combine with a large variety of compounds. The compound to be labeled with the radionuclide is so chosen because it concentrates in the organ or region of the anatomy to be studied. Detectors outside the body then record, or image, the distribution of the radioactively labeled compound. The detection can be done by a single detector (Fig. 31–17) which is moved across the body, measuring the intensity of radioactivity at a large number of points. The image represents the relative intensity of radioactivity at each point. The relative radioactivity is a diagnostic tool. For example, high or low radioactivity may represent overactivity or underactivity of an organ or part of an organ, or in another case may represent a lesion or tumor. More complex *gamma cameras* make use of many detectors which simultaneously record the radioactivity at many points. The measured intensities can be displayed on a TV or computer monitor, and allow "dynamic" studies (that is, images that change in time) to be performed.

(a)

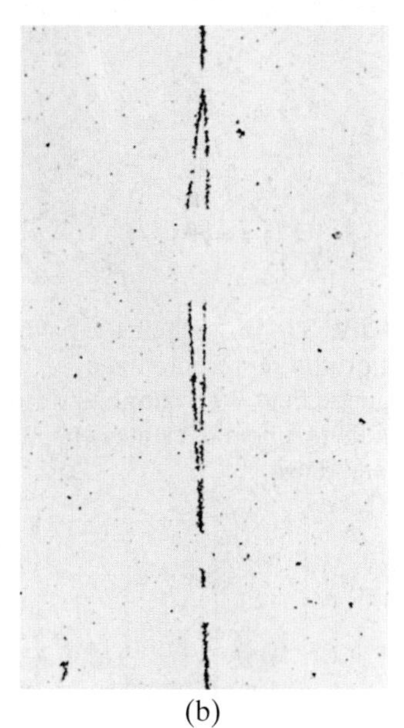

(b)

FIGURE 31–16 (a) Autoradiograph of a mature leaf of the squash plant *Cucurbita melopepo* exposed for 30 s to $^{14}CO_2$. The photosynthetic (green) tissue has become radioactive; the nonphotosynthetic tissue of the veins is free of ^{14}C and therefore does not blacken the X-ray sheet. This technique is very useful in following patterns of nutrient transport in plants. (b) An autoradiograph of a fiber of chromosomal DNA isolated from the mustard plant *Arabidopsis thaliana*. The dashed arrays of film grains show the Y-shaped growing point of replicating DNA.

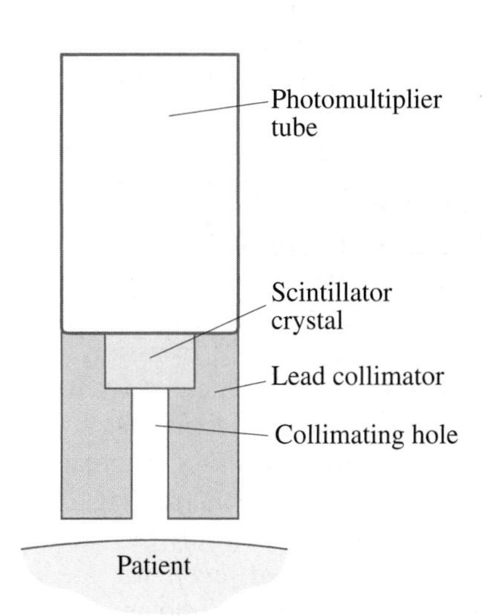

FIGURE 31–17 Collimated gamma-ray detector for scanning (moving) over a patient. The collimator is necessary to select γ rays that come in a straight line from the patient. Without the collimator, γ rays from all parts of the body could strike the scintillator, producing a very poor image.

*31–8 Emission Tomography

PHYSICS APPLIED

Medical imaging

The images formed using the standard techniques of nuclear medicine, as briefly discussed in the previous Section, are produced from radioactive tracer sources within the *volume* of the body. It is also possible to image the radioactive emissions in a single plane or slice through the body using the computed tomography techniques discussed in Section 25–12. A basic gamma camera is moved around the patient to measure the radioactive intensity from the tracer at many points and angles; the data are processed in much the same way as for X-ray CT scans (Section 25–12). This technique is referred to as **single photon emission tomography** (SPET).†

SPET

† Also known as SPECT, "single photon emission computed tomography."

PET

Another important technique is **positron emission tomography** (PET), which makes use of positron emitters such as ${}^{11}_{6}C$, ${}^{13}_{7}N$, ${}^{15}_{8}O$, and ${}^{18}_{9}F$ whose half-lives are short. These isotopes are incorporated into molecules that, when inhaled or injected, accumulate in the organ or region of the body to be studied. When such a nuclide β decays, the emitted positron travels at most a few millimeters before it collides with a normal electron. In this collision, the positron and electron are annihilated, producing two γ rays $(e^+ + e^- \rightarrow 2\gamma)$, each having an energy of 511 keV $(= m_e c^2)$. The two γ rays fly off in opposite directions $(180° \pm 0.25°)$ since they must have almost exactly equal and opposite momentum to conserve momentum (the momenta of the initial e^+ and e^- are essentially zero compared to the momenta afterwards of the γ rays). Because the photons travel along the same line in opposite directions, their detection in coincidence by rings of detectors surrounding the patient (Fig. 31–18) readily establishes the line along which the emission took place. If the difference in time of arrival of the two photons could be determined accurately, the actual position of the emitting nuclide along that line could be calculated. Present-day electronics can measure times to at best ± 300 ps, so at the γ ray's speed $(c = 3 \times 10^8\,\mathrm{m/s})$, the actual position could be determined to an accuracy on the order of about $d = vt \approx (3 \times 10^8\,\mathrm{m/s})(300 \times 10^{-12}\,\mathrm{s}) \approx 10\,\mathrm{cm}$, which is not very useful. Although there may be future potential for time-of-flight measurements to determine position, today computed tomography techniques are used instead, similar to those for X-ray CT, which can reconstruct PET images with a resolution on the order of 3–5 mm. One big advantage of PET is that no collimators are needed (as for detection of a single photon—see Fig. 31–17). Thus, fewer photons are "wasted" and lower doses can be administered to the patient with PET.

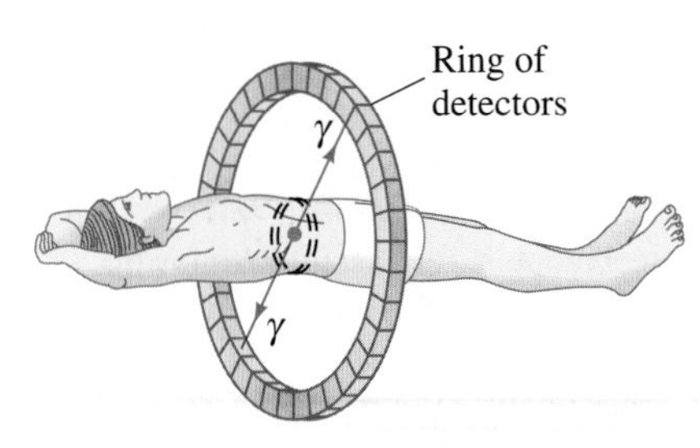

FIGURE 31–18 Positron emission tomography (PET) uses a ring of detectors, typically scintillators with photomultiplier tubes (Section 30–13), to detect the two annihilation γ rays $(e^+ + e^- \rightarrow 2\gamma)$ emitted at 180° to each other.

Both PET and SPET systems can give images that relate to biochemistry, metabolism, and function. This is to be compared to X-ray CT scans (Section 25–12), whose images reflect shape and structure—that is, the anatomy of the imaged region.

31–9 Nuclear Magnetic Resonance (NMR) and Magnetic Resonance Imaging (MRI)

NMR

Nuclear magnetic resonance (NMR) is a phenomenon that soon after its discovery in 1946 became a powerful research tool in a variety of fields from physics to chemistry and biochemistry. It is also an important medical imaging technique. We first briefly discuss the phenomenon, and then look at its applications.

We saw in Chapter 28 (Section 28–6) that when atoms are placed in a magnetic field, atomic energy levels split into several closely spaced levels (see Fig. 28–8). Nuclei, too, exhibit these magnetic properties. We will examine only the simplest, the hydrogen (H) nucleus, since it is the one most used, even for medical imaging. The ${}^{1}_{1}H$ nucleus consists of a single proton. Its spin angular momentum (and its magnetic moment), like that of the electron, can take on only two values when placed in a magnetic field: we call these "spin up" (parallel to the field) and "spin down" (antiparallel to the field), as suggested in Fig. 31–19. When a magnetic field is present, the energy of the nucleus splits into two levels as shown in Fig. 31–20, with the spin up (parallel to field) having the lower energy. (This is like the Zeeman effect for atomic levels, Fig. 28–8.)

FIGURE 31–19 Schematic picture of a proton represented in a magnetic field $\vec{\mathbf{B}}$ (pointing upward) with its two possible states of spin, up and down.

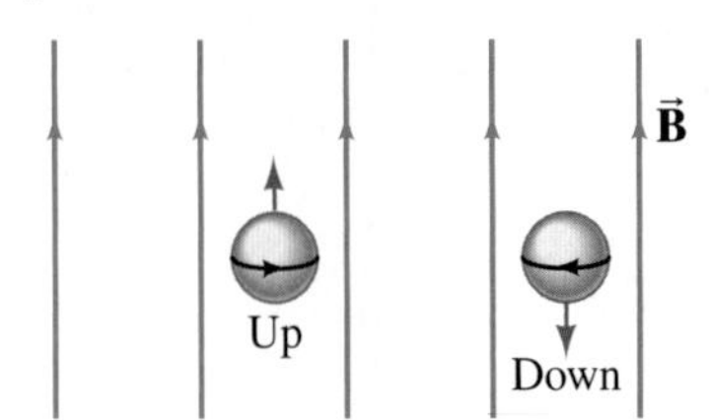

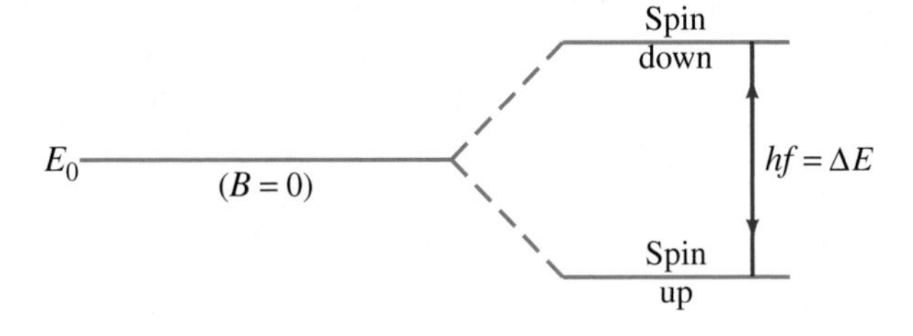

FIGURE 31–20 Energy E_0 in the absence of a magnetic field splits into two levels in the presence of a magnetic field.

The difference in energy, ΔE, between these two levels is proportional to the total magnetic field, B_T, at the nucleus:

$$\Delta E = kB_T,$$

where k is a proportionality constant that is different for different nuclides.

In a standard **nuclear magnetic resonance** (NMR) setup, the sample to be examined is placed in a static magnetic field. A radiofrequency (RF) pulse of electromagnetic radiation (that is, photons) is applied to the sample. If the frequency, f, of this pulse corresponds precisely to the energy difference between the two energy levels (Fig. 31–20), so that

$$hf = \Delta E = kB_T, \tag{31–11}$$

then the photons of the RF beam will be absorbed, exciting many of the nuclei from the lower state to the upper state. This is a resonance phenomenon, whose photons can be detected, since there is significant absorption only if f is very near $f = kB_T/h$. Hence the name "nuclear magnetic resonance." For free ${}^1_1H$ nuclei, the frequency is 42.58 MHz for a magnetic field $B_T = 1.0\,T$. If the H atoms are bound in a molecule the total magnetic field B_T at the H nuclei will be the sum of the external applied field (B_{ext}) plus the local magnetic field (B_{local}) due to electrons and nuclei of neighboring atoms. Since f is proportional to B_T, the value of f for a given external field will be slightly different for the bound H atoms than for free atoms:

$$hf = k(B_{ext} + B_{local}).$$

This small change in frequency can be measured, and is called the "chemical shift." A great deal has been learned about the structure of molecules and bonds using such NMR measurements.

PHYSICS APPLIED

NMR imaging (MRI)

For producing medically useful NMR images—now commonly called MRI, or **magnetic resonance imaging**—the element most used is hydrogen since it is the commonest element in the human body and gives the strongest NMR signals. The experimental apparatus is shown in Fig. 31–21. The large coils set up the static magnetic field, and the RF coils produce the RF pulse of electromagnetic waves (photons) that cause the nuclei to jump from the lower state to the upper one (Fig. 31–20). These same coils (or another coil) can detect the absorption of energy or the emitted radiation (also of frequency $f = \Delta E/h$, Eq. 31–11) when the nuclei jump back down to the lower state.

FIGURE 31–21 Typical NMR imaging setup: (a) diagram; (b) photograph.

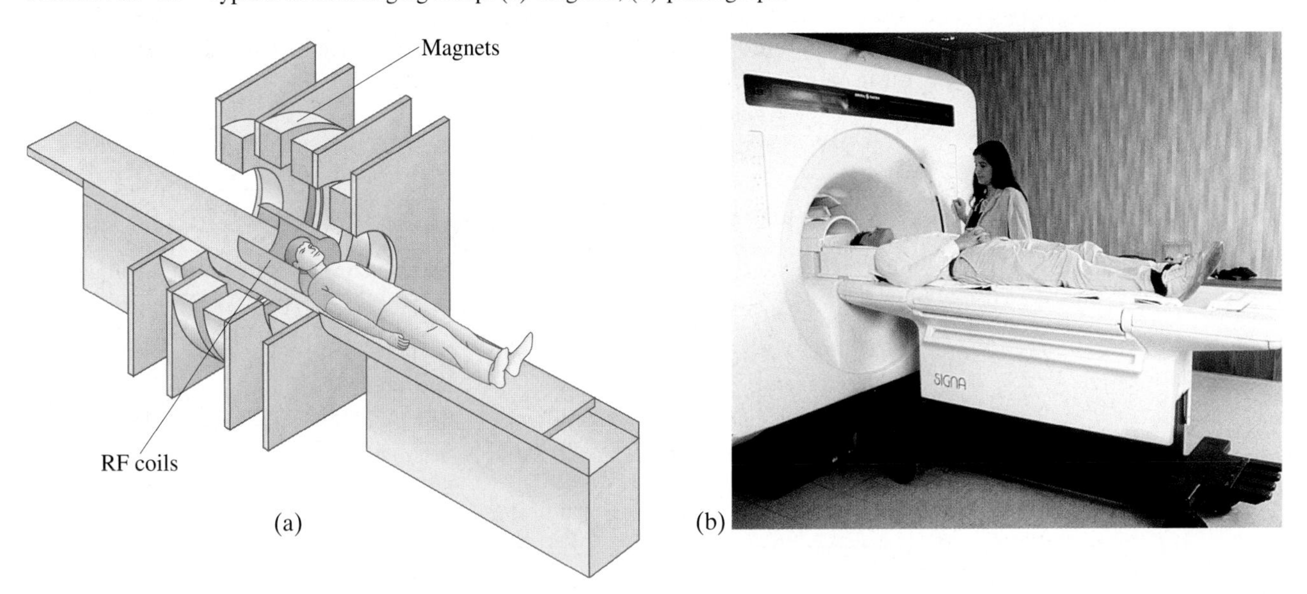

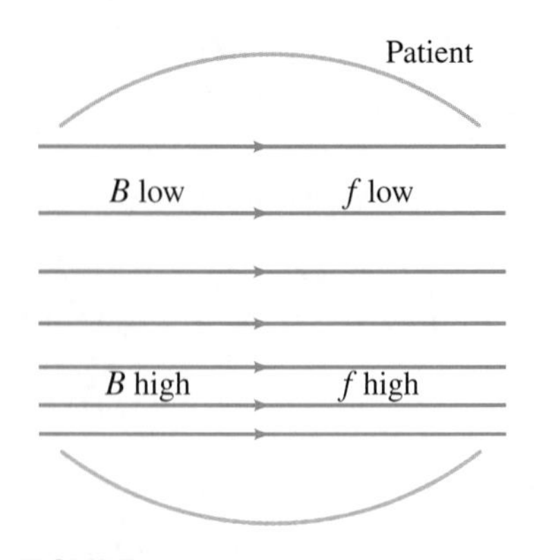

FIGURE 31–22 A static field that is stronger at the bottom than at the top. The frequency of absorbed or emitted radiation is proportional to B in NMR.

FIGURE 31–23 False-color NMR image (MRI) of a vertical section through the head showing structures in the normal brain.

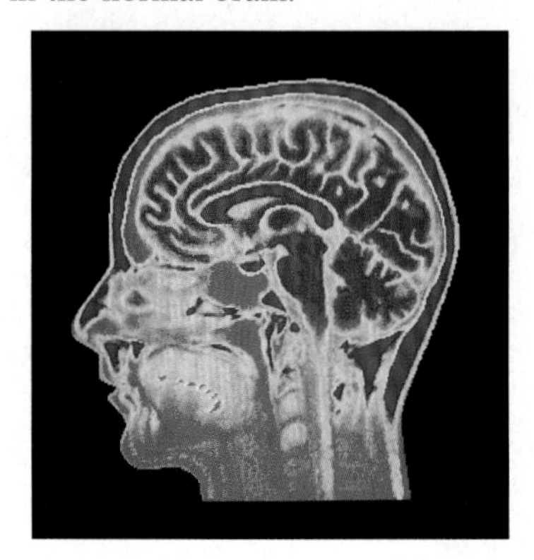

The formation of a two-dimensional or three-dimensional image can be done using techniques similar to those for computed tomography (Section 25–12). The simplest thing to measure for creating an image is the intensity of absorbed and/or reemitted radiation from many different points of the body, and this would be a measure of the density of H atoms at each point. But how do we determine from what part of the body a given photon comes? One technique is to give the static magnetic field a gradient; that is, instead of applying a uniform magnetic field, B_T, the field is made to vary with position across the width of the sample (or patient). Since the frequency absorbed by the H nuclei is proportional to B_T (Eq. 31–11), only one plane within the body will have the proper value of B_T to absorb photons of a particular frequency f. By varying f, absorption by different planes can be measured. Alternately, if the field gradient is applied *after* the RF pulse, the frequency of the emitted photons will be a measure of where they were emitted. See Fig. 31–22. If a magnetic field gradient in one direction is applied during excitation (absorption of photons) and photons of a single frequency are transmitted, only H nuclei in one thin slice will be excited. By applying a gradient in a different direction, perpendicular to the first, during reemission, the frequency f of the reemitted radiation will represent depth in that slice. Other ways of varying the magnetic field throughout the volume of the body can be used in order to correlate NMR frequency with position.

A reconstructed image based on the density of H atoms (that is, the intensity of absorbed or emitted radiation) is not very interesting. More useful are images based on the rate at which the nuclei decay back to the ground state, and such images can produce resolution of 1 mm or better. This NMR technique (sometimes called **spin-echo**) produces images of great diagnostic value, both in the delineation of structure (anatomy) and in the study of metabolic processes. An NMR image is shown in Fig. 31–23.

NMR imaging is considered to be noninvasive. We can calculate the energy of the photons involved: as mentioned above, in a 1.0-T magnetic field, $f = 42.58\,\mathrm{MHz}$ for ${}^{1}_{1}\mathrm{H}$. This corresponds to an energy of $hf = (6.6 \times 10^{-34}\,\mathrm{J\cdot s})(43 \times 10^{6}\,\mathrm{Hz}) \approx 3 \times 10^{-26}\,\mathrm{J}$ or about $10^{-7}\,\mathrm{eV}$. Since molecular bonds are on the order of 1 eV, it is clear that the RF photons can cause little cellular disruption. This should be compared to X- or γ rays, whose energies are 10^4 to 10^6 eV and thus can cause significant damage. The static magnetic fields, though often large (≈ 0.1 to 1 T), are believed to be harmless (except for people wearing heart pacemakers).

Table 31–2 lists the recently developed techniques we have discussed for imaging the interior of the body, along with the optimum resolution attainable today. Of course, resolution is only one factor that must be considered; it must be remembered that the different imaging techniques provide different types of information, useful for different types of diagnosis.

TABLE 31–2 Medical Imaging Techniques

Technique	Where Discussed in This Book	Resolution
Conventional X-ray	Section 25–12	$\frac{1}{2}$ mm
CT scan, X-ray	Section 25–12	$\frac{1}{2}$ mm
Nuclear medicine (tracers)	Section 31–7	1 cm
SPET (single photon emission)	Section 31–8	1 cm
PET (positron emission)	Section 31–8	3–5 mm
NMR	Section 31–9	$\frac{1}{2}$–1 mm
Ultrasound	Section 12–9	2 mm

Summary

A **nuclear reaction** occurs when two nuclei collide and two or more other nuclei (or particles) are produced. In this process, as in radioactivity, **transmutation** (change) of elements occurs.

The **reaction energy** or ***Q*-value** of a reaction $a + X \rightarrow Y + b$ is

$$Q = (M_a + M_X - M_b - M_Y)c^2 \quad \textbf{(31–1)}$$

$$= K_b + K_Y - K_a - K_X. \quad \textbf{(31–2)}$$

In **fission** a heavy nucleus such as uranium splits into two intermediate-sized nuclei after being struck by a neutron. $^{235}_{92}\text{U}$ is fissionable by slow neutrons, whereas some fissionable nuclei require fast neutrons. Much energy is released in fission because the binding energy per nucleon is lower for heavy nuclei than it is for intermediate-sized nuclei, so the mass of a heavy nucleus is greater than the total mass of its fission products. The fission process releases neutrons, so that a **chain reaction** is possible. The **critical mass** is the minimum mass of fuel needed to sustain a chain reaction. In a **nuclear reactor** or nuclear bomb, a **moderator** is needed to slow down the released neutrons.

The **fusion** process, in which small nuclei combine to form larger ones, also releases energy. The energy from our Sun is believed to originate in the fusion reactions known as the **proton–proton cycle** in which four protons fuse to form a $^{4}_{2}\text{He}$ nucleus producing over 25 MeV of energy. A useful fusion reactor for power generation has not yet proved possible because of the difficulty in containing the fuel (e.g., deuterium) long enough at the high temperature required.

Radiation can cause damage to materials, including biological tissue. Quantifying amounts of radiation is the subject of **dosimetry**. The **curie** (Ci) and the **becquerel** (Bq) are units that measure the **source activity** or rate of decay of a sample: $1\,\text{Ci} = 3.70 \times 10^{10}$ disintegrations per second, whereas $1\,\text{Bq} = 1$ disintegration/s. The **absorbed dose**, often specified in **rads**, measures the amount of energy deposited per unit mass of absorbing material: 1 rad is the amount of radiation that deposits energy at the rate of $10^{-2}\,\text{J/kg}$ of material. The SI unit of absorbed dose is the **gray**: $1\,\text{Gy} = 1\,\text{J/kg} = 100\,\text{rad}$. The **effective dose** is often specified by the **rem** = rad × QF, where QF is the "quality factor" of a given type of radiation; 1 rem of any type of radiation does approximately the same amount of biological damage. The average dose received per person per year in the United States is about 0.36 rem. The SI unit for effective dose is the **sievert**: $1\,\text{Sv} = 10^2\,\text{rem}$.

[*Nuclear radiation is used in medicine as therapy and for imaging of biological processes, as well as several types of tomographic **imaging** of the human body: PET, SPET, and MRI; the latter makes use of **nuclear magnetic resonance** (NMR).]

Questions

(NOTE: Masses are found in Appendix B.)

1. Fill in the missing particles or nuclei:
(*a*) $n + {}^{137}_{56}\text{Ba} \rightarrow ? + \gamma$; (*b*) $n + {}^{137}_{56}\text{Ba} \rightarrow {}^{137}_{55}\text{Cs} + ?$;
(*c*) $d + {}^{2}_{1}\text{H} \rightarrow {}^{4}_{2}\text{He} + ?$; (*d*) $\alpha + {}^{197}_{79}\text{Au} \rightarrow ? + d$
where d stands for deuterium.

2. The isotope: $^{32}_{15}\text{P}$ is produced by the reaction: $n + ? \rightarrow {}^{32}_{15}\text{P} + p$. What must be the target nucleus?

3. When $^{22}_{11}\text{Na}$ is bombarded by deuterons ($^{2}_{1}\text{H}$), an α particle is emitted. What is the resulting nuclide?

4. Why are neutrons such good projectiles for producing nuclear reactions?

5. A proton strikes a $^{20}_{10}\text{Ne}$ nucleus, and an α particle is observed to emerge. What is the residual nucleus? Write down the reaction equation.

6. Are fission fragments β^+ or β^- emitters? Explain.

7. If $^{235}_{92}\text{U}$ released only 1.5 neutrons per fission on the average, would a chain reaction be possible? If so, what would be different?

8. $^{238}_{92}\text{U}$ releases an average of 2.5 neutrons per fission compared to 2.9 for $^{239}_{94}\text{Pu}$. Pure samples of which of these two nuclei do you think would have the smaller critical mass? Explain.

9. The energy from nuclear fission appears in the form of thermal energy—but the thermal energy of what?

10. Why can't uranium be enriched by chemical means?

11. How can a neutron, with practically no kinetic energy, excite a nucleus to the extent shown in Fig. 31–2?

12. Why would a porous block of uranium be more likely to explode if kept under water rather than in air?

13. A reactor that uses highly enriched uranium can use ordinary water (instead of heavy water) as a moderator and still have a self-sustaining chain reaction. Explain.

14. Why must the fission process release neutrons if it is to be useful?

15. Discuss the relative merits and disadvantages, including pollution and safety, of power generation by fossil fuels, nuclear fission, and nuclear fusion.

16. What is the reason for the "secondary system" in a nuclear reactor, Fig. 31–7? That is, why is the water heated by the fuel in a nuclear reactor not used directly to drive the turbines?

17. Why are neutrons released in a fission reaction?

18. Why do gamma particles penetrate matter more easily than beta particles do?

19. A higher temperature is required for deuterium–deuterium ignition than for deuterium–tritium. Explain.

20. Light energy emitted by the Sun and stars comes from the fusion process. What conditions in the interior of stars make this possible?

21. How do stars, and our Sun, maintain confinement of the plasma for fusion?

22. What is the basic difference between fission and fusion?

23. People who work around metals that emit alpha particles are trained that there is little danger from proximity or even touching the material, but that they must take extreme precautions against ingesting it. Hence, there are strong rules against eating and drinking while working, and against machining the metal. Why?

The frequency, f, of the applied voltage must be equal to that of the circulating protons. When ions of charge q are circulating *within* the hollow dees, the net force F on each is due to the magnetic field B, so $F = qvB$, where v is the speed of the ion at a given moment (Eq. 20–4). The magnetic force is perpendicular to both $\vec{\mathbf{v}}$ and $\vec{\mathbf{B}}$ and causes the ions to move in circles; the acceleration inside the dees is thus centripetal and equals v^2/r, where r is the radius of the ion's path at a given moment. We use Newton's second law, $F = ma$, and find that

Cyclotron frequency

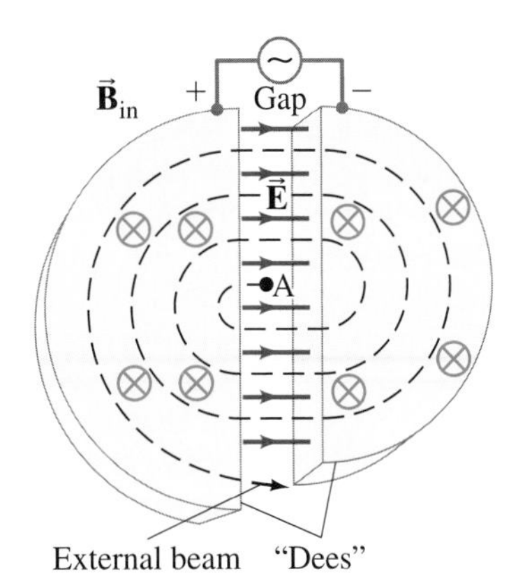

FIGURE 32–2 (repeated) Diagram of a cyclotron.

$$F = ma$$

$$qvB = \frac{mv^2}{r}$$

when the protons are within the dees (not the gap), so

$$v = \frac{qBr}{m}.$$

The time required for a complete revolution is the period T and is equal to

$$T = \frac{\text{distance}}{\text{speed}} = \frac{2\pi r}{qBr/m} = \frac{2\pi m}{qB}.$$

Hence the frequency of revolution f is

$$f = \frac{1}{T} = \frac{qB}{2\pi m}. \qquad \textbf{(32–2)}$$

This is known as the **cyclotron frequency**.

EXAMPLE 32–2 Cyclotron. A small cyclotron of maximum radius $R = 0.25\,\text{m}$ accelerates protons in a 1.7-T magnetic field. Calculate (*a*) the frequency needed for the applied alternating voltage, and (*b*) the kinetic energy of protons when they leave the cyclotron.

APPROACH The frequency of the protons revolving within the dees (Eq. 32–2) must equal the frequency of the voltage applied across the gap if the protons are going to increase in speed.

SOLUTION (*a*) From Eq. 32–2,

$$f = \frac{qB}{2\pi m} = \frac{(1.6\times10^{-19}\,\text{C})(1.7\,\text{T})}{(6.28)(1.67\times10^{-27}\,\text{kg})} = 2.6\times10^{7}\,\text{Hz} = 26\,\text{MHz},$$

which is in the radio-wave region of the EM spectrum (Fig. 22–8).

(*b*) The protons leave the cyclotron at $r = R = 0.25\,\text{m}$. From $qvB = mv^2/r$ (see above), we have $v = qBr/m$, so

$$\text{KE} = \frac{1}{2}mv^2 = \frac{1}{2}m\frac{q^2B^2R^2}{m^2} = \frac{q^2B^2R^2}{2m} = \frac{(1.6\times10^{-19}\,\text{C})^2(1.7\,\text{T})^2(0.25\,\text{m})^2}{(2)(1.67\times10^{-27}\,\text{kg})} = 1.4\times10^{-12}\,\text{J} = 8.7\,\text{MeV}.$$

The KE is much less than the rest energy of the proton (938 MeV), so relativity is not needed.

NOTE The magnitude of the voltage applied to the dees does not affect the final energy. But the higher this voltage, the fewer the revolutions required to bring the protons to full energy.

An important aspect of the cyclotron is that the frequency of the applied voltage, as given by Eq. 32–2, does not depend on the radius r of the particle's path. Thus the frequency does not have to be changed as the protons or ions start from the source and are accelerated to paths of larger and larger radii. But this is only true at nonrelativistic energies. At higher speeds, the momentum (Eq. 26–4) is $p = m_0 v/\sqrt{1 - v^2/c^2}$, so m in Eq. 32–2 has to be replaced by γm_0 and the cyclotron frequency f (Eq. 32–2) depends on speed v. To keep the particles in synch, machines called **synchrocyclotrons** reduce the frequency in time, in parallel with the mass increase, as a packet of charged particles increases in speed and mass at larger orbits.

Synchrotron

Another way to accelerate relativistic particles is to increase the magnetic field B in time so as to keep f (Eq. 32–2) constant as the particles speed up. Such devices are called **synchrotrons**, and today they can be enormous. At the European Center for Nuclear Research (CERN) in Geneva, Switzerland, the new (2007) Large Hadron Collider (LHC) will be 4.3 km in radius and accelerate protons to 7 TeV. The *Tevatron* accelerator at Fermilab (the Fermi National Accelerator Laboratory) at Batavia, Illinois, has a radius of 1.0 km. The Tevatron uses superconducting magnets to accelerate protons to about 1000 GeV = 1 TeV (hence its name); 1 TeV = 10^{12} eV. These large synchrotrons use a narrow ring of magnets (see Fig. 32–3) with each magnet placed at the same radius from the center of the circle. The magnets are interrupted by gaps where high voltage accelerates the particles; another way to describe the acceleration is to say the particles "surf" on a traveling electromagnetic wave within radiofrequency (RF) cavities.

Synchrotron

FIGURE 32–3 (a) Aerial view of Fermilab at Batavia, Illinois; the main accelerator is a circular ring 1.0 km in radius. (b) The interior of the tunnel of the main accelerator at Fermilab. The upper (rectangular-shaped) ring of magnets is for the older 500-GeV accelerator. Below it is the ring of superconducting magnets for the 1-TeV Tevatron.

(a)

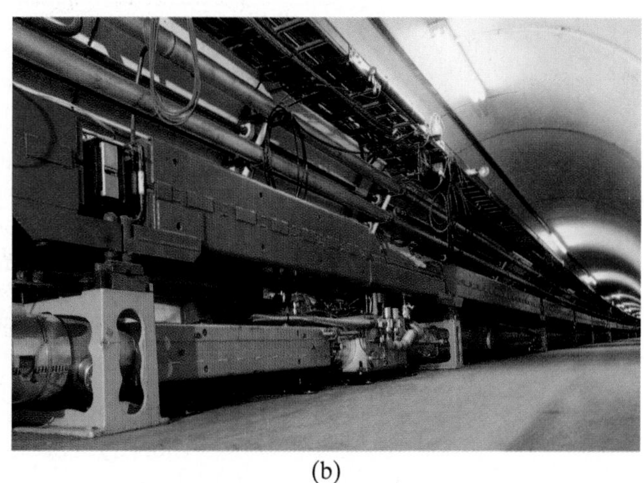

(b)

Once charged particles are injected, they must move in a circle of constant radius. This is accomplished by giving them considerable energy initially in a smaller accelerator (the injector), and then slowly increasing the magnetic field as they speed up in the large synchrotron.

One problem of any accelerator is that accelerating electric charges radiate electromagnetic energy (see Chapter 22). Since ions or electrons are accelerated in an accelerator, we can expect considerable energy to be lost by radiation. The effect increases with energy and is especially important in circular machines where centripetal acceleration is present, such as synchrotrons, and hence is called **synchrotron radiation**. Synchrotron radiation can be useful, however. Intense beams of photons are sometimes needed, and they are usually obtained from an electron synchrotron.

Synchrotron radiation

Linear Accelerators

Linac

In a **linear accelerator** (linac), electrons or ions are accelerated along a straight-line path, Fig. 32–4a, passing through a series of tubular conductors. Voltage applied to the tubes is alternating so that when electrons (say) reach a gap, the tube in front of them is positive and the one they just left is negative. At low speeds, the particles cover less distance in the same amount of time, so the tubes are shorter at first. Electrons, with their small mass, get close to the speed of light quickly, $v \approx c$, and the tubes are nearly equal in length. Linear accelerators are particularly important for accelerating electrons because of the absence of synchrotron radiation. The largest electron linear accelerator is at Stanford (Stanford Linear Accelerator Center, or SLAC), Fig. 32–4b. It is about 3 km (2 mi) long and can accelerate electrons to 50 GeV. Many hospitals have 10-MeV electron linacs that strike a metal foil to produce γ ray photons to irradiate tumors.

FIGURE 32–4 (a) Diagram of a simple linear accelerator. (b) Photo of the Stanford Linear Accelerator (SLAC) in California.

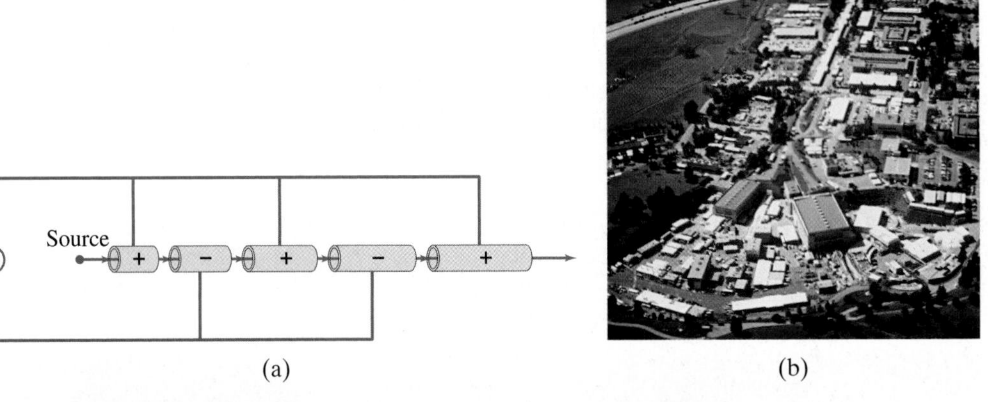

Colliding Beams

Colliders

High-energy physics experiments can be done by allowing a beam of particles from an accelerator to strike a stationary target. But to obtain the maximum possible collision energy from a given accelerator, two beams of particles are accelerated to very high energy and are steered so that they collide head-on. One way to accomplish such **colliding beams** with a single accelerator is through the use of **storage rings**, in which oppositely revolving beams can be repeatedly brought into collision with one another at particular points. For example, in the experiments that provided strong evidence for the top quark (see Chapter opening photo and Section 32–9), the Fermilab Tevatron accelerated protons and antiprotons each to 900 GeV, so that the combined energy of head-on collisions was 1.8 TeV.

The largest collider will soon be the Large Hadron Collider (LHC) at CERN, with a circumference of 26.7 km (Fig. 32–5), scheduled to be completed about 2007. The two colliding beams will each carry 7-TeV protons for a total interaction energy of 14 TeV.

FIGURE 32–5 The large circle represents the position of the tunnel, about 100 m below the ground at CERN (near Geneva) on the French-Swiss border, which will house the LHC. The smaller circle shows the position of the Super Proton Synchrotron that will be used for accelerating protons prior to injection into the LHC.

EXAMPLE 32–3 **Speed of a 1.0-TeV proton.** What is the speed of a 1.0-TeV proton produced at Fermilab?

APPROACH KE $= 1.0\,\text{TeV} = 1.0 \times 10^{12}\,\text{eV}$ is much greater than the rest mass of the proton, $0.938 \times 10^{9}\,\text{eV}$, so relativistic calculations must be used. In particular, we use Eq. 26–6:

$$\text{KE} = (\gamma - 1)m_0c^2 = \frac{m_0c^2}{\sqrt{1 - v^2/c^2}} - m_0c^2.$$

SOLUTION Compared to the KE of $1.0 \times 10^{12}\,\text{eV}$, the rest energy ($\approx 10^{-3}\,\text{TeV}$) can be neglected, so we write

$$\text{KE} = \frac{m_0c^2}{\sqrt{1 - v^2/c^2}}.$$

Then

$$1 - \frac{v^2}{c^2} = \left(\frac{m_0c^2}{\text{KE}}\right)^2$$

or

$$\frac{v}{c} = \sqrt{1 - \left(\frac{m_0c^2}{\text{KE}}\right)^2} = \sqrt{1 - \left(\frac{938 \times 10^6\,\text{eV}}{1.0 \times 10^{12}\,\text{eV}}\right)^2}$$

$$v = 0.9999996c.$$

The proton is traveling at a speed extremely close to c, the speed of light.

32–2 Beginnings of Elementary Particle Physics—Particle Exchange

The accepted model for elementary particles today views *quarks* and *leptons* as the basic constituents of ordinary matter. To understand our present-day view we need to begin with the ideas leading up to its formulation.†

Elementary particle physics might be said to have begun in 1935 when the Japanese physicist Hideki Yukawa (1907–1981) predicted the existence of a new particle that would in some way mediate the strong nuclear force. To understand Yukawa's idea, we first consider the electromagnetic force. When we first discussed electricity, we saw that the electric force acts over a distance, without contact. To better perceive how a force can act over a distance, we used the idea of a **field**. The force that one charged particle exerts on a second can be said to be due to the electric field set up by the first. Similarly, the magnetic field can be said to carry the magnetic force. Later (Chapter 22), we saw that electromagnetic (EM) fields can travel through space as waves. Finally, in Chapter 27, we saw that electromagnetic radiation (light) can be considered as either a wave or as a collection of particles called *photons*. Because of this wave–particle duality, it is possible to imagine that the electromagnetic force between charged particles is due to

(1) the EM field set up by one charged particle and felt by the other, or
(2) an exchange of photons (γ particles) between them.

It is (2) that we want to concentrate on here, and a crude analogy for how an exchange of particles could give rise to a force is suggested in Fig. 32–6. In part (a), two children start throwing heavy pillows at each other; each throw and each catch results in the child being pushed backward by the impulse. This is the equivalent of a repulsive force. On the other hand, if the two children exchange pillows by grabbing them out of the other person's hand, they will be pulled toward each other, as when an attractive force acts.

FIGURE 32–6 Forces equivalent to particle exchange. (a) Repulsive force (children throwing pillows at each other). (b) Attractive force (children grabbing pillows from each other's hands).

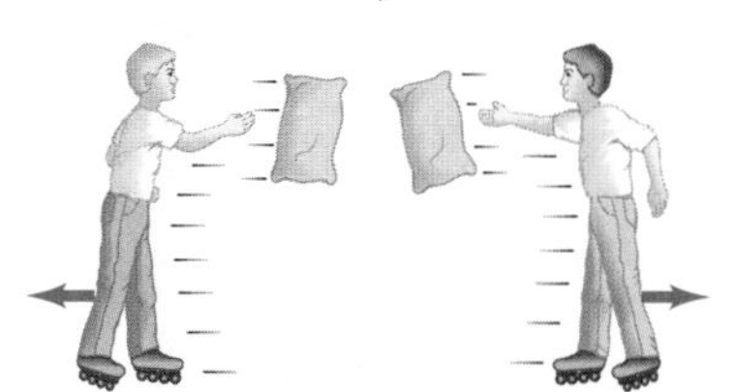
(a) Repulsive force (children throwing pillows)

(b) Attractive force (children grabbing pillows from each other's hands)

†Just telling you how it is today would not be a scientific discussion; nor would it give understanding—see footnote on page 769.

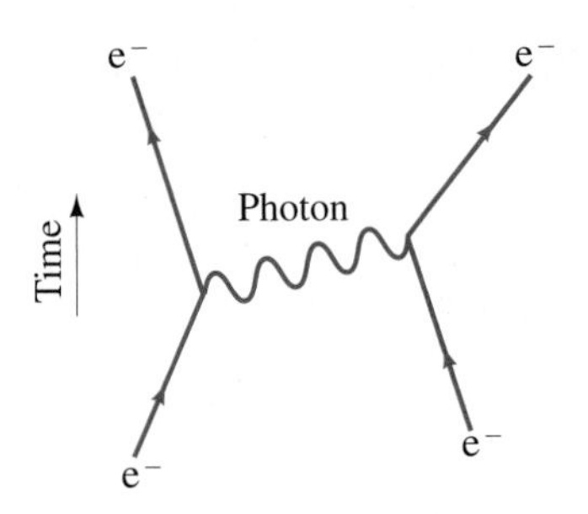

FIGURE 32–7 Feynman diagram showing a photon acting as the carrier of the electromagnetic force between two electrons. This is sort of an x vs. t graph, with t increasing upward. Starting at the bottom, two electrons approach each other (the distance between them decreases in time). As they get close, momentum and energy get transferred from one to the other, carried by a photon (or, perhaps, by more than one), and the two electrons bounce apart.

For the electromagnetic force, it is photons that are exchanged between two charged particles that give rise to the force between them. A simple diagram describing this photon exchange is shown in Fig. 32–7. Such a diagram, called a **Feynman diagram** after its inventor, the American physicist Richard Feynman (1918–1988), is based on the theory of **quantum electrodynamics** (QED).

Figure 32–7 represents the simplest case in QED, in which a single photon is exchanged. One of the charged particles emits the photon and recoils somewhat as a result; and the second particle absorbs the photon. In any collision or *interaction*, energy and momentum are transferred from one charged particle to the other, carried by the photon. The photon is absorbed by the second particle very shortly after it is emitted by the first and is not observable; hence it is referred to as a *virtual* photon, in contrast to one that is free and can be detected by instruments. The photon is said to *mediate*, or *carry*, the electromagnetic force.

Particles that mediate or "carry" forces

By analogy with photon exchange that mediates the electromagnetic force, Yukawa argued in this early theory that there ought to be a particle that mediates the strong nuclear force—the force that holds nucleons together in the nucleus. Yukawa called this predicted particle a **meson** (meaning "medium mass"). Figure 32–8 is a Feynman diagram showing meson exchange: a meson carrying the strong force between a neutron and a proton.

Mass estimate of exchange particle

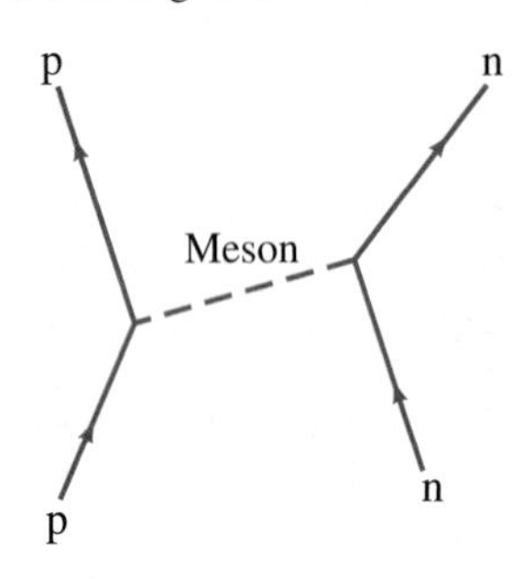

FIGURE 32–8 Meson exchange when a proton and neutron interact via the strong nuclear force.

We can make a rough estimate of the mass of the meson as follows. Suppose the proton on the left in Fig. 32–8 is at rest. For it to emit a meson would require energy (to make the meson's mass) which, coming from nowhere, would violate conservation of energy. But the uncertainty principle allows nonconservation of energy by an amount ΔE if it occurs only for a time Δt given by $(\Delta E)(\Delta t) \approx h/2\pi$. We set ΔE equal to the energy needed to create the mass m of the meson: $\Delta E = mc^2$. Conservation of energy is violated only as long as the meson exists, which is the time Δt required for the meson to pass from one nucleon to the other, where it is absorbed and disappears. If we assume the meson travels at relativistic speed, close to the speed of light c, then Δt need be at most about $\Delta t = d/c$, where d is the maximum distance that can separate the interacting nucleons. Thus we can write

$$\Delta E\,\Delta t \approx \frac{h}{2\pi}$$

$$mc^2\left(\frac{d}{c}\right) \approx \frac{h}{2\pi}$$

or

Mass of exchange particle

$$mc^2 \approx \frac{hc}{2\pi d}. \tag{32–3}$$

The range of the strong nuclear force (the maximum distance away it can be felt), is small—not much more than the size of a nucleon or small nucleus (see Eq. 30–1)—so let us take $d \approx 1.5 \times 10^{-15}$ m. Then from Eq. 32–3,

$$mc^2 \approx \frac{hc}{2\pi d} = \frac{(6.6 \times 10^{-34}\,\mathrm{J\cdot s})(3.0 \times 10^{8}\,\mathrm{m/s})}{(6.28)(1.5 \times 10^{-15}\,\mathrm{m})} \approx 2.1 \times 10^{-11}\,\mathrm{J} = 130\,\mathrm{MeV}.$$

The mass of the predicted meson, roughly $130\,\mathrm{MeV}/c^2$, is about 250 times the electron mass of $0.51\,\mathrm{MeV}/c^2$.†

†Note that since the electromagnetic force has infinite range, Eq. 32–3 with $d = \infty$ tells us that the exchanged particle for the electromagnetic force, the photon, will have zero rest mass, which it does.

The particle predicted by Yukawa was discovered in cosmic rays by C. F. Powell and G. Occhialini in 1947, and is called the "π" or pi meson, or simply the **pion**. It comes in three charge states: +, −, or 0. The π^+ and π^- have mass of $139.6\,\mathrm{MeV}/c^2$ and the π^0 a mass of $135.0\,\mathrm{MeV}/c^2$, all close to Yukawa's prediction. All three interact strongly with matter. Reactions observed in the laboratory, using a particle accelerator, include

Pion

$$\begin{aligned} \mathrm{p} + \mathrm{p} &\rightarrow \mathrm{p} + \mathrm{p} + \pi^0, \\ \mathrm{p} + \mathrm{p} &\rightarrow \mathrm{p} + \mathrm{n} + \pi^+. \end{aligned} \qquad \textbf{(32–4)}$$

The incident proton from the accelerator must have sufficient energy to produce the additional mass of the free pion.

Yukawa's theory of pion exchange as carrier of the strong force is now out of date, and has been replaced by *quantum chromodynamics* in which the basic entities are *quarks*, and the basic carriers of the strong force are *gluons*, as we shall discuss shortly. But the basic idea of the earlier theory, that forces can be understood as the exchange of particles, remains valid.

Forces carried by particles even in standard model

There are four known types of force—or interaction—in nature. The electromagnetic force is carried by the photon, the strong nuclear force by gluons. What about the other two: the weak nuclear force and gravity? These too are believed to be mediated by particles. The particles that transmit the weak force are referred to as the W^+, W^-, and Z^0, and were detected in 1983 (Fig. 32–9).

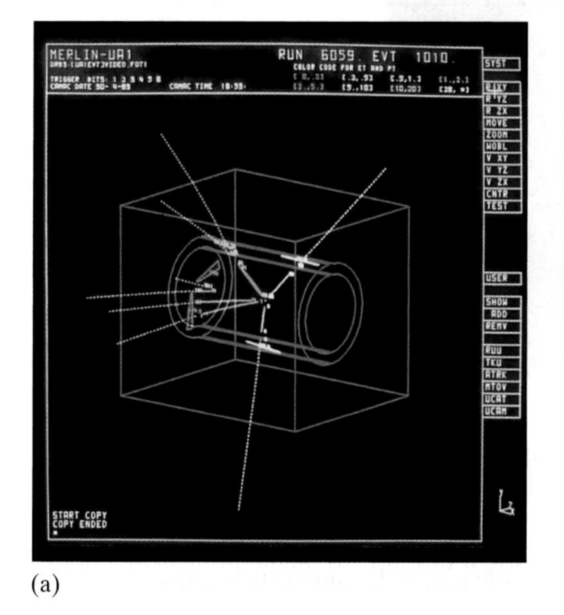

(a)

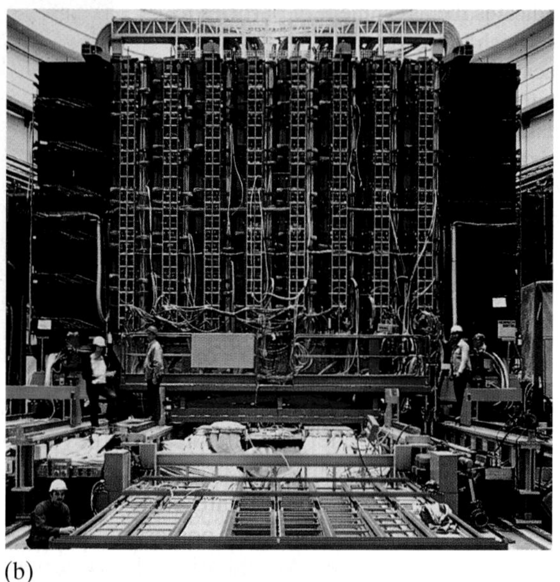
(b)

FIGURE 32–9 (a) Computer reconstruction of a Z-particle decay into an electron and a positron ($Z^0 \rightarrow e^+ + e^-$) whose tracks are shown in white, which took place in the UA1 detector at CERN. (b) Photo of the UA1 detector at CERN as it was being built.

The quantum (or carrier) of the gravitational force is called the **graviton**, and if it exists it has not yet been observed. A comparison of the four forces is given in Table 32–1, where they are listed according to their (approximate) relative strengths. Notice that although gravity may be the most obvious force in daily life (because of the huge mass of the Earth), on a nuclear scale it is the weakest of the four forces, and its effect at the particle level can nearly always be ignored.

Graviton

TABLE 32–1 The Four Forces in Nature

Type	Relative Strength (approx., for 2 protons in nucleus)	Field Particle
Strong nuclear	1	Gluons[†] (mesons)
Electromagnetic	10^{-2}	Photon
Weak nuclear	10^{-6}	$W^{\pm}$ and Z^0
Gravitational	10^{-38}	Graviton (?)

[†] Until the 1970s, thought to be mesons, but now gluons (see Section 32–10).

Theoreticians have wondered why the W and Z have large masses rather than being massless like the photon. Electroweak theory suggests an explanation by means of a new **Higgs field** and its particle, the **Higgs boson**, which interact with the W and Z to "slow them down." In being forced to go slower than the speed of light, they must acquire mass. The search for the Higgs boson will be a priority for experimental particle physicists when CERN's Large Hadron Collider (Section 32–1) starts running. So far, searches have excluded a Higgs lighter than 115 GeV/c^2. Yet it is expected to have a mass no larger than 200 GeV/c^2. We are narrowing in on it.

Higgs

32–11 Grand Unified Theories

With the success of the unified electroweak theory, attempts are being made to incorporate it and QCD for the strong (color) force into a so-called **grand unified theory** (GUT). One type of such a grand unified theory of the electromagnetic, weak, and strong forces has been worked out in which there is only one class of particle—leptons and quarks belong to the same family and are able to change freely from one type to the other—and the three forces are different aspects of a single underlying force. The unity is predicted to occur, however, only on a scale of less than about 10^{-32} m corresponding to an extremely high energy of about 10^{16} GeV. If two elementary particles (leptons or quarks) approach each other to within this **unification scale**, the apparently fundamental distinction between them would not exist at this level, and a quark could readily change to a lepton, or vice versa. Baryon and lepton numbers would not be conserved. The weak, electromagnetic, and strong (color) force would blend to a force of a single strength.

GUT

Unification of forces

What happens between the unification distance of 10^{-32} m and more normal (larger) distances is referred to as **symmetry breaking**. As an analogy, consider an atom in a crystal. Deep within the atom, there is much symmetry—in the innermost regions the electron cloud is spherically symmetric (Chapter 28). Farther out, this symmetry breaks down—the electron clouds are distributed preferentially along the lines (bonds) joining the atoms in the crystal. In a similar way, at 10^{-32} m the force between elementary particles is theorized to be a single force—it is symmetrical and does not single out one type of "charge" over another. But at larger distances, that symmetry is broken and we see three distinct forces. (In the "standard model" of electroweak interactions, Section 32–10, the symmetry breaking between the electromagnetic and the weak interactions occurs at about 10^{-18} m.)

Symmetry breaking

CONCEPTUAL EXAMPLE 32–9 **Symmetry.** The table in Fig. 32–15 has four identical place settings. Four people sit down to eat. Describe the symmetry of this table and what happens to it when someone starts the meal.

RESPONSE The table has several kinds of symmetry. It is symmetric to rotations of 90°: that is, the table will look the same if everyone moved one chair to the left or to the right. It is also north–south symmetric and east–west symmetric, so that swaps across the table don't affect the way the table looks. It also doesn't matter whether any person picks up the fork to the left of the plate or the fork to the right. But once that first person picks up either fork, the choice is set for all the rest at the table as well. The symmetry has been *broken*. The underlying symmetry is still there—the blue glasses could still be chosen either way—but some choice must get made and at that moment the symmetry of the diners is broken.

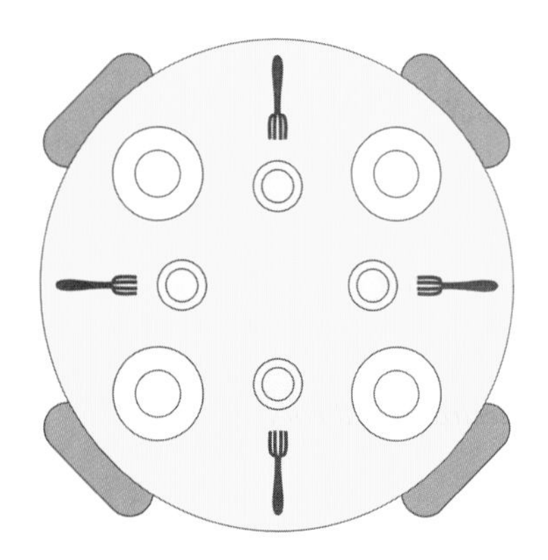

FIGURE 32–15 Symmetry around a table. Example 32–9.

Since unification occurs at such tiny distances and huge energies, the theory is difficult to test experimentally. But it is not completely impossible. One testable prediction is the idea that the proton might decay (via, for example, $p \rightarrow \pi^0 + e^+$) and violate conservation of baryon number. This could happen if two quarks approached to within 10^{-31} m of each other. But it is very unlikely at normal temperature and energy, so the decay of a proton can only be an unlikely process. In the simplest form of GUT, the theoretical estimate of the proton lifetime for the decay mode $p \rightarrow \pi^0 + e^+$ is about 10^{31} yr, and this has just come within the realm of testability.† Proton decays have still not been seen, and experiments put the lower limit on the proton lifetime for the above mode to be about 10^{33} yr, somewhat greater than this prediction. This may seem a disappointment, but on the other hand, it presents a challenge. Indeed more complex GUTs are not affected by this result.

Proton decay?

EXAMPLE 32–10 ESTIMATE Proton decay. An experiment uses 3300 tons of water waiting to see a proton decay of the type $p \rightarrow \pi^0 + e^+$. If the experiment is run for 4 years without detecting a decay, estimate the lower limit on the proton half-life.

APPROACH As with radioactive decay, the number of decays is proportional to the number of parent species (N), the time interval (Δt), and the decay constant (λ) which is related to the half-life $T_{\frac{1}{2}}$ by (see Eqs. 30–3 and 30–6):

$$\Delta N = -\lambda N \,\Delta t = -\frac{\ln 2}{T_{\frac{1}{2}}} N \,\Delta t.$$

SOLUTION Dealing only with magnitudes, we solve for $T_{\frac{1}{2}}$:

$$T_{\frac{1}{2}} = \frac{N}{\Delta N} \Delta t \ln 2.$$

Thus for $\Delta N < 1$ over the four-year trial,

$$T_{\frac{1}{2}} > N(4\,\text{yr})(0.693),$$

where N is the number of protons in 3300 tons of water. To determine N, we note that each molecule of H_2O contains $(2 + 8 =)10$ protons. So one mole of water (18 g, 6×10^{23} molecules) contains $10 \times 6 \times 10^{23}$ protons in 18 g of water, or about 3×10^{26} protons per kilogram. One ton is 10^3 kg, so the chamber contains $(3.3 \times 10^6\,\text{kg})(3 \times 10^{26}\,\text{protons/kg}) \approx 1 \times 10^{33}$ protons. Then our very rough estimate for a lower limit on the proton half-life is $T_{\frac{1}{2}} > (10^{33})(4\,\text{yr})(0.7) \approx 3 \times 10^{33}$ yr.

An interesting prediction of unified theories relates to cosmology (Chapter 33). It is thought that during the first 10^{-35} s after the theorized Big Bang that created the universe, the temperature was so extremely high that particles had energies corresponding to the unification scale. Baryon number would not have been conserved then, perhaps allowing an imbalance that might account for the observed predominance of matter $(B > 0)$ over antimatter $(B < 0)$ in the universe.

Connection with cosmology

This last example is interesting, for it illustrates a deep connection between investigations at either end of the size scale: theories about the tiniest objects (elementary particles) have a strong bearing on the understanding of the universe on a large scale. We will look at this more in the next Chapter.

†This is much larger than the age of the universe $(\approx 14 \times 10^9\,\text{yr})$. But we don't have to wait 10^{31} yr to see. Instead we can wait for one decay among 10^{31} protons over a year (see Eqs. 30–3 and 30–6, $\Delta N = \lambda N \,\Delta t = 0.693 N \,\Delta t / T_{\frac{1}{2}}$).

32–12 Strings and Supersymmetry

Even more ambitious than grand unified theories are attempts to also incorporate gravity, and thus unify all four forces in nature into a single theory. (Such theories are sometimes referred to misleadingly as **theories of everything**.) There are consistent theories that attempt to unify all four forces called **string theories**, in which the elementary particles (Table 32–5) are imagined not as points but as one-dimensional strings perhaps 10^{-35} m long.

String theory

Supersymmetry

A related idea is **supersymmetry**, which applied to strings is known as **superstring theory**. Supersymmetry predicts that interactions exist that would change fermions into bosons and vice versa, and that all known fermions have supersymmetric boson partners. Thus, for each quark we know (a fermion), there would be a *squark* (a *boson*) or "supersymmetric" quark. For every lepton there would be a *slepton*. Likewise, for every known boson (photons and gluons, for example), there would be a supersymmetric fermion (*photinos* and *gluinos*). Supersymmetry predicts also that a *graviton*, which transmits the gravity force, has a partner, the gravitino. Supersymmetric particles are sometimes called "SUSYs" for short, and may be a candidate for the "dark matter" of the universe (discussed in Chapter 33). But why hasn't this "missing part" of the universe ever been detected? The best guess is that supersymmetric particles might be heavier than their conventional counterparts, perhaps too heavy to have been produced in today's accelerators. Until a supersymmetric particle is found, and it may be possible at CERN's new LHC, supersymmetry is just an elegant guess.

The world of elementary particles is opening new vistas. What happens in the future is bound to be exciting.

Summary

Particle accelerators are used to accelerate charged particles, such as electrons and protons, to very high energy. High-energy particles have short wavelength and so can be used to probe the structure of matter at very small distances in great detail. High kinetic energy also allows the creation of new particles through collision (via $E = mc^2$).

Cyclotrons and **synchrotrons** use a magnetic field to keep the particles in a circular path and accelerate them at intervals by high voltage. **Linear accelerators** accelerate particles along a line. **Colliding beams** allow higher interaction energy.

An **antiparticle** has the same mass as a particle but opposite charge. Certain other properties may also be opposite: for example, the antiproton has **baryon number** (nucleon number) opposite to that for the proton.

In all nuclear and particle reactions, the following conservation laws hold: momentum, angular momentum, mass–energy, electric charge, baryon number, and **lepton numbers**.

Certain particles have a property called **strangeness**, which is conserved by the strong force but not by the weak force. The properties **charm**, **bottomness**, and **topness** also are conserved by the strong force but not by the weak.

Just as the electromagnetic force can be said to be due to an exchange of photons, the strong nuclear force was first thought to be carried by *mesons* that have rest mass, but recent theory says the force is carried by massless **gluons**. The W and Z particles carry the weak force. These fundamental force carriers (photon, W and Z, gluons) are called **gauge bosons**.

Other particles can be classified as either *leptons* or *hadrons*. **Leptons** participate in the weak and electrically charged electromagnetic interactions. **Hadrons**, which today are considered to be made up of **quarks**, participate in the strong interaction as well. The hadrons can be classified as **mesons**, with baryon number zero, and **baryons**, with nonzero baryon number.

All particles, except for the photon, electron, neutrinos, and proton, decay with measurable half-lives varying from 10^{-25} s to 10^{3} s. The half-life depends on which force is predominant. Weak decays usually have half-lives greater than about 10^{-13} s. Electromagnetic decays have half-lives on the order of 10^{-16} to 10^{-19} s. The shortest lived particles, called **resonances**, decay via the strong interaction and live typically for only about 10^{-23} s.

Today's standard model of elementary particles considers **quarks** as the basic building blocks of the hadrons. The six quark "flavors" are called **up**, **down**, **strange**, **charmed**, **bottom**, and **top**. It is expected that there are the same number of quarks as leptons (six of each), and that quarks and leptons are the truly elementary particles along with the gauge bosons (γ, W, Z, gluons). Quarks are said to have **color**, and, according to **quantum chromodynamics** (QCD), the strong color force acts between their color charges and is transmitted by **gluons**. **Electroweak theory** views the weak and electromagnetic forces as two aspects of a single underlying interaction. QCD plus the electroweak theory are referred to as the **Standard Model**.

Grand unified theories of forces suggest that at very short distances $\left(10^{-32}\text{ m}\right)$ and very high energy, the weak, electromagnetic, and strong forces appear as a single force, and the fundamental difference between quarks and leptons disappears.

Questions

1. Give a reaction between two nucleons, similar to Eq. 32–4, that could produce a π^-.
2. If a proton is moving at very high speed, so that its kinetic energy is much greater than its rest energy (m_0c^2), can it then decay via $p \rightarrow n + \pi^+$?
3. What would an "antiatom," made up of the antiparticles to the constituents of normal atoms, consist of? What might happen if *antimatter*, made of such antiatoms, came in contact with our normal world of matter?
4. What particle in a decay signals the electromagnetic interaction?
5. Does the presence of a neutrino among the decay products of a particle necessarily mean that the decay occurs via the weak interaction? Do all decays via the weak interaction produce a neutrino? Explain.
6. Why is it that a neutron decays via the weak interaction even though the neutron and one of its decay products (proton) are strongly interacting?
7. Which of the four interactions (strong, electromagnetic, weak, gravitational) does an electron take part in? A neutrino? A proton?
8. Check that charge and baryon number are conserved in each of the decays in Table 32–2.
9. Which of the particle decays in Table 32–2 occur via the electromagnetic interaction?
10. Which of the particle decays in Table 32–2 occur by the weak interaction?
11. By what interaction, and why, does $\Sigma^\pm$ decay to Λ^0? What about Σ^0 decaying to Λ^0? ?
12. The Δ baryon has spin $\frac{3}{2}$, baryon number 1, and charge $Q = +2, +1, 0,$ or -1. Why is there no charge state $Q = -2$?
13. Which of the particle decays in Table 32–4 occur via the electromagnetic interaction?
14. Which of the particle decays in Table 32–4 occur by the weak interaction?
15. Quarks have spin $\frac{1}{2}$. How do you account for the fact that baryons have spin $\frac{1}{2}$ or $\frac{3}{2}$, and mesons have spin 0 or 1?
16. Suppose there were a kind of "neutrinolet" that was massless, had no color charge or electrical charge, and did not feel the weak force. Could you say that this particle even exists?
17. Is it possible for a particle to be both (*a*) a lepton and a baryon? (*b*) a baryon and a hadron? (*c*) a meson and a quark? (*d*) a hadron and a lepton? Explain.
18. Using the ideas of quantum chromodynamics, would it be possible to find particles made up of two quarks and no antiquarks? What about two quarks and two antiquarks?
19. Why do neutrons decay when they are free but not when they are inside the nucleus?
20. Is the reaction $e^- + p \rightarrow n + \bar{\nu}_e$ possible? Explain.
21. Occasionally, the Λ will decay by the following reaction: $\Lambda^0 \rightarrow p^+ + e^- + \bar{\nu}_e$. Which of the four forces in nature is responsible for this decay? How do you know?

Problems

32–1 Particles and Accelerators

1. (I) What is the total energy of a proton whose kinetic energy is 6.35 GeV?
2. (I) Calculate the wavelength of 35-GeV electrons.
3. (I) What strength of magnetic field is used in a cyclotron in which protons make 2.8×10^7 revolutions per second?
4. (I) What is the time for one complete revolution for a very high-energy proton in the 1.0-km-radius Fermilab accelerator?
5. (I) If α particles are accelerated by the cyclotron of Example 32–2, what must be the frequency of the voltage applied to the dees?
6. (II) (*a*) If the cyclotron of Example 32–2 accelerated α particles, what maximum energy could they attain? What would their speed be? (*b*) Repeat for deuterons $({}^2_1H)$. (*c*) In each case, what frequency of voltage is required?
7. (II) Which is better for picking out details of the nucleus: 30-MeV alpha particles or 30-MeV protons? Compare each of their wavelengths with the size of a nucleon in a nucleus.
8. (II) The voltage across the dees of a cyclotron is 55 kV. How many revolutions do protons make to reach a kinetic energy of 25 MeV?
9. (II) What is the wavelength (= maximum resolvable distance) of 7.0-TeV protons?
10. (II) A cyclotron with a radius of 1.0 m is to accelerate deuterons $({}^2_1H)$ to an energy of 12 MeV. (*a*) What is the required magnetic field? (*b*) What frequency is needed for the voltage between the dees? (*c*) If the potential difference between the dees averages 22 kV, how many revolutions will the particles make before exiting? (*d*) How much time does it take for one deuteron to go from start to exit. (*e*) Estimate how far it travels during this time.
11. (II) The 4.25-km-radius tunnel that will be used to house the magnets for the Large Hadron Collider (LHC) calls for proton beams of energy 7.0 TeV. What magnetic field will be required?
12. (II) The 1.0-km radius Fermilab Tevatron takes about 20 seconds to bring the energies of the stored protons from 150 GeV to 1.0 TeV. The acceleration is done once per turn. Estimate the energy given to the protons on each turn. (You can assume that the speed of the protons is essentially c the whole time.)
13. (III) Show that the energy of a particle (charge e) in a synchrotron, in the relativistic limit $(v \approx c)$, is given by $E\,(\text{in eV}) = Brc$, where B is magnetic field strength and r the radius of the orbit (SI units).
14. (III) What magnetic field intensity is needed at the 1.0-km-radius Fermilab synchrotron for 1.0-TeV protons?

32–2 to 32–6 Particle Interactions, Particle Exchange

15. (I) How much energy is released in the decay
$$\pi^+ \rightarrow \mu^+ + \nu_\mu?$$
See Table 32–2.

16. (I) About how much energy is released when a Λ^0 decays to $n + \pi^0$? (See Table 32–2.)

17. (I) How much energy is required to produce a neutron–antineutron pair?

18. (I) Estimate the range of the strong force if the mediating particle were the kaon instead of the pion.

19. (II) Two protons are heading toward each other with equal speeds. What minimum kinetic energy must each have if a π^0 meson is to be created in the process? (See Table 32–2.)

20. (II) What minimum kinetic energy must two neutrons each have if they are traveling at the same speed toward each other, collide, and produce a K^+K^-pair in addition to themselves? (See Table 32–2.)

21. (II) Estimate the range of the weak force using Eq. 32–3, given the masses of the W and Z particles as about 80 to 90 GeV/c^2.

22. (II) What are the wavelengths of the two photons produced when a proton and antiproton at rest annihilate?

23. (II) The Λ cannot decay by the following reactions. What conservation law is violated in each of the reactions?
(*a*) $\Lambda^0 \rightarrow n + \pi^-$
(*b*) $\Lambda^0 \rightarrow p + K^-$
(*c*) $\Lambda^0 \rightarrow \pi^+ + \pi^-$

24. (II) For the decay $\Lambda^0 \rightarrow p + \pi^-$, calculate (*a*) the Q-value (energy released), and (*b*) the kinetic energy of the p and π^-, assuming the Λ^0 decays from rest. (Use relativistic formulas.)

25. (II) (*a*) Show, by conserving momentum and energy, that it is impossible for an isolated electron to radiate only a single photon. (*b*) With this result in mind, how can you defend the photon exchange diagram in Fig. 32–7?

26. (II) What would be the wavelengths of the two photons produced when an electron and a positron, each with 420 keV of kinetic energy, annihilate head on?

27. (II) In the rare decay $\pi^+ \rightarrow e^+ + \nu_e$, what is the kinetic energy of the positron? Assume the π^+ decays from rest.

28. (II) Which of the following reactions and decays are possible? For those forbidden, explain what laws are violated.
(*a*) $\pi^- + p \rightarrow n + \eta^0$
(*b*) $\pi^+ + p \rightarrow n + \pi^0$
(*c*) $\pi^+ + p \rightarrow p + e^+$
(*d*) $p \rightarrow e^+ + \nu_e$
(*e*) $\mu^+ \rightarrow e^+ + \bar{\nu}_\mu$
(*f*) $p \rightarrow n + e^+ + \nu_e$

29. (II) Calculate the kinetic energy of each of the two products in the decay $\Xi^- \rightarrow \Lambda^0 + \pi^-$. Assume the Ξ^- decays from rest.

30. (III) Could a π^+ meson be produced if a 100-MeV proton struck a proton at rest? What minimum kinetic energy must the incoming proton have?

31. (III) Calculate the maximum kinetic energy of the electron in the decay $\mu^- \rightarrow e^- + \bar{\nu}_e + \nu_\mu$. [*Hint*: in what direction do the two neutrinos move relative to the electron in order to give the latter the maximum kinetic energy? Both energy and momentum are conserved; use relativistic formulas.]

32–7 to 32–11 Resonances, Standard Model, Quarks, QCD, GUT

32. (I) Use Fig. 32–11 to estimate the energy width and then the lifetime of the Δ resonance using the uncertainty principle.

33. (I) The measured width of the J/ψ meson is 88 keV. Estimate its lifetime.

34. (I) The measured width of the ψ (3685) meson is 277 keV. Estimate its lifetime.

35. (I) What is the energy width (or uncertainty) of (*a*) η^0, and (*b*) Σ^0? See Table 32–2.

36. (I) The B^- meson is a $b\bar{u}$ quark combination. (*a*) Show that this is consistent for all quantum numbers. (*b*) What are the quark combinations for B^+, B^0, $\bar{B}^0$?

37. (II) Which of the following decays are possible? For those that are forbidden, explain which laws are violated.
(*a*) $\Xi^0 \rightarrow \Sigma^+ + \pi^-$
(*b*) $\Omega^- \rightarrow \Sigma^0 + \pi^- + \nu$
(*c*) $\Sigma^0 \rightarrow \Lambda^0 + \gamma + \gamma$

38. (II) What are the quark combinations that can form (*a*) a neutron, (*b*) an antineutron, (*c*) a Λ^0, (*d*) a $\bar{\Sigma}^0$?

39. (II) What particles do the following quark combinations produce: (*a*) uud, (*b*) $\bar{u}\bar{u}\bar{s}$, (*c*) $\bar{u}s$, (*d*) $d\bar{u}$, (*e*) $\bar{c}s$?

40. (II) What is the quark combination needed to produce a D^0 meson ($Q = B = S = 0, c = +1$)?

41. (II) The D_S^+ meson has $S = c = +1,\ B = 0$. What quark combination would produce it?

42. (II) Draw a possible Feynman diagram using quarks (as in Fig. 32–13c) for the reaction $\pi^- + p \rightarrow \pi^0 + n$.

43. (II) Draw a Feynman diagram for the reaction $n + \nu_\mu \rightarrow p + \mu^-$.

General Problems

44. What is the total energy of a proton whose kinetic energy is 25 GeV? What is its wavelength?

45. Assume there are 5.0×10^{13} protons at 1.0 TeV stored in the 1.0-km-radius ring of the Tevatron. (*a*) How much current (amperes) is carried by this beam? (*b*) How fast would a 1500-kg car have to move to carry the same kinetic energy as this beam?

46. Protons are injected into the 1.0-km-radius Fermilab Tevatron with an energy of 150 GeV. If they are accelerated by 2.5 MV each revolution, how far do they travel and approximately how long does it take for them to reach 1.0 TeV?

47. (*a*) How much energy is released when an electron and a positron annihilate each other? (*b*) How much energy is released when a proton and an antiproton annihilate each other? (All particles KE $\approx$ 0.)

48. Which of the following reactions are possible, and by what interaction could they occur? For those forbidden, explain why.
(*a*) $\pi^- + \mathrm{p} \rightarrow \mathrm{K}^+ + \Sigma^-$
(*b*) $\pi^+ + \mathrm{p} \rightarrow \mathrm{K}^+ + \Sigma^+$
(*c*) $\pi^- + \mathrm{p} \rightarrow \Lambda^0 + \mathrm{K}^0 + \pi^0$
(*d*) $\pi^+ + \mathrm{p} \rightarrow \Sigma^0 + \pi^0$
(*e*) $\pi^- + \mathrm{p} \rightarrow \mathrm{p} + \mathrm{e}^- + \bar{\nu}_\mathrm{e}$

49. Which of the following reactions are possible, and by what interaction could they occur? For those forbidden, explain why.
(*a*) $\pi^- + \mathrm{p} \rightarrow \mathrm{K}^0 + \mathrm{p} + \pi^0$
(*b*) $\mathrm{K}^- + \mathrm{p} \rightarrow \Lambda^0 + \pi^0$
(*c*) $\mathrm{K}^+ + \mathrm{n} \rightarrow \Sigma^+ + \pi^0 + \gamma$
(*d*) $\mathrm{K}^+ \rightarrow \pi^0 + \pi^0 + \pi^+$
(*e*) $\pi^+ \rightarrow \mathrm{e}^+ + \nu_\mathrm{e}$

50. One decay mode for a π^+ is $\pi^+ \rightarrow \mu^+ + \nu_\mu$. What would be the equivalent decay for a π^-? Check conservation rules.

51. Symmetry breaking occurs in the electroweak theory at about 10^{-18} m. Show that this corresponds to an energy that is on the order of the mass of the $\mathrm{W}^\pm$.

52. The mass of a π^0 can be measured by observing the reaction $\pi^- + \mathrm{p} \rightarrow \pi^0 + \mathrm{n}$ at very low incident π^- kinetic energy (assume it is zero). The neutron is observed to be emitted with a kinetic energy of 0.60 MeV. Use conservation of energy and momentum to determine the π^0 mass.

53. Calculate the Q-value for each of the reactions, Eq. 32–4, for producing a pion.

54. Calculate the Q-value for the reaction $\pi^- + \mathrm{p} \rightarrow \Lambda^0 + \mathrm{K}^0$, when negative pions strike stationary protons. Estimate the minimum pion kinetic energy needed to produce this reaction. [*Hint*: assume Λ^0 and K^0 move off with the same velocity.]

55. How many fundamental fermions are there in a water molecule?

56. A proton and an antiproton annihilate each other at rest and produce two pions, π^- and π^+. What is the kinetic energy of each pion?

57. (*a*) Show that the so-called unification distance of 10^{-32} m in grand unified theory is equivalent to an energy of about 10^{16} GeV. Use the uncertainty principle, and also de Broglie's wavelength formula, and explain how they apply. (*b*) Calculate the temperature corresponding to 10^{16} GeV.

58. For the reaction $\mathrm{p} + \mathrm{p} \rightarrow 3\mathrm{p} + \bar{\mathrm{p}}$, where one of the initial protons is at rest, use relativistic formulas to show that the threshold energy is $6m_\mathrm{p}c^2$, equal to three times the magnitude of the Q-value of the reaction, where m_p is the proton mass. [*Hint*: assume all final particles have the same velocity.]

59. The lifetimes listed in Table 32–2 are in terms of *proper time*, measured in a reference frame where the particle is at rest. If a tau lepton is created with a kinetic energy of 450 MeV, how long would its track be as measured in the lab, on average, ignoring any collisions?

60. Identify the missing particle in the following reactions.
(*a*) $\mathrm{p} + \mathrm{p} \rightarrow \mathrm{p} + \mathrm{n} + \pi^+ + ?$
(*b*) $\mathrm{p} + ? \rightarrow \mathrm{n} + \mu^+$

61. Use the quark model to describe the reaction
$$\bar{\mathrm{p}} + \mathrm{n} \rightarrow \pi^- + \pi^0.$$

62. What fraction of the speed of light c is the speed of a 7.0-TeV proton?

Answers to Exercises

A: 1.24×10^{-18} m.

B: $\mathrm{s\bar{u}}$.

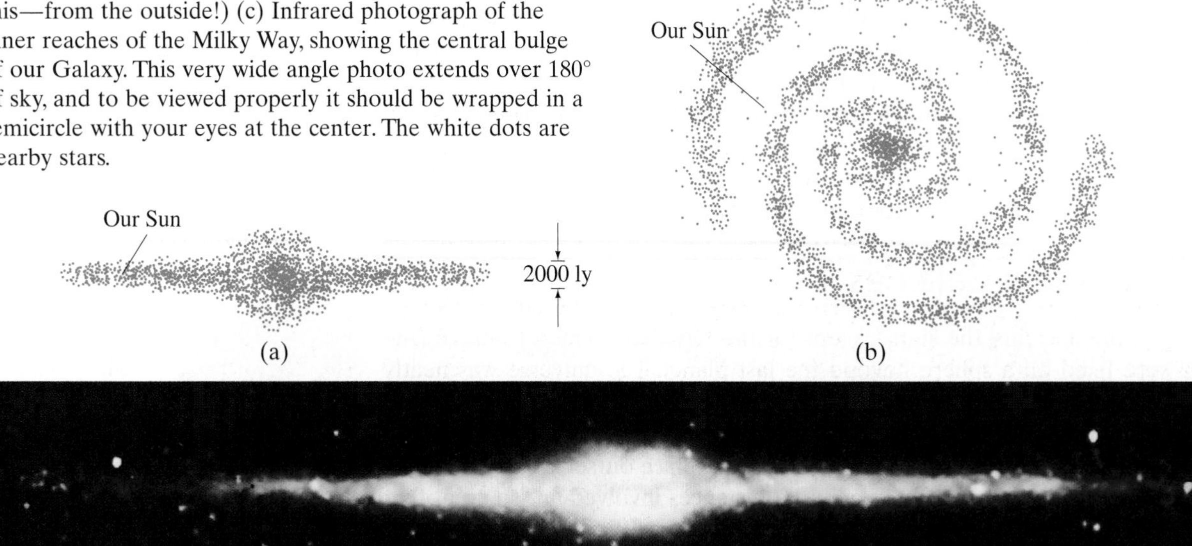

FIGURE 33–2 Our Galaxy, as it would appear from the outside: (a) "edge view," in the plane of the disc; (b) "top view," looking down on the disc. (If only we could see it like this—from the outside!) (c) Infrared photograph of the inner reaches of the Milky Way, showing the central bulge of our Galaxy. This very wide angle photo extends over 180° of sky, and to be viewed properly it should be wrapped in a semicircle with your eyes at the center. The white dots are nearby stars.

Our Galaxy has a diameter of almost 100,000 light-years and a thickness of roughly 2000 ly. It has a bulging central "nucleus" and spiral arms (Fig. 33–2). Our Sun, which seems to be just another star, is located about halfway from the galactic center to the edge, some 26,000 ly from the center. Our Galaxy contains roughly 100 billion (10^{11}) stars. The Sun orbits the galactic center approximately once every 250 million years or so, so its speed is about 200 km/s relative to the center of the Galaxy. The total mass of all the stars in our Galaxy is estimated to be about 3×10^{41} kg, which is ordinary matter. In addition, there is strong evidence that our Galaxy is surrounded by an invisible "halo" of "dark matter," which we discuss in Section 33–8.

EXAMPLE 33–1 ESTIMATE Our Galaxy's mass. Estimate the total mass of our Galaxy using the orbital data of the Sun (including our solar system) about the center of the Galaxy. Assume that most of the mass of the Galaxy is concentrated near the center of the Galaxy.

APPROACH We assume that the Sun and the solar system (total mass m) move in a circular orbit about the center of the Galaxy (total mass M), and that the mass M can be considered as being located at the center of the Galaxy. We then apply Newton's second law, $F = ma$, with a being the centripetal acceleration, $a = v^2/r$, and F being the universal law of gravitation (Chapter 5).

SOLUTION Our Sun and solar system orbit the center of the Galaxy, according to the best measurements as mentioned above, with a speed of about $v = 200\,\text{km/s}$ at a distance from the Galaxy center of about $r = 26{,}000\,\text{ly}$. We use Newton's second law:

$$F = ma$$

$$G\frac{Mm}{r^2} = m\frac{v^2}{r}$$

where M is the mass of the Galaxy and m is the mass of our Sun and solar system. Solving this, we find

$$M = \frac{rv^2}{G} \approx \frac{(26{,}000\,\text{ly})(10^{16}\,\text{m/ly})(2 \times 10^5\,\text{m/s})^2}{6.67 \times 10^{-11}\,\text{N}\cdot\text{m}^2/\text{kg}^2} \approx 2 \times 10^{41}\,\text{kg}.$$

NOTE In terms of *numbers* of stars, if they are like our Sun $(m = 2.0 \times 10^{30}\,\text{kg})$, there would be about $(2 \times 10^{41}\,\text{kg})/(2 \times 10^{30}\,\text{kg}) \approx 10^{11}$ or about 100 billion stars.

In addition to stars both within and outside the Milky Way, we can see by telescope many faint cloudy patches in the sky which were all referred to once as "nebulae" (Latin for "clouds"). A few of these, such as those in the constellations Andromeda and Orion, can actually be discerned with the naked eye on a clear night. Some are **star clusters** (Fig. 33–3), groups of stars that are so numerous they appear to be a cloud. Others are glowing clouds of gas or dust (Fig. 33–4), and it is for these that we now mainly reserve the word **nebula**. Most fascinating are those that belong to a third category: they often have fairly regular elliptical shapes and seem to be a great distance beyond our Galaxy. Immanuel Kant (about 1755) seems to have been the first to suggest that these latter might be circular discs, but appear elliptical because we see them at an angle, and are faint because they are so distant. At first it was not universally accepted that these objects were **extragalactic**—that is, outside our Galaxy. The very large telescopes constructed in the twentieth century revealed that individual stars could be resolved within these extragalactic objects and that many contain spiral arms. Edwin Hubble (1889–1953) did much of this observational work in the 1920s using the 2.5-m (100-inch) telescope† on Mt. Wilson near Los Angeles, California, then the world's largest. Hubble demonstrated that these objects were indeed extragalactic because of their great distances. The distance to our nearest large galaxy,‡ Andromeda, is over 2 million light-years, a distance 20 times greater than the diameter of our Galaxy. It seemed logical that these nebulae must be **galaxies** similar to ours. (Note that it is usual to capitalize the word "galaxy" only when it refers to our own.) Today it is thought there are roughly 40×10^9 galaxies in the observable universe—that is, roughly as many galaxies as there are stars in a galaxy. See Fig. 33–5.

FIGURE 33–3 This globular star cluster is located in the constellation Hercules.

FIGURE 33–4 This gaseous nebula, found in the constellation Carina, is about 9000 light-years from us.

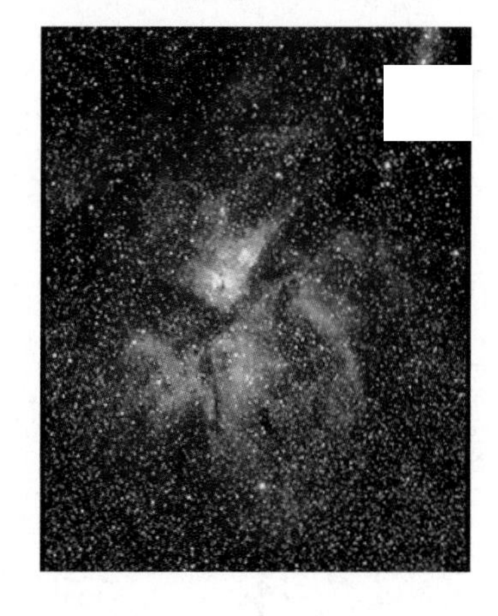

Many galaxies tend to be grouped in **galaxy clusters** held together by their mutual gravitational attraction. There may be anywhere from a few to many thousands

†2.5 m (= 100 inches) refers to the diameter of the curved objective mirror. The bigger the mirror, the more light it collects (greater intensity) and the less diffraction there is (better resolution), so more and fainter stars can be seen. See Chapter 25. Until recently, photographic films or plates were used to take long time exposures. Now large solid-state CCD sensors (Section 25–1) are available containing 100 million pixels (compared to 5 or 6 megapixels in a good-quality digital camera).

‡The *Magellanic clouds* are much closer than Andromeda, but are small and are usually considered small satellite galaxies of our own Galaxy.

FIGURE 33–5 Photographs of galaxies. (a) Spiral galaxy in the constellation Hydra. (b) Two galaxies: the larger and more dramatic one is known as the Whirlpool galaxy. (c) A false-color infrared image of the same galaxies as in (b), here showing the arms of the spiral as being more regular than in the visible light photo (b); the different colors correspond to different light intensities. Visible light is scattered and absorbed by interstellar dust much more than infrared is, so the latter gives us a clearer image.

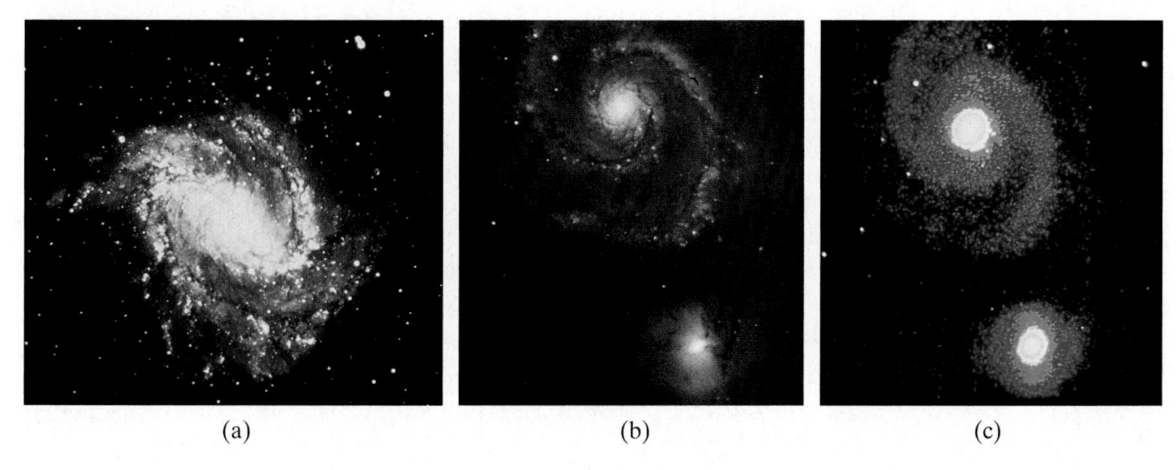

(a) (b) (c)

TABLE 33–1 Heavenly Distances

Object	Approx. Distance from Earth (ly)
Moon	4×10^{-8}
Sun	1.6×10^{-5}
Size of solar system (distance to Pluto)	6×10^{-4}
Nearest star (Proxima Centauri)	4.3
Center of our Galaxy	2.6×10^{4}
Nearest large galaxy	2.4×10^{6}
Farthest galaxies	10^{10}

of galaxies in each cluster. Furthermore, clusters themselves seem to be organized into even larger aggregates: clusters of clusters of galaxies, or **superclusters**. The farthest detectable galaxies are on the order of 10^{10} ly distant. See Table 33–1.

CONCEPTUAL EXAMPLE 33–2 **Looking back in time.** Astronomers often think of their telescopes as time machines, looking back toward the origin of the universe. How far back do they look?

RESPONSE The distance in light-years measures precisely how long in years the light has been traveling to reach us, so Table 33–1 tells us also how far back in time we are looking. For example, if we saw Proxima Centauri explode into a supernova today, then the event would have really occurred 4.3 years ago. The most distant galaxies, 10^{10} ly away, emitted the light we see now 10^{10} years ago; so what we see was how they were then, 10^{10} yr ago, close to the beginning of the universe.

EXERCISE A Suppose we could place a huge mirror 1 light-year away from us. What would we see in this mirror if it is facing us on Earth? When did it take place? (This might be called a "time machine.")

Besides the usual stars, clusters of stars, galaxies, and clusters and superclusters of galaxies, the universe contains a number of other interesting objects. Among these are stars known as *red giants*, *white dwarfs*, *neutron stars*, exploding stars called *novae* and *supernovae*, and *black holes* (very probably) whose gravity is so strong even light can not escape them. In addition, there is electromagnetic radiation that reaches the Earth but does not emanate from the bright pointlike objects we call stars: particularly important is the microwave background radiation that arrives nearly uniformly from all directions in the universe. We will discuss all these phenomena.

Quasar (QSO)

Finally, there are "active galactic nuclei" (AGN), which are very luminous pointlike sources of light in the centers of distant galaxies. The most dramatic examples of AGN are *quasars* ("quasistellar objects" or QSOs), which are so luminous that the surrounding starlight of the galaxy is drowned out. Their luminosity is thought to come from matter starting to fall into a giant black hole at a galaxy's center.

33–2 Stellar Evolution: The Birth and Death of Stars

The stars appear unchanging. Night after night the heavens reveal no significant variations. Indeed, on a human time scale, the vast majority of stars change very little (except for novae, supernovae, and certain variable stars). Although stars *seem* fixed in relation to each other, many move sufficiently for the motion to be detected. Speeds of stars relative to neighboring stars can be hundreds of km/s, but at their great distance from us, this motion is detectable only by careful measurement. Furthermore, there is a great range of brightness among stars. The differences in brightness are due to differences in the amount of light stars emit as well as to their different distances from us.

Luminosity and Brightness of Stars

A useful parameter for a star or galaxy is its **luminosity** (or "absolute luminosity"), L, by which we mean the total power radiated in watts. Also important is the **apparent brightness**, l, defined as the power crossing unit area at the Earth perpendicular to the path of the light. Given that energy is conserved, and ignoring any absorption in space, the total emitted power L when it reaches a distance d from the star will be spread over a sphere of surface area $4\pi d^2$. If d is the distance from the star to the Earth, then L must be equal to $4\pi d^2$ times l (power per unit area at Earth). That is,

$$l = \frac{L}{4\pi d^2}. \qquad \textbf{(33–1)}$$

EXAMPLE 33–3 Apparent brightness. Suppose a particular star has absolute luminosity equal to that of the Sun but is 10 ly away from Earth. By what factor will it appear dimmer than the Sun?

APPROACH The luminosity L is the same for both stars, so the apparent brightness depends only on their relative distances. We use the inverse square law as stated in Eq. 33–1 to determine the relative brightness.

SOLUTION Using Eq. 33–1, we find that the star appears dimmer by a factor

$$\frac{l_{\text{star}}}{l_{\text{Sun}}} = \frac{d_{\text{Sun}}^2}{d_{\text{star}}^2} = \frac{(1.5 \times 10^8\,\text{km})^2}{(10\,\text{ly})^2(10^{13}\,\text{km/ly})^2} \approx 2 \times 10^{-12}.$$

Careful study of nearby stars has shown that the absolute luminosity for most stars depends on the mass:† *the more massive the star, the greater its luminosity.* Another important parameter of a star is its surface temperature, which can be determined from the spectrum of electromagnetic frequencies it emits, just as for a blackbody (Section 27–2). As we saw in Chapter 27, as the temperature of a body increases, the spectrum shifts from predominantly lower frequencies (and longer wavelengths, such as red) to higher frequencies (and shorter wavelengths such as blue). Quantitatively, the relation is given by Wien's law (Eq. 27–2): the peak wavelength λ_{P} in the spectrum of light emitted by a blackbody (and stars are fairly good approximations to blackbodies) is inversely proportional to its kelvin temperature T; that is, $\lambda_{\text{P}} T = 2.90 \times 10^{-3}\,\text{m}\cdot\text{K}$. The surface temperatures of stars typically range from about 3500 K (reddish) to perhaps 50,000 K (UV).

Luminosity increases with star's mass

EXAMPLE 33–4 Determining star temperature and star size. Suppose that the distances from Earth to two nearby stars can be reasonably estimated, and that their measured apparent brightnesses suggest the two stars have about the same absolute luminosity, L. The spectrum of one of the stars peaks at about 700 nm (so it is reddish). The spectrum of the other peaks at about 350 nm (bluish). Use Wien's law (Eq. 27–2) and the Stefan-Boltzmann equation (Section 14–8) to determine (*a*) the surface temperature of each star, and (*b*) how much larger one star is than the other.

APPROACH We determine the surface temperature T for each star using Wien's law and each star's peak wavelength. Then, using the Stefan-Boltzmann equation (power output or luminosity $\propto AT^4$), we can find the surface area ratio and relative sizes of the two stars.

SOLUTION (*a*) Wien's law (Eq. 27–2) states that $\lambda_{\text{P}} T = 2.90 \times 10^{-3}\,\text{m}\cdot\text{K}$. So the temperature of the reddish star is

$$T_{\text{r}} = \frac{2.90 \times 10^{-3}\,\text{m}\cdot\text{K}}{700 \times 10^{-9}\,\text{m}} = 4140\,\text{K}.$$

The temperature of the bluish star will be double this since its peak wavelength is half (350 nm vs. 700 nm); just to check:

$$T_{\text{b}} = \frac{2.90 \times 10^{-3}\,\text{m}\cdot\text{K}}{350 \times 10^{-9}\,\text{m}} = 8280\,\text{K}.$$

(*b*) The Stefan-Boltzmann equation, which we discussed in Chapter 14 (see Eq. 14–5), states that the power radiated *per unit area* of surface from a body is proportional to the fourth power of the kelvin temperature, T^4. Now the temperature of the bluish star is double that of the reddish star, so the bluish one must radiate $(2^4) = 16$ times as much energy per unit area. But we are given that they have the same luminosity (the same total power output); so the surface area of the blue star must be $\frac{1}{16}$ that of the red one. Since the surface area of a sphere is $4\pi r^2$, we conclude that the radius of the reddish star is $\sqrt{16} = 4$ times larger than the radius of the bluish star (or $4^3 = 64$ times the volume).

† Applies to "main-sequence" stars (see next page). The mass of a star can be determined by observing its gravitational effects. Many stars are part of a cluster, the simplest being a binary star in which two stars orbit around each other, allowing their masses to be determined using rotational mechanics.

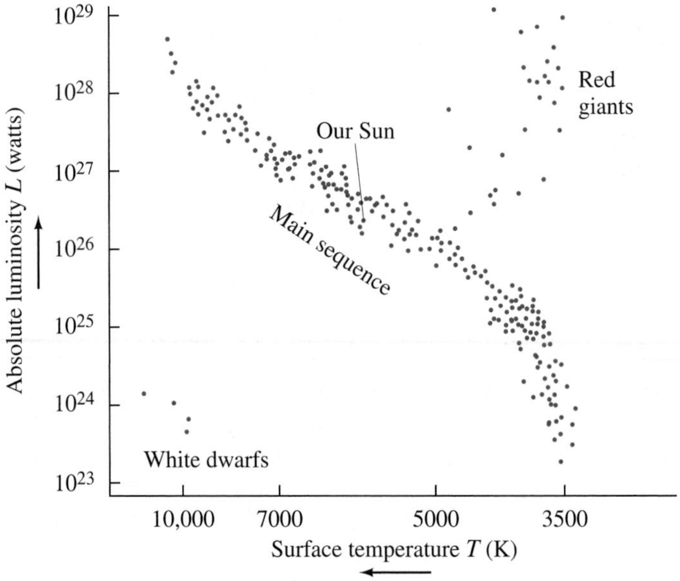

FIGURE 33–6 Hertzsprung–Russell (H–R) diagram. Note that the temperature T increases to the left.

H–R Diagram

H–R diagram

Main-sequence stars

Red giants

White dwarfs

An important astronomical discovery, made around 1900, was that for most stars, the color is related to the absolute luminosity and therefore to the mass. A useful way to present this relationship is by the so-called Hertzsprung–Russell (H–R) diagram. On the H–R diagram, the horizontal axis shows the temperature T whereas the vertical axis is the luminosity L; each star is represented by a point on the diagram, Fig. 33–6. Most stars fall along the diagonal band termed the **main sequence**. Starting at the lower right we find the coolest stars, reddish in color; they are the least luminous and therefore of low mass. Farther up toward the left we find hotter and more luminous stars that are whitish, like our Sun. Still farther up we find still more massive and more luminous stars, bluish in color. Stars that fall on this diagonal band are called *main-sequence stars*. There are also stars that fall outside the main sequence. Above and to the right we find extremely large stars, with high luminosities but with low (reddish) color temperature: these are called **red giants**. At the lower left, there are a few stars of low luminosity but with high temperature: these are the **white dwarfs**.

EXAMPLE 33–5 ESTIMATE Distance to a star using H–R and color. Suppose that detailed study of a certain star suggests that it most likely fits on the main sequence of an H–R diagram. Its measured apparent brightness is $l = 1.0 \times 10^{-12}\,\mathrm{W/m^2}$, and the peak wavelength of its spectrum is $\lambda_P \approx 600\,\mathrm{nm}$. Estimate its distance from us.

APPROACH We find the temperature using Wien's law, Eq. 27–2. The absolute luminosity is estimated for a main sequence star on the H–R diagram of Fig. 33–6, and then the distance is found using Eq. 33–1.

SOLUTION The star's temperature, from Wien's law (Eq. 27–2), is

$$T \approx \frac{2.90 \times 10^{-3}\,\mathrm{m\cdot K}}{600 \times 10^{-9}\,\mathrm{m}} \approx 4800\,\mathrm{K}.$$

A star on the main sequence of an H–R diagram at this temperature has absolute luminosity of about $L \approx 1 \times 10^{26}\,\mathrm{W}$, read off of Fig. 33–6. Then, from Eq. 33–1,

$$d = \sqrt{\frac{L}{4\pi l}} \approx \sqrt{\frac{1 \times 10^{26}\,\mathrm{W}}{4(3.14)(1.0 \times 10^{-12}\,\mathrm{W/m^2})}} \approx 3 \times 10^{18}\,\mathrm{m}.$$

Its distance from us in light-years is

$$d = \frac{3 \times 10^{18}\,\mathrm{m}}{10^{16}\,\mathrm{m/ly}} \approx 300\,\mathrm{ly}.$$

EXERCISE B Estimate the distance to a 6000 K star with an apparent brightness of $2.0 \times 10^{-12}\,\mathrm{W/m^2}$.

Stellar Evolution; Nucleosynthesis

Why are there different types of stars, such as red giants and white dwarfs, as well as main-sequence stars? Were they all born this way, in the beginning? Or might each different type represent a different age in the life cycle of a star? Astronomers and astrophysicists today believe the latter is the case. Note, however, that we cannot actually follow any but the tiniest part of the life cycle of any given star since they live for ages vastly greater than ours, on the order of millions or billions of years. Nonetheless, let us follow the process of **stellar evolution** from the birth to the death of a star, as astrophysicists have theoretically reconstructed it today.

Birth of a star

Stars are born, it is believed, when gaseous clouds (mostly hydrogen) contract due to the pull of gravity. A huge gas cloud might fragment into numerous contracting masses, each mass centered in an area where the density was only slightly greater than that at nearby points. Once such "globules" formed, gravity would cause each to contract in toward its center of mass. As the particles of such a *protostar* accelerate inward, their kinetic energy increases. When the kinetic energy is sufficiently high, the Coulomb repulsion between the positive charges is not strong enough to keep the hydrogen nuclei apart, and nuclear fusion can take place. In a star like our Sun, the "burning" of hydrogen[†] (that is, fusion) occurs via the *proton–proton cycle* (Section 31–3, Eqs. 31–6), in which four protons fuse to form a ${}^4_2$He nucleus with the release of γ rays, positrons, and neutrinos: $4\,{}^1_1\mathrm{H} \rightarrow {}^4_2\mathrm{He} + 2\,e^+ + 2\,\nu_e + 2\,\gamma$. These reactions require a temperature of about 10^7 K, corresponding to an average kinetic energy (kT) of about 1 keV (Eq. 13–8). In more massive stars, the carbon cycle produces the same net effect: Four ${}^1_1$H produce a ${}^4_2$He—see Section 31–3. The fusion reactions take place primarily in the core of a star, where T is sufficiently high. (The surface temperature is, of course, much lower—on the order of a few thousand kelvins.) The tremendous release of energy in these fusion reactions produces an outward pressure sufficient to halt the inward gravitational contraction; and our protostar, now really a young *star*, stabilizes on the main sequence. Exactly where the star falls along the main sequence depends on its mass. The more massive the star, the farther up (and to the left) it falls on the H–R diagram of Fig. 33–6. To reach the main sequence requires perhaps 30 million years, if it is a star like our Sun, and it is expected to remain there[‡] about 10 billion years (10^{10} yr). Although most stars are billions of years old, there is evidence that stars are actually being born at this moment.

Contraction due to gravity

Fusion begins when T (and $\overline{\mathrm{KE}}$) is large enough

Proton–proton cycle

Carbon cycle

Reaching the main sequence

As hydrogen fuses to form helium, the helium that is formed is denser and tends to accumulate in the central core where it was formed. As the core of helium grows, hydrogen continues to fuse in a shell around it: see Fig. 33–7. When much of the hydrogen within the core has been consumed, the production of energy decreases at the center and is no longer sufficient to prevent the huge gravitational forces from once again causing the core to contract and heat up. The hydrogen in the shell around the core then fuses even more fiercely because of this rise in temperature, causing the outer envelope of the star to expand and to cool. The surface temperature, thus reduced, produces a spectrum of light that peaks at longer wavelength (reddish).

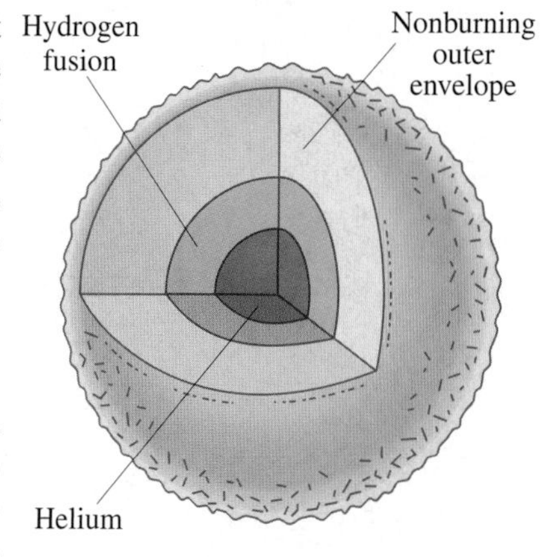

FIGURE 33–7 A shell of "burning" hydrogen (fusing to become helium) surrounds the core where the newly formed helium gravitates.

[†]The word "burn" is put in quotation marks because these high-temperature fusion reactions occur via a *nuclear* process, and must not be confused with ordinary burning (of, say, paper, wood, or coal) in air, which is a *chemical* reaction, occurring at the *atomic* level (and at a much lower temperature).

[‡]More massive stars, since they are hotter and the Coulomb repulsion is more easily overcome, "burn" much more quickly, and so use up their fuel faster, resulting in shorter lives. A star 10 times more massive than our Sun, for example, will remain on the main sequence only for about 10^7 years. Stars less massive than our Sun live much longer than our Sun's 10^{10} yr.

By this time the star has left the main sequence. It has become redder, and as it has grown in size, it has become more luminous. So it will have moved to the right and upward on the H–R diagram, as shown in Fig. 33–8. As it moves upward, it enters the **red giant** stage. Thus, theory explains the origin of red giants as a natural step in a star's evolution. Our Sun, for example, has been on the main sequence for about $4\frac{1}{2}$ billion years. It will probably remain there another 4 or 5 billion years. When our Sun leaves the main sequence, it is expected to grow in size (as it becomes a red giant) until it occupies all the volume out to roughly the present orbit of the planet Mercury.

Red giants

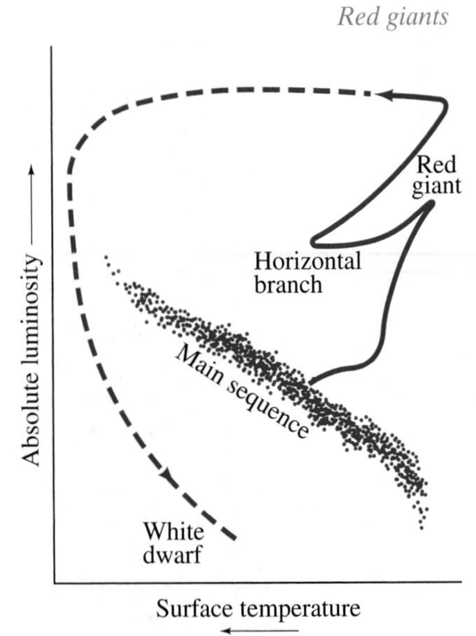

FIGURE 33–8 Evolutionary "track" of a star like our Sun represented on an H–R diagram.

If the star is like our Sun, or larger, further fusion can occur. As the star's outer envelope expands, its core is shrinking and heating up. When the temperature reaches about 10^8 K, even helium nuclei, in spite of their greater charge and hence greater electrical repulsion, can then reach each other and undergo fusion. The reactions are

$$
\begin{aligned}
{}^4_2\mathrm{He} + {}^4_2\mathrm{He} &\longrightarrow {}^8_4\mathrm{Be} \\
{}^4_2\mathrm{He} + {}^8_4\mathrm{Be} &\longrightarrow {}^{12}_6\mathrm{C}
\end{aligned}
\qquad \textbf{(33–2)}
$$

with the emission of two γ rays. These two reactions must occur in quick succession (because ${}^8_4$Be is very unstable), and the net effect is

$$3\,{}^4_2\mathrm{He} \longrightarrow {}^{12}_6\mathrm{C}. \qquad (Q = 7.3\,\mathrm{MeV})$$

This fusion of helium causes a change in the star which moves rapidly to the "horizontal branch" on the H–R diagram (Fig. 33–8). Further fusion reactions are possible, with ${}^4_2$He fusing with ${}^{12}_6$C to form ${}^{16}_8$O. In very massive stars, higher Z elements like ${}^{20}_{10}$Ne or ${}^{24}_{12}$Mg can be made. This process of creating heavier nuclei from lighter ones (or by absorption of neutrons at higher Z) is called **nucleosynthesis**.

Nucleosynthesis

The final fate of a star depends on its mass. Stars can lose mass as parts of their envelope drift off into space. Stars born with a mass less than about 8 (or perhaps 10) solar masses eventually end up with a residual mass less than about 1.4 solar masses, which is known as the *Chandrasekhar limit*. For them, no further fusion energy can be obtained. The core of such a "low mass" star (original mass $\lesssim$ 8 solar masses) contracts under gravity; the outer envelope expands again and the star becomes an even larger red giant. Eventually the outer layers escape into space, the core shrinks, the star cools, and typically follows the dashed route shown in Fig. 33–8, descending downward, becoming a **white dwarf**. A white dwarf with a mass equal to that of the Sun would be about the size of the Earth. A white dwarf contracts to the point at which the electron clouds start to overlap, but collapses no further because, as the Pauli exclusion principle claims, no two electrons can be in the same quantum state. Arriving at this point is called *electron degeneracy*. A white dwarf continues to lose internal energy by radiation, decreasing in temperature and becoming dimmer until its light goes out. It has then become a cold dark chunk of ash.

White dwarfs

Stars whose residual mass is greater than the Chandrasekhar limit of 1.4 solar masses (original mass greater than about 8 or 10 solar masses) are thought to follow a quite different scenario. A star with this great a mass can contract under gravity and heat up even further. In the range $T = 2.5\text{–}5 \times 10^9$ K, nuclei as heavy as ${}^{56}_{26}$Fe and ${}^{56}_{28}$Ni can be made. But here the formation of heavy nuclei from lighter ones by fusion, ends. As we saw in Fig. 30–1, the average binding energy per nucleon begins to decrease for A greater than about 60. Further fusions would *require* energy, rather than release it.

Production of heavy elements

Elements heavier than Ni are thought to form mainly by neutron capture, particularly in supernova explosions. Large numbers of free neutrons, resulting from nuclear reactions, are present inside highly evolved stars and they can readily combine with, say, a ${}^{56}_{26}$Fe nucleus to form (if three are captured) ${}^{59}_{26}$Fe, which decays to ${}^{59}_{27}$Co. The ${}^{59}_{27}$Co can capture neutrons, also becoming neutron rich and decaying by β^- to the next higher Z element, and so on. The highest Z elements are thought to form by such neutron capture during supernova explosions when hordes of neutrons are available.

Yet at these extremely high temperatures, well above 10^9 K, the kinetic energy of the nuclei is so high that fusion of elements heavier than iron is still possible even though the reactions require energy input. But the high-energy collisions can also cause the breaking apart of iron and nickel nuclei into He nuclei, and eventually into protons and neutrons:

$$^{56}_{26}\text{Fe} \rightarrow 13\,^{4}_{2}\text{He} + 4\text{n}$$

$$^{4}_{2}\text{He} \rightarrow 2\text{p} + 2\text{n}.$$

These are energy-requiring (endothermic) reactions, but at such extremely high temperature and pressure there is plenty of energy available, enough even to force electrons and protons together to form neutrons in inverse β decay:

$$\text{e}^- + \text{p} \rightarrow \text{n} + \nu.$$

As the core collapses under the huge gravitational forces, the tremendous mass becomes essentially an enormous nucleus made up almost exclusively of neutrons. The size of the star is no longer limited by the exclusion principle applied to electrons, but rather applied to neutrons (*neutron degeneracy*), and the star begins to contract rapidly toward forming an enormously dense **neutron star**. The contraction of the core would mean a great reduction in gravitational potential energy. Somehow this energy would have to be released. Indeed, it was suggested in the 1930s that the final core collapse to a neutron star may be accompanied by a catastrophic explosion whose tremendous energy could form virtually all elements of the periodic table and blow away the entire outer envelope of the star (Fig. 33–9), spreading its contents into interstellar space. Such explosions are believed to produce some of the observed *supernovae*. The presence of heavy elements on Earth and in our solar system suggests that our solar system formed from the debris of supernovae.

Neutron stars

Supernovae

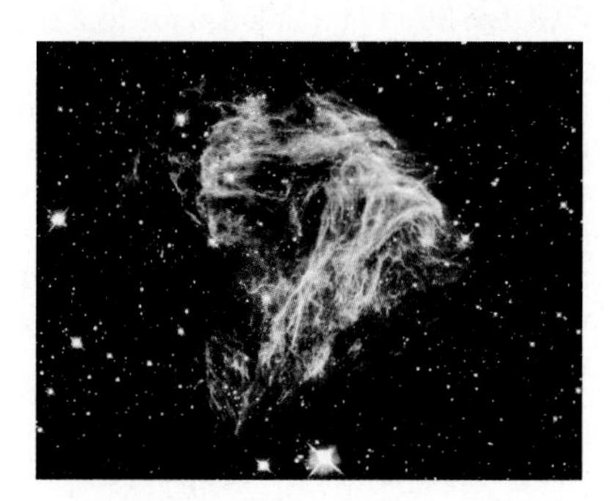

FIGURE 33–9 These glowing filaments, observed by the Hubble Space Telescope, are remnants of a supernova whose light would have reached Earth thousands of years ago. Inside is a powerful rotating neutron star called a *pulsar*.

The core of a neutron star contracts to the point at which all neutrons are as close together as they are in a nucleus. That is, the density of a neutron star is on the order of 10^{14} times greater than normal solids and liquids on Earth. A cupful of such dense matter would weigh billion of tons. A neutron star that has a mass 1.5 times that of our Sun would have a diameter of only about 20 km.

If the final mass of a neutron star is less than about two or three solar masses, its subsequent evolution is thought to be similar to that of a white dwarf. If the mass is greater than this, the star collapses under gravity, overcoming even the neutron exclusion principle. Gravity would then be so strong that even light emitted from it could not escape—it would be pulled back in by the force of gravity. Since no radiation could escape from such a star, we could not see it—it would be black. An object may pass by it and be deflected by its gravitational field, but if it came too close it would be swallowed up, never to escape. This is a **black hole**.

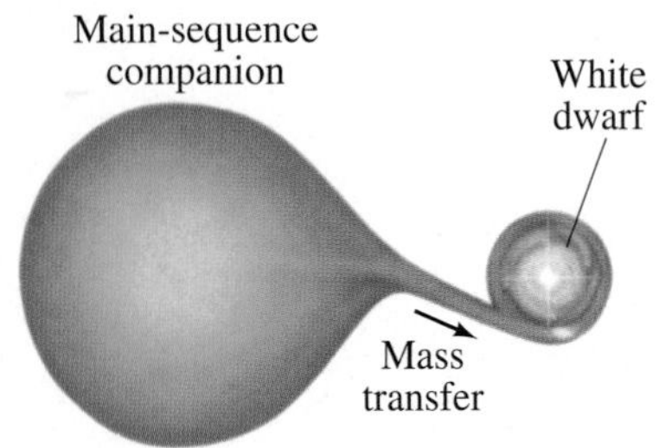

FIGURE 33–10 Hypothetical model for novae and type Ia supernovae, showing how a white dwarf could pull mass from its normal companion.

Novae and Supernovae

Novae (singular is *nova*, meaning "new" in Latin) are faint stars that have suddenly increased in brightness by as much as a factor of 10^4 and last for a month or two before fading. Novae are thought to be faint white dwarfs that have pulled mass from a nearby companion (they make up a *binary* system), as illustrated in Fig. 33–10. The captured mass of hydrogen fuses into helium at a high rate for a few weeks. Many novae (maybe all) are *recurrent*—they repeat their bright glow years later.

Supernovae are also brief explosive events, but release millions of times more energy than novae, up to 10^{10} times more luminous than our Sun. The peak of brightness may equal that of the entire galaxy in which they are located, but lasts only a few days. They remain bright but slowly fade over a few months. Many supernovae form by core collapse to a neutron star as described above.

Type Ia supernovae

Type Ia supernovae are different. They all seem to have very nearly the same luminosity. They are believed to be binary stars, one of which is a white dwarf that pulls mass from its companion, much like for a nova, Fig. 33–10. The mass is higher, and as mass is captured and the total mass reaches the Chandrasekhar limit of 1.4 solar masses, the star begins to collapse and then explodes as a supernova.

33–3 Distance Measurements

How astronomical distances are measured

We have talked about the vast distances of objects in the universe. But how do we measure these distances? One basic technique employs simple geometry to measure the **parallax** of a star. By parallax we mean the apparent motion of a star, against the background of more distant stars, due to the Earth's motion about the Sun. As shown in Fig. 33–11, the sighting angle of a star relative to the plane of Earth's orbit (angle θ) can be determined at different times of the year. Since we know the distance d from Earth to Sun, we can reconstruct the right triangles shown in Fig. 33–11 and can determine† the distance D to the star.

†This is essentially the way the heights of mountains are determined, by "triangulation." See Example 1–9.

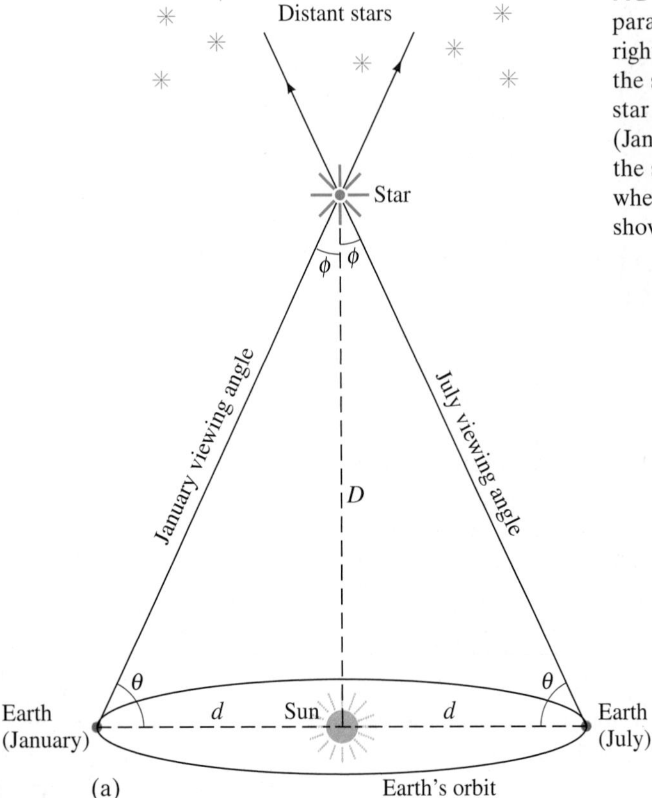

FIGURE 33–11 (a) Distance to a star determined by parallax (not to scale!). The imaginary triangles are right triangles and ϕ is a very small angle. (b) View of the sky showing the apparent position of the "nearby" star relative to more distant stars, at two different times (January and July). The viewing angle in January puts the star more to the right relative to distant stars, whereas in July it is more to the left (dashed circle shows January location).

EXAMPLE 33–6 ESTIMATE Distance to a star using parallax. Estimate the distance D to a star if the angle θ in Fig. 33–11 is measured to be 89.99994°.

APPROACH From trigonometry, $\tan\phi = d/D$ in Fig. 33–11. The Sun–Earth distance is $d = 1.5 \times 10^8$ km.

SOLUTION The angle $\phi = 90° - 89.99994° = 0.00006°$, or about 1.0×10^{-6} radians. We can use $\tan\phi \approx \phi$ since ϕ is very small. We solve for D in $\tan\phi = d/D$. The distance D to the star is

$$D = \frac{d}{\tan\phi} \approx \frac{d}{\phi} = \frac{1.5 \times 10^8\,\text{km}}{1.0 \times 10^{-6}\,\text{rad}} = 1.5 \times 10^{14}\,\text{km},$$

or about 15 ly.

Distances to stars are often specified in terms of parallax angle given in seconds of arc: 1 second (1″) is $\frac{1}{60}$ of a minute (′) of arc, which is $\frac{1}{60}$ of a degree, so $1'' = \frac{1}{3600}$ of a degree. The distance is then specified in parsecs (meaning *par*allax angle in *sec*onds of arc), where the **parsec** (pc) is defined as $1/\phi$ with ϕ in seconds. In Example 33–6, $\phi = (6 \times 10^{-5})°(3600) = 0.22''$ of arc, so we would say the star is at a distance of $1/0.22'' = 4.5$ pc. It is easy to show that the parsec is given by

Parsec (unit)

$$\begin{aligned} 1\,\text{pc} &= 3.26\,\text{ly} \\ &= (3.26\,\text{ly})(9.46 \times 10^{15}\,\text{m/ly}) = 3.08 \times 10^{16}\,\text{m}. \end{aligned}$$

Parallax can be used to determine the distance to stars as far away as about 100 light-years (≈ 30 parsecs) from Earth, and from an orbiting satellite perhaps 5 to 10 times farther. Beyond that distance, parallax angles are too small to measure. For greater distances, more subtle techniques must be employed. We might compare the apparent brightnesses of two galaxies and use the inverse square law (intensity drops off as the square of the distance) to roughly estimate their relative distances. We can't expect this technique to be very precise because we don't expect all galaxies to have the same intrinsic luminosity. A perhaps better estimate assumes the brightest stars in all galaxies (or the brightest galaxies in galaxy clusters) are similar and have about the same absolute luminosity. Consequently, their *apparent* brightness would be a measure of how far away they were.

Another technique makes use of the H–R diagram. Measurement of a star's surface temperature (from its spectra) places it at a certain point (within 20%) on the H–R diagram, assuming it is a main-sequence star, and then its luminosity can be estimated off the vertical axis (Fig. 33–6). Its apparent brightness and Eq. 33–1 give its approximate distance; see Example 33–5.

Cepheid variables

A better estimate comes from comparing *variable stars*, such as *Cepheid variables* whose averaged intrinsic luminosity (varying in time) has been found to be correlated to their periods.

SNIa as standard candles

The largest distances are estimated by comparing the apparent brightnesses of type Ia supernovae (SNIa). Type Ia supernovae all have a similar origin (they collapse to a neutron star at 1.4 solar masses, as described on the previous page and Fig. 33–10), and their brief explosive burst of light is expected to be of nearly the same total luminosity. They are thus sometimes referred to as "standard candles."

Another important technique for estimating distance is from the "redshift" in the line spectra of elements and compounds. The redshift is related to the expansion of the universe, as discussed in Section 33–5. It is useful for objects further than 10^7 to 10^8 ly away.

As we look farther and farther away, the measurement techniques are less and less reliable, so there is more and more uncertainty in the measurements of large distances.

33–4 General Relativity: Gravity and the Curvature of Space

We have seen that the force of gravity plays an important role in the processes that occur in stars. Gravity too is important for the evolution of the universe as a whole. The reasons gravity plays the dominant role in the universe, and not one of the other of the four forces in nature, are (1) it is long-range and (2) it is always attractive. The strong and weak nuclear forces act over very short distances only, on the order of the size of a nucleus; hence they do not act over astronomical distances (they do act between nuclei and nucleons in stars to produce nuclear reactions). The electromagnetic force, like gravity, acts over great distances. But it can be either attractive or repulsive. And since the universe does not seem to contain large areas of net electric charge, a large net force does not occur. But gravity acts as an attractive force between *all* masses, and there are large accumulations in the universe of only the one "sign" of mass (not + and − as with electric charge). However, the force of gravity as Newton described it in his law of universal gravitation shows discrepancies on a cosmological scale. Einstein, in his general theory of relativity, developed a theory of gravity that now forms the basis of cosmological dynamics.

In the *special theory of relativity* (Chapter 26), Einstein concluded that there is no way for an observer to determine whether a given frame of reference is at rest or is moving at constant velocity in a straight line. Thus the laws of physics must be the same in different inertial reference frames. But what about the more general case of motion where reference frames can be *accelerating*?

Einstein tackled the problem of accelerating reference frames in his **general theory of relativity** and in it also developed a theory of gravity. The mathematics of General Relativity is complex, so our discussion will be mainly qualitative.

FIGURE 33–12 A freely falling elevator. The released book hovers next to the owner's hand; (b) is a few moments after (a).

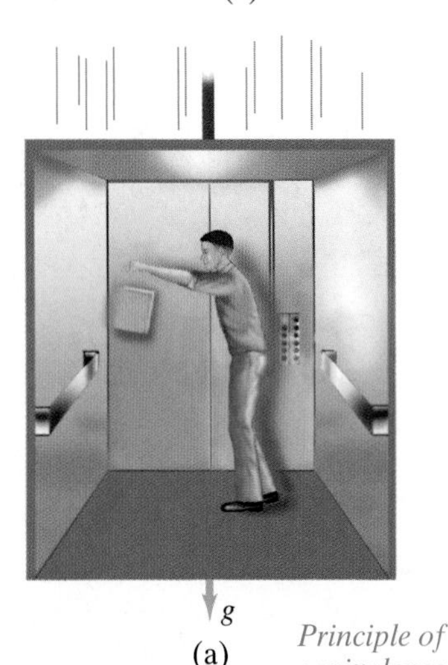

We begin with Einstein's **principle of equivalence**, which states that

Principle of equivalence

no experiment can be performed that could distinguish between a uniform gravitational field and an equivalent uniform acceleration.

If observers sensed that they were accelerating (as in a vehicle speeding around a sharp curve), they could not prove by any experiment that in fact they weren't simply experiencing the pull of a gravitational field. Conversely, we might think we are being pulled by gravity when in fact we are undergoing an "inertial" acceleration having nothing to do with gravity.

As a thought experiment, consider a person in a freely falling elevator near the Earth's surface. If our observer held out a book and let go of it, what would happen? Gravity would pull it downward toward the Earth, but at the same rate $(g = 9.8\,\mathrm{m/s^2})$ at which the person and elevator were falling. So the book would hover right next to the person's hand (Fig. 33–12). The effect is exactly the same as if this reference frame was at rest and *no* forces were acting. On the other hand, if the elevator was out in space where the gravitational field is essentially zero, the released book would float, just as it does in Fig. 33–12. Next, if the elevator (out in space) is accelerating upward at an acceleration of $9.8\,\mathrm{m/s^2}$, the book as seen by our observer would fall to the floor with an acceleration of $9.8\,\mathrm{m/s^2}$, just as if it were falling due to gravity at the surface of the Earth. According to the principle of equivalence, the observer could not determine whether the book fell because the elevator was accelerating upward, or because a gravitational field was acting downward and the elevator was at rest. The two descriptions are equivalent.

The principle of equivalence is related to the concept that there are two types of mass. Newton's second law, $F = ma$, uses **inertial mass**. We might say that inertial mass represents "resistance" to any type of force. The second type of mass is **gravitational mass**. When one body attracts another by the gravitational force (Newton's law of universal gravitation, $F = Gm_1 m_2/r^2$, Chapter 5), the strength of the force is proportional to the product of the *gravitational masses* of the two bodies. This is much like the electric force

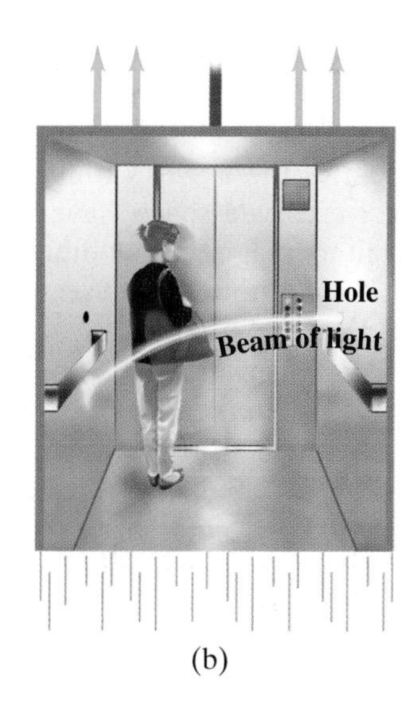

FIGURE 33–13 (a) Light beam goes straight across an elevator not accelerating. (b) The light beam bends (exaggerated) in an elevator accelerating in an upward direction.

between two bodies which is proportional to the product of their electric charges. The electric charge of a body is not related to its inertial mass; so why should we expect that a body's gravitational mass (call it gravitational charge if you like) be related to its inertial mass? All along we have assumed they were the same. Why? Because no experiment—not even of high precision—has been able to discern any measurable difference between inertial mass and gravitational mass. This is another way to state the equivalence principle: *gravitational mass is equivalent to inertial mass.*

The principle of equivalence can be used to show that light ought to be deflected due to the gravitational force of a massive body. Consider another thought experiment, in which an elevator is in free space where virtually no gravity acts. If a light beam enters a hole in the side of the elevator, the beam travels straight across the elevator and makes a spot on the opposite side if the elevator is at rest (Fig. 33–13a). If the elevator is accelerating upward, as in Fig. 33–13b, the light beam still travels straight across in a reference frame at rest. In the upwardly accelerating elevator, however, the beam is observed to curve downward. Why? Because during the time the light travels from one side of the elevator to the other, the elevator is moving upward at ever-increasing speed. Next we note that according to the equivalence principle, an upwardly accelerating reference frame is equivalent to a downward gravitational field. Hence, we can picture the curved light path in Fig. 33–13b as being the effect of a gravitational field. Thus, from the principle of equivalence, we expect gravity to exert a force on a beam of light and to bend it out of a straight-line path!

Gravity bends light

That light is affected by gravity is an important prediction of Einstein's general theory of relativity. And it can be tested. The amount a light beam would be deflected from a straight-line path must be small even when passing a massive body. (For example, light near the Earth's surface after traveling 1 km is predicted to drop only about 10^{-10} m, which is equal to the diameter of a small atom and not detectable.) The most massive body near us is the Sun, and it was calculated that light from a distant star would be deflected by 1.75″ of arc (tiny but detectable) as it passed near the Sun (Fig. 33–14). However, such a measurement could be made only during a total eclipse of the Sun, so that the Sun's tremendous brightness would not overwhelm the starlight passing near its edge. An opportune eclipse occurred in 1919, and scientists journeyed to the South Atlantic to observe it. Their photos of stars around the Sun revealed shifts in accordance with Einstein's prediction.

FIGURE 33–14 (a) Three stars in the sky. (b) If the light from one of these stars passes very near the Sun, whose gravity bends the rays, the star will appear higher than it actually is.

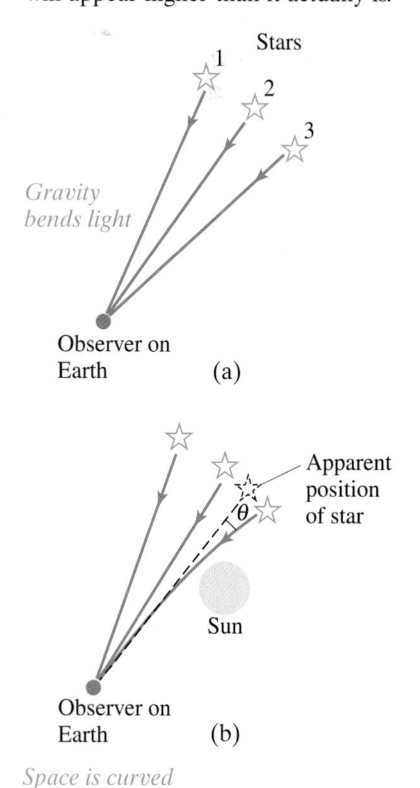

Space is curved

If a light beam can follow a curved path, as discussed above, then perhaps we can say that *space itself is curved* and that it is the gravitational mass that causes the curvature. Indeed, the curvature of space—or rather, of four-dimensional space-time—is a basic aspect of Einstein's General Relativity (GR).

What is meant by **curved space**? To understand, recall that our normal method of viewing the world is via Euclidean plane geometry. In Euclidean geometry, there are many axioms and theorems we take for granted, such as that the sum of the angles of any triangle is 180°. Non-Euclidean geometries, which involve curved space, have also been imagined by mathematicians. It is hard enough to imagine three-dimensional curved space, much less curved four-dimensional space-time. So let us try to understand the idea of curved space by using two-dimensional surfaces.

Consider, for example, the two-dimensional surface of a sphere. It is clearly curved, Fig. 33–15, at least to us who view it from the outside—from our three-dimensional world. But how would hypothetical two-dimensional creatures determine whether their two-dimensional space were flat (a plane) or curved? One way would be to measure the sum of the angles of a triangle. If the surface is a plane, the sum of the angles is 180°, as we learn in plane geometry. But if the space is curved, and a sufficiently large triangle is constructed, the sum of the angles will *not* be 180°. To construct a triangle on a curved surface, say the sphere of Fig. 33–15, we must use the equivalent of a straight line: that is, the shortest distance between two points, which is called a **geodesic**. On a sphere, a geodesic is an arc of a great circle (an arc in a plane passing through the center of the sphere) such as the Earth's equator and the Earth's longitude lines. Consider, for example, the large triangle of Fig. 33–15: its sides are two longitude lines passing from the north pole to the equator, and the third side is a section of the equator as shown. The two longitude lines make 90° angles with the equator (look at a world globe to see this more clearly). They make an angle with each other at the north pole, which could be, say, 90° as shown; the sum of these angles is $90^\circ + 90^\circ + 90^\circ = 270^\circ$. This is clearly *not* a Euclidean space. Note, however, that if the triangle is small in comparison to the radius of the sphere, the angles will add up to nearly 180°, and the triangle (and space) will seem flat.

Geodesic

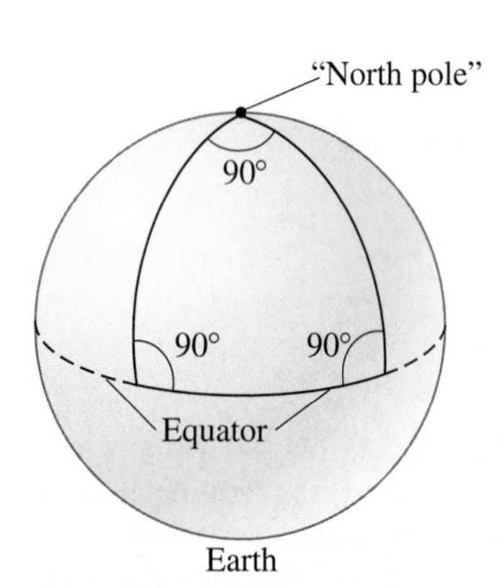

FIGURE 33–15 On a two-dimensional curved surface, the sum of the angles of a triangle may not be 180°.

Another way to test the curvature of space is to measure the radius r and circumference C of a large circle. On a plane surface, $C = 2\pi r$. But on a two-dimensional spherical surface, C is *less* than $2\pi r$, as can be seen in Fig. 33–16. The proportionality between C and r is *less* than 2π. Such a surface is said to have *positive curvature*. On the saddlelike surface of Fig. 33–17, the circumference of a circle is greater than $2\pi r$, and the sum of the angles of a triangle is less than 180°. Such a surface is said to have a *negative curvature*.

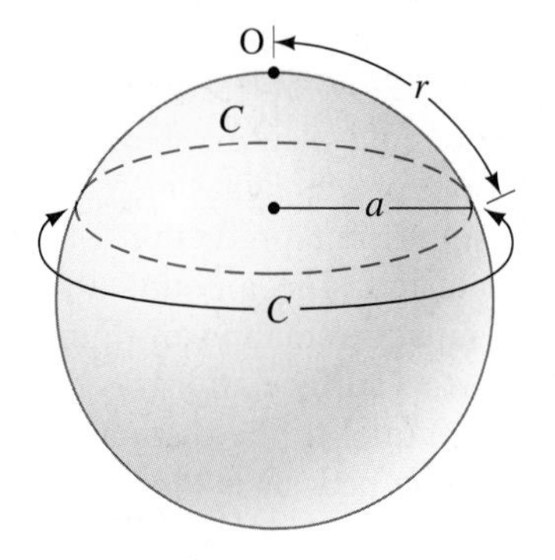

FIGURE 33–16 On a spherical surface (a two-dimensional world) a circle of circumference C is drawn about point O as the center. The radius of the circle (not the sphere) is the distance r along the surface. (Note that in our three-dimensional view, we can tell that $2\pi a = C$; since $r > a$, then $2\pi r > C$.)

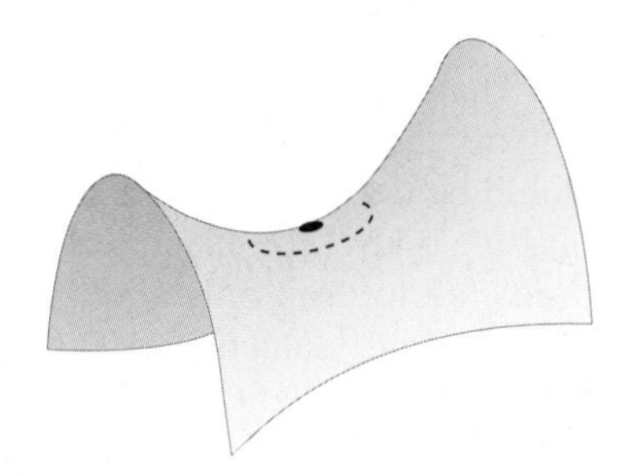

FIGURE 33–17 Example of a two-dimensional surface with negative curvature.

Curvature of the Universe

Now, what about our universe? On a large scale (not just near a large mass), what is the overall curvature of the universe? Does it have positive curvature, negative curvature, or is it flat (zero curvature)?

The universe: open or closed?

If the universe had a positive curvature, the universe would be *closed*, or *finite* in volume. This would *not* mean that the stars and galaxies extended out to a certain boundary, beyond which there is empty space. There is no boundary or edge in such a universe. If a particle were to move in a straight line in a particular direction, it would eventually return to the starting point—perhaps eons of time later.

On the other hand, if the curvature of space was zero or negative, the universe would be *open*. It could just go on forever. An open universe could be *infinite*, but not necessarily according to recent research.

Today the evidence is very strong that the universe on a large scale is very close to being flat. Indeed, it is so close to being flat that we can't tell if it might be very slightly positive or very slightly negative.

Black Holes

According to Einstein's theory, space-time is curved near massive bodies. We might think of space as being like a thin rubber sheet: if a heavy weight is hung from it, it curves as shown in Fig. 33–18. The weight corresponds to a huge mass that causes space (space itself!) to curve. Thus, in Einstein's theory† we do not speak of the "force" of gravity acting on bodies. Instead we say that bodies and light rays move as they do because space-time is curved. A body at rest or moving slowly near the great mass of Fig. 33–18 would follow a geodesic (the equivalent of a straight line in plane geometry) toward that body.

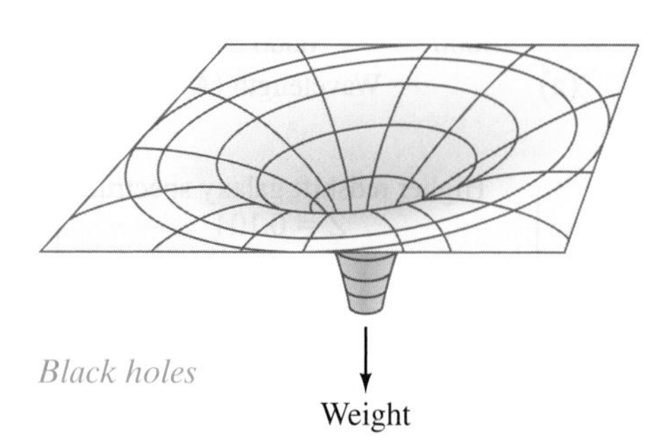

FIGURE 33–18 Rubber-sheet analogy for space-time curved by matter.

Black holes

The extreme curvature of space-time shown in Fig. 33–18 could be produced by a **black hole**. A black hole, as we saw in the Section 33–2, is so dense that even light cannot escape from it. To become a black hole, a body of mass M must undergo **gravitational collapse**, contracting by gravitational self-attraction to within a radius called the **Schwarzschild radius**:

$$R = \frac{2GM}{c^2},$$

where G is the gravitational constant and c the speed of light.

Event horizon

The Schwarzschild radius also represents the event horizon of a black hole. By **event horizon** we mean the surface beyond which no signals can ever reach us, and thus inform us of events that happen. As a star collapses toward a black hole, the light it emits is pulled harder and harder by gravity, but we can still see it. Once the matter passes within the event horizon the emitted light cannot escape, but is pulled back in by gravity.

All we can know about a black hole is its mass, its angular momentum (there could be rotating black holes), and its electric charge. No other information, no details of its structure or the kind of matter it was formed of, can be known because no information can escape.

How might we observe black holes? We cannot see them because no light can escape from them. They would be black objects against a black sky. But they do exert a gravitational force on nearby bodies. The black hole believed to be at the center of our Galaxy was discovered by examining the motion of matter in its vicinity. Another technique is to examine stars which appear to be rotating as if they were members of a *binary system* (two stars rotating about their common center of mass), although the companion is invisible. If the unseen star is a black hole, it might be expected to pull off gaseous material from its visible companion (as in Fig. 33–10). As this matter approached the black hole, it would be highly accelerated and should emit X-rays of a characteristic type before plunging inside the event horizon. Such X-rays, plus a sufficiently high mass estimate from the rotational motion, can provide evidence for a black hole. One of the many candidates for a black hole is in the binary-star system Cygnus X-1.

† Alexander Pope (1688–1744) wrote an epitaph for Newton:

> "Nature, and Nature's laws lay hid in night:
> God said, *Let Newton be!* and all was light."

Sir John Squire (1884–1958), perhaps uncomfortable with Einstein's profound thoughts, added:

> "It did not last: the Devil howling '*Ho!*
> *Let Einstein be!*' restored the status quo."

What does it mean that distant galaxies are all moving away from us, and with ever greater speed the farther they are from us? It seems to suggest some kind of explosive expansion that started at some very distant time in the past. And at first sight we seem to be in the middle of it all. But we aren't. The expansion appears the same from any other point in the universe. To understand why, see Fig. 33–20. In Fig. 33–20a we have the view from Earth (or from our Galaxy). The velocities of surrounding galaxies are indicated by arrows, pointing away from us, and the arrows are longer for galaxies more distant from us. Now, what if we were on the galaxy labeled A in Fig. 33–20a? From Earth, galaxy A appears to be moving to the right at a velocity, call it $\vec{\mathbf{v}}_A$, represented by the arrow pointing to the right. If we were *on* galaxy A, Earth would appear to be moving to the left at velocity $-\vec{\mathbf{v}}_A$. To determine the velocities of other galaxies relative to A, we vectorially add the velocity vector, $-\vec{\mathbf{v}}_A$, to all the velocity arrows shown in Fig. 33–20a. This yields Fig. 33–20b, where we see clearly that the universe is expanding away from galaxy A as well; and the velocities of galaxies receding from A are proportional to their distance from A.

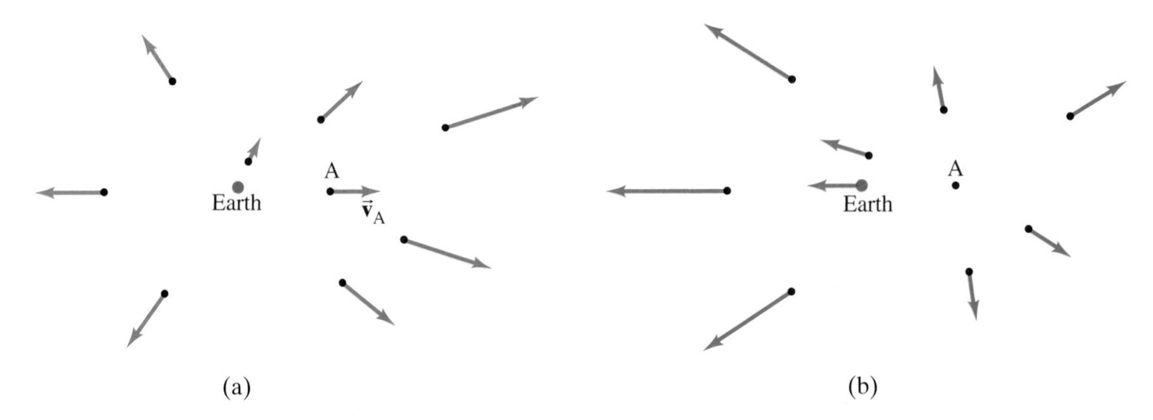

FIGURE 33–20 Expansion of the universe looks the same from any point in the universe.

Thus the expansion of the universe can be stated as follows: All galaxies are racing away from *each other* at an average rate of about 71 km/s per megaparsec of distance between them. The ramifications of this idea are profound, and we discuss them in a moment.

Cosmological principle

A basic assumption in cosmology has been that on a large scale, the universe would look the same to observers at different places at the same time. In other words, the universe is both *isotropic* (looks the same in all directions) and *homogeneous* (would look the same if we were located elsewhere, say in another galaxy). This assumption is called the **cosmological principle**. On a local scale, say in our solar system or within our Galaxy, it clearly does not apply (the sky looks different in different directions). But it has long been thought to be valid if we look on a large enough scale, so that the average population density of galaxies and clusters of galaxies ought to be the same in different areas of the sky. This seems to be valid on distances greater than about 200 Mpc (700 Mly). The expansion of the universe (Fig. 33–20) is consistent with the cosmological principle; and the near uniformity of the cosmic microwave background radiation (discussed in Section 33–6) supports it.

The expansion of the universe, as described by Hubble's law, strongly suggests that galaxies must have been closer together in the past than they are now. This is, in fact, the basis of the *Big Bang* theory of the origin of the universe, which pictures the universe as a relentless expansion starting from a very hot and compressed beginning. We discuss the Big Bang in detail shortly, but first let us see what can be said about the age of the universe.

One way to estimate the age of the universe uses the Hubble parameter. With $H \approx 22\,\mathrm{km/s}$ per 10^6 light-years, the time required for the galaxies to arrive at their present separations would be approximately (starting with $v = d/t$ and using Hubble's law, Eq. 33–6),

$$t = \frac{d}{v} = \frac{d}{Hd} = \frac{1}{H} \approx \frac{(10^6\,\mathrm{ly})(0.95 \times 10^{13}\,\mathrm{km/ly})}{(22\,\mathrm{km/s})(3.16 \times 10^7\,\mathrm{s/yr})} \approx 13.7 \times 10^9\,\mathrm{yr},$$

or 13.7 billion years. The age of the universe calculated in this way is called the *characteristic expansion time* or "Hubble age." It is a rough estimate and assumes the rate of expansion of the universe was constant (which today we are quite sure is not true). Recent precise measurements (2003) give the age as $13.7 \times 10^9\,\mathrm{yr}$, in remarkable agreement with the rough Hubble age estimate.

Age of the universe

* Steady-State Model

Before discussing the Big Bang in detail, we mention one alternative to the Big Bang—the **steady-state model**—which assumed that the universe is infinitely old and on average looks the same now as it always has. (This assumed uniformity in time as well as space was called the *perfect cosmological principle.*) According to the steady-state model, no large-scale changes have taken place in the universe as a whole, particularly no Big Bang. To maintain this view in the face of the recession of galaxies away from each other, mass–energy conservation must be violated. That is, matter must be created continuously to maintain the assumption of uniformity. The rate of mass creation required is very small—about one nucleon per cubic meter every 10^9 years.

The steady-state model provided the Big Bang model with healthy competition in the mid-twentieth century. But the discovery of the cosmic microwave background radiation (next Section), as well as the observed expansion of the universe, has made the Big Bang model almost universally accepted.

33–6 The Big Bang and the Cosmic Microwave Background

The expansion of the universe seems to suggest that typical objects in the universe were once much closer together than they are now. This is the basis for the idea that the universe began about 13.7 billion years ago as an expansion from a state of very high density and temperature known affectionately as the **Big Bang**.

The Big Bang

The Big Bang was not an explosion, because an explosion blows pieces out into the surrounding space. Instead, the Big Bang was the start of an expansion of space itself. The volume of the observable universe was very small at the start and has been expanding ever since. The initial tiny volume of extremely dense matter is not to be thought of as a concentrated mass in the midst of a much larger space around it. The initial tiny but dense volume *was* the universe—the entire universe. There wouldn't have been anything else. When we say that the universe was once smaller than it is now, we mean that the average separation between galaxies (or other objects) was less. Thus, it is the *size of the universe itself* that has increased since the Big Bang.

FIGURE 33–21 Robert Wilson (left) and Arno Penzias, and behind them their "horn antenna."

A major piece of evidence supporting the Big Bang is the **cosmic microwave background** radiation (or CMB) whose discovery came about as follows.

In 1964, Arno Penzias and Robert Wilson were experiencing difficulty with what they assumed to be background noise, or "static," in their radio telescope (a large antenna device for detecting radio waves from the heavens, Fig. 33–21). Eventually, they became convinced that it was real and that it was coming from outside our Galaxy. They made precise measurements at a wavelength $\lambda = 7.35\,\mathrm{cm}$, in the microwave region of the electromagnetic spectrum (Fig. 22–8). The intensity of this radiation was found initially not to vary by day or night or time of year, nor to depend on direction. It came from all directions in the universe with equal intensity, to a precision of better than 1%. It could only be concluded that this radiation came from the universe as a whole.

The 2.73-K cosmic microwave background radiation

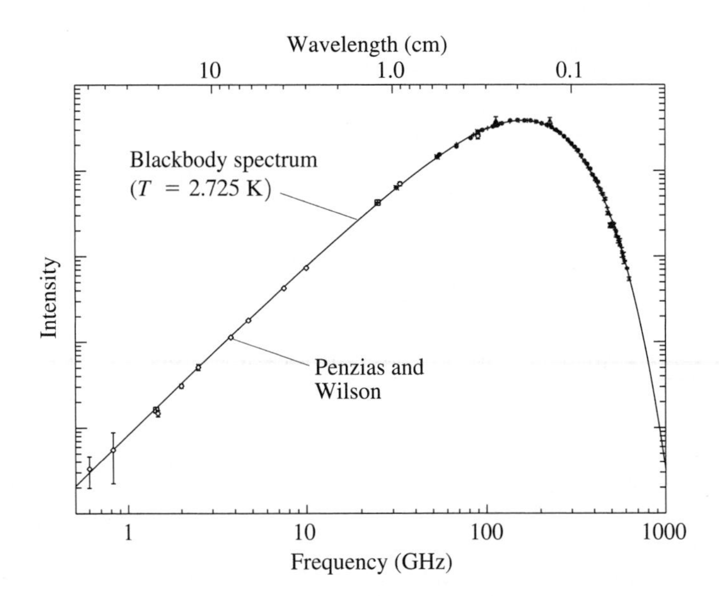

FIGURE 33–22 Spectrum of cosmic microwave background radiation, showing blackbody curve and experimental measurements including that of Penzias and Wilson. (Thanks to G. F. Smoot and D. Scott. The vertical bars represent the experimental uncertainty in a measurement.)

The intensity of this cosmic microwave background radiation as measured at $\lambda = 7.35\,\text{cm}$ corresponds to blackbody radiation (see Section 27–2) at a temperature of about 3 K. When radiation at other wavelengths was measured, the intensities were found to fall on a blackbody curve as shown in Fig. 33–22, corresponding to a temperature of 2.725 K.

The remarkable uniformity of the cosmic microwave background radiation was in accordance with the cosmological principle. But theorists felt that there needed to be some small inhomogeneities, or "anisotropies," in the CMB that would have provided "seeds" around which galaxy formation could have started. Small areas of slightly higher density and temperature, which could have contracted under gravity to form stars and galaxies, were indeed found. These tiny inhomogeneities were detected first by the COBE (Cosmic Background Explorer) satellite experiment (1992) and by subsequent experiments with greater detail, culminating in 2003 with the WMAP (Wilkinson Microwave Anisotropy Probe) results. See Fig. 33–23. WMAP gives the CMB temperature as $2.725 \pm 0.002\,\text{K}$.

Importance of CMB: the Big Bang

The CMB provides strong evidence in support of the Big Bang, and gives us information about conditions in the very early universe. In fact, in the late 1940s, George Gamow and his collaborators calculated that a Big Bang origin of the universe should have generated just such a microwave background radiation.

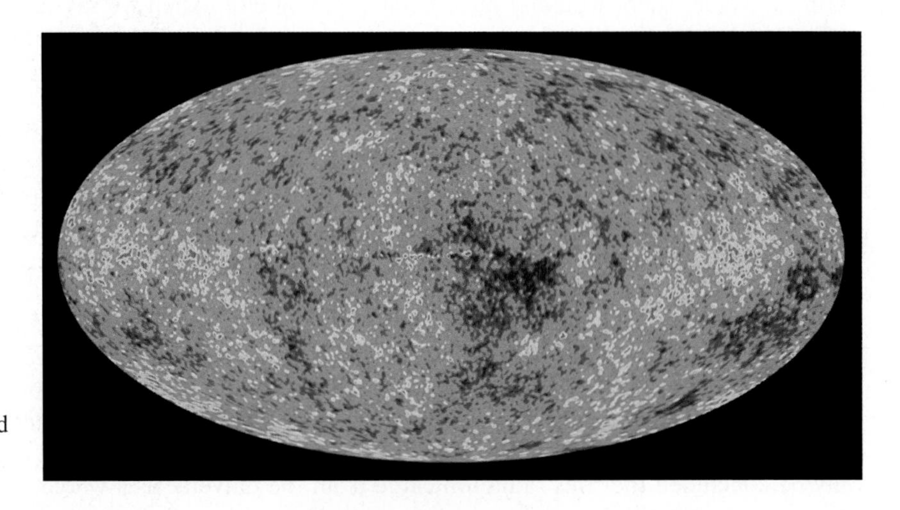

FIGURE 33–23 The cosmic microwave background radiation over the entire sky, color-coded to represent differences in temperature from the average 2.725 K: the color scale ranges from $+200\,\mu\text{K}$ (red) to $-200\,\mu\text{K}$ (dark blue), representing slightly hotter and colder spots (and also variations in density). Results are from the WMAP satellite in 2003: the angular resolution is 0.2°. The larger version of WMAP at the start of this Chapter was done at a specific frequency band and includes our Galaxy in the foreground (red stripe) which here has been subtracted out.

To understand why, let us look at what a Big Bang might have been like. The temperature must have been extremely high at the start, so high that there could not have been any atoms in the very early stages of the universe. Instead, the universe would have consisted solely of radiation (photons) and a plasma of charged electrons and other elementary particles. The universe would have been opaque—the photons in a sense "trapped," traveling very short distances before being scattered again, primarily by electrons. Indeed, the details of the microwave background radiation is strong evidence that matter and radiation were once in equilibrium at a very high temperature. As the universe expanded, the energy spread out over an increasingly larger volume and the temperature dropped. Only when the temperature had fallen to about 3000 K, some 380,000 years later, could nuclei and electrons combine together as atoms. With the disappearance of free electrons, as they combined with nuclei to form atoms, the radiation would have been freed—**decoupled** from matter, we say. The universe became *transparent* because photons were now free to travel nearly unimpeded straight through the universe.

Photons decoupled

As the universe expanded, so too the wavelengths of the radiation lengthened (you might think of standing waves, Section 11–13), thus redshifting to longer wavelengths that correspond to lower temperature (recall Wien's law, $\lambda_P T$ = constant, Section 27–2), until they would have reached the 2.7-K background radiation we observe today.

* Looking Back toward the Big Bang—Lookback Time

Figure 33–24 shows our Earth point of view, looking out in all directions back toward the Big Bang and the brief (380,000 year long) period when radiation was trapped in the early plasma (yellow band). The time it takes light to reach us from an event (say 5×10^9 yr ago) is called its **lookback time**. The "close-up" insert in Fig. 33–24 shows a photon scattering repeatedly inside the plasma and then exiting the plasma in a straight line. No matter what direction we look, our view of the very early universe is blocked by this wall of plasma—we can see only as far as its surface, called the "surface of last scattering," but not into it. Wavelengths from there are redshifted by $z \approx 1000$. Time $\Delta t'$ in Fig. 33–24 is the lookback time (not real time that goes forward).

Lookback time

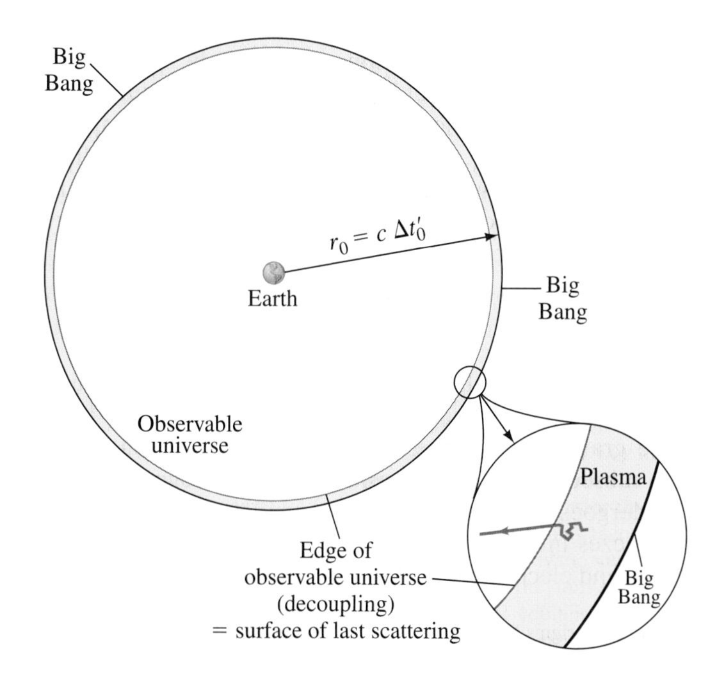

FIGURE 33–24 When we look out from the Earth, we look back in time. Any other observer in the universe would see the same thing. The farther an object is from us, the earlier in time the light we see left it. We cannot see quite as far as the Big Bang; we can see only as far as the "surface of last scattering," which represents the CMB. The blowup shows the earliest 380,000 years of the universe when it was opaque: a photon is shown scattering many times and then (at decoupling, 380,000 yr after the Big Bang) becoming free to travel in a straight line. If this photon wasn't heading our way when "liberated," many others were. Galaxies are not shown, but would be concentrated close to Earth in this diagram. *Note:* This diagram is not a normal map. Maps show a section of the world as might be seen all *at a given time*. This diagram shows space (like a map), but each point is *not* at the same time. The light coming from a point a distance r from Earth took a time $\Delta t' = r/c$ to reach Earth, and thus shows an event that took place long ago, a time $\Delta t' = r/c$ in the past, which we call its "lookback time." The Big Bang happened $\Delta t_0' = 13.7$ Gyr ago.

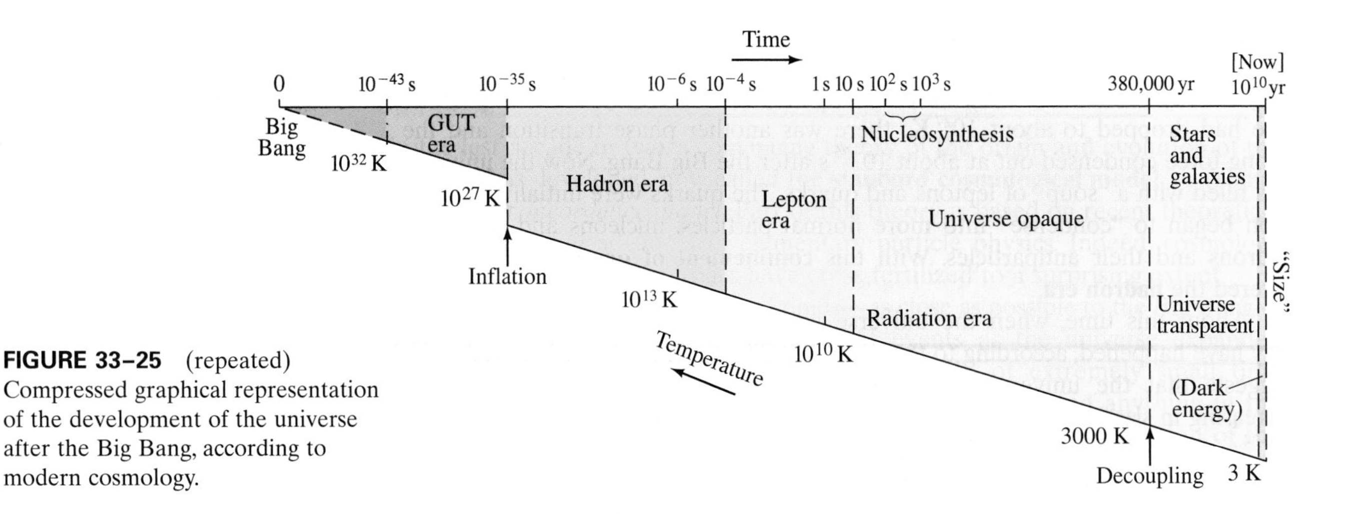

FIGURE 33–25 (repeated) Compressed graphical representation of the development of the universe after the Big Bang, according to modern cosmology.

have resulted in a slight excess of nucleons over antinucleons. And it is these "leftover" nucleons that we are made of today. The excess of nucleons over antinucleons was about one part in 10^9. Earlier, during the hadron era, there should have been about as many nucleons as photons. After it ended, the "left-over" nucleons thus numbered only about one nucleon per 10^9 photons, and this ratio has persisted to this day. Protons, neutrons, and all other heavier particles were thus tremendously reduced in number by about 10^{-6} s after the Big Bang. The lightest hadrons, the pions, disappeared as the nucleons had; because they are the lightest mass hadrons (140 MeV), they were the last hadrons to go, about 10^{-4} s after the Big Bang. Lighter particles, including electrons and neutrinos, were the dominant form of matter, and the universe entered the **lepton era**.

Lepton era

By the time the first full second had passed (clearly the most eventful second in history!), the universe had cooled to about 10 billion degrees, 10^{10} K. The average kinetic energy was about 1 MeV. This was still sufficient energy to create electrons and positrons and balance their annihilation reactions, since their masses correspond to about 0.5 MeV. So there were about as many e^+ and e^- as there were photons. But within a few more seconds, the temperature had dropped sufficiently so that e^+ and e^- could no longer be formed. Annihilation ($e^+ + e^- \rightarrow$ photons) continued. And, like nucleons before them, electrons and positrons all but disappeared from the universe—except for a slight excess of electrons over positrons (later to join with nuclei to form atoms). Thus, about $t = 10$ s after the Big Bang, the universe entered the **radiation era**. Its major constituents were photons and neutrinos. But the neutrinos, partaking only in the weak force, rarely interacted. So the universe, until then experiencing significant amounts of energy in matter and in radiation, now became **radiation-dominated**: much more energy was contained in radiation than in matter, a situation that would last tens of thousands of years (Fig. 33–25).

Radiation-dominated universe

Making He nuclei

Meanwhile, during the next few minutes, crucial events were taking place. Beginning about 2 or 3 minutes after the Big Bang, nuclear fusion began to occur. The temperature had dropped to about 10^9 K, corresponding to an average kinetic energy $\overline{\text{KE}} \approx 100$ keV, where nucleons could strike each other and be able to fuse (Section 31–3), but now cool enough so newly formed nuclei would not be immediately broken apart by subsequent collisions. Deuterium, helium, and very tiny amounts of lithium nuclei were probably made. But the universe was cooling too quickly, and larger nuclei were not made. After only a few minutes, probably not even a quarter of an hour after the Big Bang, the temperature dropped far enough that nucleosynthesis stopped, not to start again for millions of years (in stars). Thus, after the first hour or so of the universe, matter consisted mainly of bare nuclei of hydrogen (about 75%) and helium (about 25%)[†] and electrons. But radiation (photons) continued to dominate.

Our story is almost complete. The next important event is presumed to have occurred 380,000 years later. The universe had expanded to about $\frac{1}{1000}$ of its present size, and the temperature had cooled to about 3000 K. The average kinetic energy of nuclei, electrons, and photons was less than an electron volt. Since ionization energies of atoms are on the order of eV, then as the temperature dropped below this point, electrons could orbit the bare nuclei and remain there (without being ejected by collisions), thus forming atoms. With the birth of atoms, the photons—which had been continually scattering from the free electrons—now became free to spread nearly unhindered throughout the universe. As mentioned in the previous Section, the photons became **decoupled** from matter. The total energy contained in radiation had been decreasing (lengthening in wavelength as the universe expanded), and even before decoupling (at about $t = 56{,}000$ yr) the total energy contained in matter became dominant. The universe was said to have become **matter-dominated.** As the universe continued to expand, the electromagnetic radiation cooled further, to 2.7 K today, forming the cosmic microwave background radiation we detect from everywhere in the universe.

Birth of stable atoms

Matter-dominated universe

After the birth of atoms, then stars and galaxies could begin to form—presumably by self-gravitation around mass concentrations (inhomogeneities). Stars began to form about 200 million years after the Big Bang, galaxies after almost 10^9 years. The universe continued to evolve until today, some 13.7 billion years later.

* * *

This scenario is by no means "proven." But it does provide a workable picture, for the first time, of how the universe may have begun and evolved.

A major event, and something only discovered very recently, is that when the universe was about half as old as it is now (5–7 Gyr ago), its expansion began to accelerate. This was a big surprise because it was assumed the expansion of the universe would slow down due to gravitational attraction of all objects to each other. Indeed, another major recent discovery is that ordinary matter makes up very little of the total mass–energy of the universe ($\approx 4\%$). Instead, as we discuss in the next Section, the major contributors to the energy density of the universe are *dark matter* and *dark energy*. On the right in Fig. 33–25 is a narrow vertical strip that represents the most recent 5 to 7 billion years of the universe, during which *dark energy* seems to have dominated.

33–8 Dark Matter and Dark Energy

According to the standard Big Bang model, the universe is evolving and changing. Individual stars are being created, evolving, and dying as white dwarfs, neutron stars, black holes. At the same time, the universe as a whole is expanding. One important question is whether the universe will continue to expand forever. Until the late 1990s, the universe was thought to be dominated by matter which interacts by gravity, and this question was connected to the curvature of space-time (Section 33–4). If the universe had *negative* curvature, the expansion of the universe would never stop, although the rate of expansion would decrease due to the gravitational attraction of its parts. Such a universe would be *open* and infinite. If the universe is *flat* (no curvature), it would still be open and infinite but its expansion would slowly approach a zero rate. Finally, if the universe had *positive* curvature, it would be *closed* and finite; the effect of gravity would be strong enough that the expansion would eventually stop and the universe would begin to contract, collapsing back onto itself in a **big crunch**.

[†]This standard model prediction of a 25% primordial production of helium agrees with what we observe today—the universe *does* contain about 25% He—and it is strong evidence in support of the standard Big Bang model. Furthermore, the theory says that 25% He abundance is fully consistent with there being three neutrino types, which is the number we observe. And it sets an upper limit of four to the maximum number of possible neutrino types. Here we have a situation where cosmology actually makes a specific prediction about fundamental physics.

Critical Density

According to the above scenario (which does not include the recently discovered acceleration of the universe), the fate of the universe would depend on the average mass density in the universe. For an average mass density greater than a critical value known as the **critical density**, estimated to be about

Critical density of the universe

$$\rho_c \approx 10^{-26}\,\mathrm{kg/m^3}$$

(i.e., a few nucleons/m^3 on average throughout the universe), gravity would prevent expansion from continuing forever. Eventually (if $\rho > \rho_c$) gravity would pull the universe back into a big crunch and space-time would have a positive curvature. If instead the actual density was equal to the critical density, $\rho = \rho_c$, the universe would be flat and open. If the actual density was less than the critical density, $\rho < \rho_c$, the universe would have negative curvature. See Fig. 33–26. Today we believe the universe is very close to flat. But recent evidence suggests the universe is expanding at an *accelerating* rate, as discussed below.

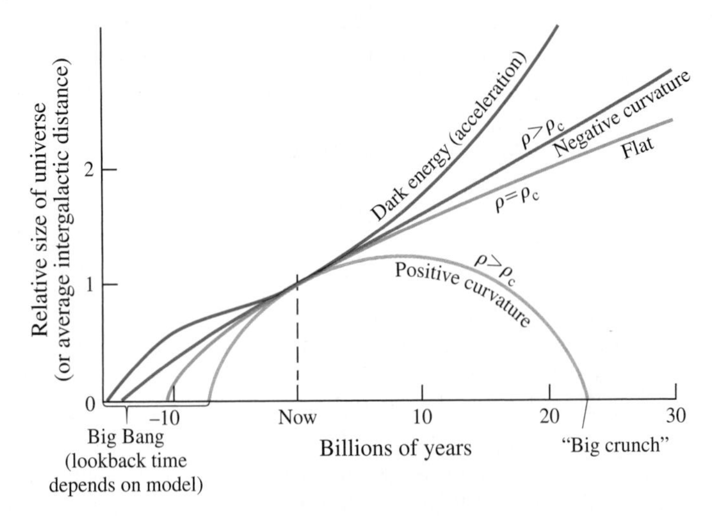

FIGURE 33–26 Three future possibilities for the universe, depending on the density ρ of ordinary matter, plus a fourth possibility that includes dark energy. Note that all curves have been chosen to have the same slope (= H, the Hubble parameter) right now. Looking back in time, the Big Bang occurs where each curve touches the horizontal (time) axis.

Dark Matter

WMAP and other experiments have convinced scientists that the universe is flat and $\rho = \rho_c$. But this ρ cannot be only normal baryonic matter (atoms are 99.9% baryons—protons and neutrons—by weight). These recent experiments put the amount of normal baryonic matter in the universe at only 4% of the critical density. What is the other 96%? There is strong evidence for a significant amount of nonluminous matter in the universe referred to as **dark matter**. For example, observations of the rotation of galaxies suggest that they rotate as if they had considerably more mass than we can see. Recall from Chapter 5, Example 5–14 (p. 123), that for a satellite revolving around Earth (mass M)

Dark matter

$$m\frac{v^2}{r} = G\frac{mM}{r^2}$$

and hence $v = \sqrt{GM/r}$. If we apply this equation to stars in a galaxy, we see that their speed depends on galactic mass. Observations show that stars farther from the galactic center revolve much faster than expected from visible matter, suggesting a great deal of invisible matter. Similarly, observations of the motion of galaxies within clusters also suggest that they have considerably more mass than can be seen. What might this nonluminous matter in the universe be? We don't know yet. It cannot be made of ordinary (baryonic) matter, so it must consist of some other sort of elementary particle.

Dark matter makes up about 23% of the mass–energy of the universe, according to the latest experiments. Thus the total mass–energy is 23% dark matter plus 4% baryons for a total of 27%, which does not bring ρ up to ρ_c. What is the other 73%? We are not sure about that either, but we have given it the name "dark energy."

Dark Energy—Cosmic Acceleration

Just before the year 2000, cosmologists received a surprise. Gravity was assumed to be the predominant force on a large scale in the universe, and it was thought that the expansion of the universe ought to be slowing down in time because gravity acts as an attractive force between objects. But measurements on type Ia supernovae (SNIa, our best standard candles—see Section 33–3) unexpectedly showed that very distant (high z) SNIa's were dimmer than expected. That is, given their great distance d as determined from their low brightness, their speed v determined from the measured z was less than expected according to Hubble's law. This result suggests that nearer galaxies are moving away from us relatively faster than those very distant ones, meaning the expansion of the universe in more recent epochs has sped up. This **acceleration** in the expansion of the universe (in place of the expected deceleration due to gravitational attraction between masses) seems to have begun roughly 5 billion years ago (8 to 9 Gyr after the Big Bang).

Acceleration

What could be causing the universe to accelerate in its expansion, against the attractive force of gravity? Does our understanding of gravity need to be revised? We don't know the answers to these questions; many scientists say dark energy is the biggest mystery facing science today. There are several speculations. But somehow it seems to have a long-range *repulsive* effect on matter, causing objects to speed away from each other ever faster. Whatever it is, it has been given the name **dark energy**.

Dark energy

One idea is a sort of quantum field given the name "quintessence." Another possibility suggests an energy latent in space itself (vacuum energy) and relates to an aspect of General Relativity known as the **cosmological constant** (symbol Λ). When Einstein developed his equations, he found that they offered no solutions for a static universe. In those days (1917) it was thought the universe was static—unchanging and everlasting. Einstein added an arbitrary constant to his equations to provide solutions for a static universe. A decade later, when Hubble showed us an expanding universe, Einstein discarded his cosmological constant as no longer needed $(\Lambda = 0)$. But now it is being reconsidered: perhaps Λ is not zero. Theoretical attempts to calculate Λ have so far given unreal values.

Cosmological constant

There is increasing evidence that the effects of some form of dark energy are very real. The data from the WMAP survey and other recent experiments agree well with theories and computer models when they input dark energy as providing 73% of the mass–energy in the universe, and when the total mass–energy density equals the critical density ρ_c.

Today's best estimate of how the mass–energy in the universe is distributed is as follows:

Mass–energy types in the universe as % of total

73% dark energy

27% matter, subject to the known gravitational force.

Of this 27%:

23% is dark matter

4% is baryons (what atoms are made of) and only $\frac{1}{10}$ of this 4% is visible matter: stars, and galaxies (that is, 0.4% of the total).

It is remarkable that only 0.4% of all the mass–energy in the universe is visible as stars and galaxies.

The idea that the universe is dominated by a completely unknown form of energy seems bizarre. Nonetheless, the exquisite agreement between theory and the measured CMB anisotropy observations plus other experimental data (clustering of galaxies—see next Section) appears to be meaningful.

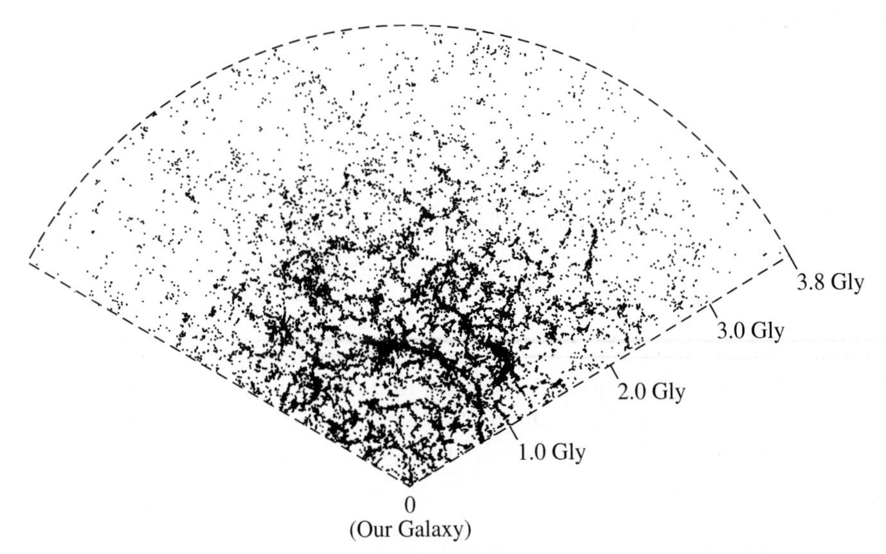

FIGURE 33–27 Distribution of some 50,000 galaxies in a 2.5° slice through almost half of the sky above the equator, as measured by the Sloan Digital Sky Survey (SDSS). Each dot represents a galaxy. The distance from us is obtained from the redshift and Hubble's law, and is given in units of 10^9 light-years (Gly). At greater distances, fewer galaxies are bright enough to be detected, thus resulting in an apparent thinning out of galaxies. The point 0 represents us, our observation point. Note the "walls" and "voids" of galaxies.

33–9 Large-Scale Structure of the Universe

The beautiful WMAP pictures of the sky (Fig. 33–23 and Chapter opening photo) show small but significant inhomogeneities in the temperature of the CMB. These anisotropies reflect compressions and expansions in the primordial plasma just before decoupling, from which stars, galaxies, and clusters of galaxies formed. Analysis of the irregularities in WMAP by mammoth computer simulations predict the distribution of clusters of galaxies and superclusters of galaxies very similar to what is seen today (Fig. 33–27). These simulations are very successful if they contain dark energy and dark matter; and the dark matter needs to be *cold* (slow speed—think of Eq. 13–8, $\frac{1}{2}m\overline{v}^2 = \frac{3}{2}kT$ where T is temperature), rather than "hot" dark matter such as neutrinos which move at or very near the speed of light. Indeed, the modern cosmological model is called the **ΛCDM** model, where lambda (Λ) stands for the cosmological constant, and CDM is **cold dark matter**.

ΛCDM *cosmological model*

Cosmologists have gained substantial confidence in this cosmological model from such a precise fit between observations and theory. They can also extract very precise values for cosmological parameters which previously were only known with low accuracy. The CMB is such an important cosmological observable that every effort is being made to extract all of the information it contains. More space missions are being prepared to observe even finer details. They could provide experimental evidence for inflation, perhaps detecting **gravity waves** as predicted by inflation models (detectable by their effect on the CMB) and also provide information about elementary particle physics at energies far beyond the reach of man-made accelerators.

33–10 Finally . . .

When we look up into the night sky, we see stars; and with the best telescopes, we see galaxies and the exotic objects we discussed earlier, including rare supernovae. But even with our best instruments we do not see the processes going on inside stars or supernovae that we hypothesized (and believe). We are dependent on brilliant theorists who come up with viable theories and ideas and verifiable models. We depend on complicated computer models whose parameters are varied until the outputs compare favorably with our observables and analyses of WMAP and other experiments. And we now have a surprisingly precise idea about some aspects of our universe: it is flat, it is 13.7 billion years old, it contains only 4% "normal" baryonic matter (for atoms), and so on. These precise results might suggest that we live at a very interesting time.

The questions raised by cosmology are difficult and profound, and may seem removed from everyday "reality." We can always say, "the Sun is shining, it's going to burn on for an unimaginably long time, all is well." Nonetheless, the questions of cosmology are deep ones that fascinate the human intellect. One aspect that is especially intriguing is this: calculations on the formation and evolution of the universe have been performed that deliberately varied the values—just slightly—of certain fundamental physical constants. The result? A universe in which life as we know it could not exist. [For example, if the difference in mass between proton and neutron were zero, or small (less than the mass of the electron, $0.511\ \mathrm{MeV}/c^2$), there would be no atoms: electrons would be captured by protons never to be freed again.] Such results have given rise to a philosophical idea called the **Anthropic principle**, which says that if the universe were even a little different than it is, we could not be here. It might even seem that the universe is exquisitely tuned, almost as if to accommodate us.

Anthropic principle

Summary

The night sky contains myriads of stars including those in the Milky Way, which is a "side view" of our **Galaxy** looking along the plane of the disc. Our Galaxy includes about 10^{11} stars. Beyond our Galaxy are billions of other galaxies.

Astronomical distances are measured in **light-years** ($1\ \mathrm{ly} \approx 10^{13}\ \mathrm{km}$). The nearest star is about 4 ly away and the nearest large galaxy is 2 million ly away. Our Galactic disc has a diameter of about 100,000 ly. Distances are often specified in **parsecs**, where $1\ \mathrm{parsec} = 3.26\ \mathrm{ly}$.

Stars are believed to begin life as collapsing masses of hydrogen gas (protostars). As they contract, they heat up (potential energy is transformed to kinetic energy). When the temperature reaches about 10 million degrees, nuclear fusion begins and forms heavier elements (**nucleosynthesis**), mainly helium at first. The energy released during these reactions heats the gas so its outward pressure balances the inward gravitational force, and the young star stabilizes as a **main-sequence** star. The tremendous luminosity of stars comes from the energy released during these thermonuclear reactions. After billions of years, as helium is collected in the core and hydrogen is used up, the core contracts and heats further. The envelope expands and cools, and the star becomes a **red giant** (larger diameter, redder color). The next stage of stellar evolution depends on the mass of the star, which may have lost much of its original mass as its outer envelope escaped into space. Stars of residual mass less than about 1.4 solar masses cool further and become **white dwarfs**, eventually fading and going out altogether. Heavier stars contract further due to their greater gravity: the density approaches nuclear density, the huge pressure forces electrons to combine with protons to form neutrons, and the star becomes essentially a huge nucleus of neutrons. This is a **neutron star**, and the energy released from its final core collapse is believed to produce **supernovae** explosions. If the star is very massive, it may contract even further and form a **black hole**, which is so dense that no matter or light can escape from it.

In the **general theory of relativity**, the **equivalence principle** states that an observer cannot distinguish acceleration from a gravitational field. Said another way, gravitational and inertial masses are the same. The theory predicts gravitational bending of light rays to a degree consistent with experiment. Gravity is treated as a curvature in space and time, the curvature being greater near massive bodies. The universe as a whole may be curved. With sufficient mass, the curvature of the universe would be positive, and the universe is *closed* and *finite*; otherwise, it would be *open* and *infinite*.

Distant galaxies display a **redshift** in their spectral lines, interpreted as a Doppler shift. The universe seems to be **expanding**, its galaxies racing away from each other at speeds (v) proportional to the distance (d) between them:

$$v = Hd, \qquad \textbf{(33–6)}$$

which is known as **Hubble's law** (H is the **Hubble parameter**). This expansion of the universe suggests an explosive origin, the **Big Bang**, which occurred about 13.7 billion years ago.

The **cosmological principle** assumes that the universe, on a large scale, is homogeneous and isotropic.

Important evidence for the Big Bang model of the universe was the discovery of the **cosmic microwave background** radiation (CMB), which conforms to a blackbody radiation curve at a temperature of 2.725 K.

The **standard model** of the Big Bang provides a possible scenario as to how the universe developed as it expanded and cooled after the Big Bang. Starting at 10^{-43} seconds after the Big Bang, according to this model, there was a series of **phase transitions** during which previously unified forces of nature "condensed out" one by one. The **inflationary scenario** assumes that during one of these phase transitions, the universe underwent a brief but rapid exponential expansion. Until about 10^{-35} s, there was no distinction between quarks and leptons. Shortly thereafter, quarks were **confined** into hadrons (the **hadron era**). About 10^{-4} s after the Big Bang, the majority of hadrons disappeared, having combined with anti-hadrons, producing photons, leptons and energy, leaving mainly photons and leptons to freely move, thus introducing the **lepton era**. By the time the universe was about 10 s old, the electrons too had mostly disappeared, having combined with their antiparticles; the universe was **radiation-dominated**. A couple of minutes later, nucleosynthesis began, but lasted only a few minutes.

It then took several hundred thousand years before the universe was cool enough for electrons to combine with nuclei to form atoms. The background radiation had expanded and cooled so much that its total energy became less than the energy in matter, and matter dominated increasingly over radiation. Then stars and galaxies formed, producing a universe not much different than it is today—some 13 billion years later.

Recent observations indicate that the universe is flat, that it contains an as-yet unknown type of **dark matter**, and that it is dominated by a mysterious **dark energy** which exerts a sort of negative gravity causing the expansion of the universe to accelerate.

Today the evidence suggests that the universe is flat and will continue to expand indefinitely. The total contributions of baryonic (normal) matter, dark matter, and dark energy sum up to the **critical density**.

Questions

1. The Milky Way was once thought to be "murky" or "milky" but is now considered to be made up of point sources. Explain.

2. A star is in equilibrium when it radiates at its surface all the energy generated at its core. What happens when it begins to generate more energy than it radiates? Less energy? Explain.

3. Describe a red giant star. List some of its properties.

4. Select a point on the H–R diagram. Mark several directions away from this point. Now describe the changes that would take place in a star moving in each of these directions.

5. Does the H–R diagram reveal anything about the core of a star?

6. Why do some stars end up as white dwarfs, and others as neutron stars or black holes?

7. Can we tell, by looking at the population on the H–R diagram, that hotter main-sequence stars have shorter lives? Explain.

8. If you were measuring star parallaxes from the Moon instead of Earth, what corrections would you have to make? What changes would occur if you were measuring parallaxes from Mars?

9. *Cepheid variable* stars change in luminosity with a typical period of several days. The period has been found to have a definite relationship with the absolute luminosity of the star. How could these stars be used to measure the distance to galaxies?

10. What is a geodesic? What is its role in General Relativity?

11. If it were discovered that the redshift of spectral lines of galaxies was due to something other than expansion, how might our view of the universe change? Would there be conflicting evidence? Discuss.

12. All galaxies appear to be moving away from us. Are we therefore at the center of the universe? Explain.

13. If you were located in a galaxy near the boundary of our observable universe, would galaxies in the direction of the Milky Way appear to be approaching you or receding from you? Explain.

14. Compare an explosion on Earth to the Big Bang. Consider such questions as: Would the debris spread at a higher speed for more distant particles, as in the Big Bang? Would the debris come to rest? What type of universe would this correspond to, open or closed?

15. If nothing, not even light, escapes from a black hole, then how can we tell if one is there?

16. What mass will give a Schwarzschild radius equal to that of the hydrogen atom in its ground state?

17. The Earth's age is often given as about 4 billion years. Find that time on Fig. 33–25. People have lived on Earth on the order of a million years. Where is that on Fig. 33–25?

18. Explain what the 2.7-K cosmic microwave background radiation is. Where does it come from? Why is its temperature now so low?

19. Why were atoms, as opposed to bare nuclei, unable to exist until hundreds of thousands of years after the Big Bang?

20. Under what circumstances would the universe eventually collapse in on itself?

Problems

33–1 to 33–3 Stars, Galaxies, Stellar Evolution, Distances

1. (I) Using the definitions of the parsec and the light-year, show that 1 pc = 3.26 ly.

2. (I) A star exhibits a parallax of 0.38 seconds of arc. How far away is it?

3. (I) The parallax angle of a star is 0.00019°. How far away is the star?

4. (I) A star is 36 pc away. What is its parallax angle? State (*a*) in seconds of arc, and (*b*) in degrees.

5. (I) What is the parallax angle for a star that is 55 ly away? How many parsecs is this?

6. (I) If one star is twice as far away from us as a second star, will the parallax angle of the farther star be greater or less than that of the nearer star? By what factor?

7. (II) A star is 35 pc away. How long does it take for its light to reach us?

8. (II) We saw earlier (Chapter 14) that the rate energy reaches the Earth from the Sun (the "solar constant") is about $1.3 \times 10^3\ \mathrm{W/m^2}$. What is (*a*) the apparent brightness l of the Sun, and (*b*) the absolute luminosity L of the Sun?

9. (II) What is the relative brightness of the Sun as seen from Jupiter as compared to its brightness from Earth? (Jupiter is 5.2 times farther from the Sun than the Earth.)

10. (II) Estimate the angular width that our Galaxy would subtend if observed from the nearest galaxy to us (Table 33–1). Compare to the angular width of the Moon from Earth.

11. (II) When our Sun becomes a red giant, what will be its average density if it expands out to the orbit of Earth (1.5×10^{11} m from the Sun)?

12. (II) When our Sun becomes a white dwarf, it is expected to be about the size of the Moon. What angular width will it subtend from the present distance to Earth?
13. (II) Calculate the density of a white dwarf whose mass is equal to the Sun's and whose radius is equal to the Earth's. How many times larger than Earth's density is this?
14. (II) A neutron star whose mass is 1.5 solar masses has a radius of about 11 km. Calculate its average density and compare to that for a white dwarf (Problem 13) and to that of nuclear matter.
15. (II) Calculate the Q-values for the He burning reactions of Eq. 33–2. (The mass of the very unstable $^8_4\mathrm{Be}$ is 8.005305 u.)
16. (II) Suppose two stars of the same apparent brightness l are also believed to be the same size. The spectrum of one star peaks at 800 nm whereas that of the other peaks at 400 nm. Use Wien's law (Section 27–2) and the Stefan-Boltzmann equation (Eq. 14–5) to estimate their relative distances from us. [*Hint*: see Examples 33–4 and 33–5.]
17. (III) Stars located in a certain cluster are assumed to be about the same distance from us. Two such stars have spectra that peak at $\lambda_1 = 500\,\mathrm{nm}$ and $\lambda_2 = 700\,\mathrm{nm}$, and the ratio of their apparent brightness is $l_1/l_2 = 0.091$. Estimate their relative sizes (give ratio of their diameters). [*Hint*: use the Stefan-Boltzmann equation, Eq. 14–5.]

33–4 General Relativity, Gravity and Curved Space

18. (I) Show that the Schwarzschild radius for a star with mass equal to that (*a*) of our Sun is 2.95 km, and (*b*) of Earth is 8.9 mm.
19. (II) What is the Schwarzschild radius for a typical galaxy (like ours)?
20. (II) Describe a triangle, drawn on the surface of a sphere, for which the sum of the angles is (*a*) 359°, and (*b*) 180°.
21. (II) What is the maximum sum-of-the-angles for a triangle on a sphere?

33–5 Redshift, Hubble's Law

22. (I) If a galaxy is traveling away from us at 1.0% of the speed of light, roughly how far away is it?
23. (I) The redshift of a galaxy indicates a velocity of 3500 km/s. How far away is it?
24. (I) Estimate the speed of a galaxy (relative to us) that is near the observable "edge" of the universe, say 12 billion light-years away.
25. (II) Estimate the observed wavelength for the 656-nm line in the Balmer series of hydrogen in the spectrum of a galaxy whose distance from us is (*a*) 1.0×10^6 ly, (*b*) 1.0×10^8 ly, (*c*) 1.0×10^{10} ly.
26. (II) Estimate the speed of a galaxy, and its distance from us, if the wavelength for the hydrogen line at 434 nm is measured on Earth as being 610 nm.
27. (II) What is the speed of a galaxy with $z = 0.60$?
28. (II) What would be the redshift parameter z for a galaxy traveling away from us at $v = 0.50c$?
29. (II) Starting from Eq. 33–3, show that the Doppler shift in wavelength is $\Delta\lambda/\lambda_0 \approx v/c$ (Eq. 33–5b) for $v \ll c$. [*Hint*: use the binomial expansion.]

33–6 to 33–8 The Big Bang, CMB, Universe Expansion

30. (I) Calculate the wavelength at the peak of the blackbody radiation distribution at 2.7 K using Wien's law.
31. (II) The critical density for closure of the universe is $\rho_c \approx 10^{-26}\,\mathrm{kg/m^3}$. State ρ_c in terms of the average number of nucleons per cubic meter.
32. (II) The scale of the universe (the average distance between galaxies) at any one moment is believed to have been inversely proportional to the absolute temperature. Estimate the size of the universe, compared to today, at (*a*) $t = 10^6$ yr, (*b*) $t = 1$ s, (*c*) $t = 10^{-6}$ s, and (*d*) $t = 10^{-35}$ s.
33. (II) At approximately what time had the universe cooled below the threshold temperature for producing (*a*) kaons $(M \approx 500\,\mathrm{MeV}/c^2)$, (*b*) Υ $(M \approx 9500\,\mathrm{MeV}/c^2)$, and (*c*) muons $(M \approx 100\,\mathrm{MeV}/c^2)$?

General Problems

34. Suppose that three main-sequence stars could undergo the three changes represented by the three arrows, A, B, and C, in the H–R diagram of Fig. 33–28. For each case, describe the changes in temperature, luminosity, and size.

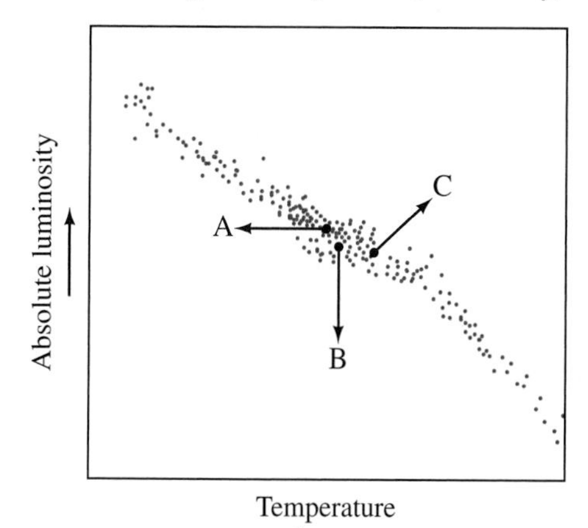

FIGURE 33–28 Problem 34.

35. Assume that the nearest stars to us have an absolute luminosity about the same as the Sun's. Their apparent brightness, however, is about 10^{11} times fainter than the Sun. From this, estimate the distance to the nearest stars. (Newton did this calculation, although he made a numerical error of a factor of 100.)
36. Use conservation of angular momentum to estimate the angular velocity of a neutron star which has collapsed to a diameter of 20 km, from a star whose radius was equal to that of our Sun (7×10^8 m), of mass 1.5 times that of the Sun, and which rotated (like our Sun) about once a month.
37. By what factor does the rotational kinetic energy change when the star in Problem 36 collapses to a neutron star?
38. A certain pulsar, believed to be a neutron star of mass 1.5 times that of the Sun, with diameter 20 km, is observed to have a rotation speed of 1.0 rev/s. If it loses rotational kinetic energy at the rate of 1 part in 10^9 per day, which is all transformed into radiation, what is the power output of the star?

39. The nearest large galaxy to our Galaxy is about 2×10^6 ly away. If both galaxies have a mass of 3×10^{41} kg, with what gravitational force does each galaxy attract the other?

40. Estimate what neutrino rest mass (in eV) would provide the critical density to close the universe. Assume the neutrino density is, like photons, about 10^9 times that of nucleons, and that nucleons make up only (*a*) 2% of the mass needed, or (*b*) 5% of the mass needed.

41. Two stars, whose spectra peak at 600 nm and 400 nm, respectively, both lie on the main sequence. Use Wien's law, the Stefan-Boltzmann equation, and the H–R diagram (Fig. 33–6) to estimate the ratio of their diameters. [*Hint*: see Examples 33–4 and 33–5.]

42. Suppose we can measure distances with parallax at 100 parsecs. What is our minimum angular resolution (in degrees), based on this information?

43. Through some coincidence, the Balmer lines from singly ionized helium in a distant star happen to overlap with the Balmer lines from hydrogen (Fig. 27–22) in the Sun. How fast is the star receding from us?

44. What is the temperature that corresponds to 1.8-TeV collisions at the Fermilab collider? To what era in cosmological history does this correspond? [*Hint*: see Fig. 33–25.]

45. Astronomers have recently measured the rotation of gas around what might be a supermassive black hole of about 2 billion solar masses at the center of a galaxy. If the radius from the galactic center to the gas clouds is 60 light-years, what Doppler shift $\Delta\lambda/\lambda_0$ do you estimate they saw?

46. A galaxy is moving away from Earth. The "blue" hydrogen line at 434 nm emitted from the galaxy is measured on Earth to be 650 nm. (*a*) How fast is the galaxy moving? (*b*) How far is it from Earth?

47. In the later stages of stellar evolution, a star (if massive enough) will begin fusing carbon nuclei to form, for example, magnesium:

$${}^{12}_{6}\mathrm{C} + {}^{12}_{6}\mathrm{C} \rightarrow {}^{24}_{12}\mathrm{Mg} + \gamma.$$

(*a*) How much energy is released in this reaction (see Appendix B). (*b*) How much kinetic energy must each carbon nucleus have (assume equal) in a head-on collision if they are just to touch (use Eq. 30–1) so that the strong force can come into play? (*c*) What temperature does this kinetic energy correspond to?

48. Consider the reaction

$${}^{16}_{8}\mathrm{O} + {}^{16}_{8}\mathrm{O} \rightarrow {}^{28}_{14}\mathrm{Si} + {}^{4}_{2}\mathrm{He},$$

and answer the same questions as in Problem 47.

49. How large would the Sun be if its density equaled the critical density of the universe, $\rho_c \approx 10^{-26}\,\mathrm{kg/m^3}$? Express your answer in light-years and compare with the Earth–Sun distance and the size of our Galaxy.

Answers to Exercises

A: Ourselves; 2 years ago.

B: 600 ly (estimating L from Fig. 33–6 as $L \approx 8 \times 10^{26}$ W; note that on a log scale, 6000 K is closer to 7000 K than it is to 5000 K).

C: 1.4.

APPENDIX A

Mathematical Review

A–1 Relationships, Proportionality, and Equations

One of the important aspects of physics is the search for relationships between different quantities—that is, determining how one quantity affects another. For example, how does temperature affect the air pressure in a tire? Or how does the net force on an object affect its acceleration? Sometimes a given quantity is affected by two or more quantities; for instance, the acceleration of an object is related to both its mass and the applied force. If you suspect that a relationship exists between two or more quantities, you can try to determine the precise nature of this relationship. This is done by varying one of the quantities and measuring how the other varies as a result. If it is likely that a particular quantity will be affected by more than one factor or quantity, only one quantity is varied at a time, while the others are held constant.†

As a simple example, the ancients found that if one circle has twice the diameter of a second circle, the first also has twice the circumference. If the diameter is three times as large, the circumference is also three times as large. In other words, an increase in the diameter results in a proportional increase in the circumference. We say that the circumference is *directly proportional to* the diameter. This can be written in symbols as $C \propto D$, where "$\propto$" means "is proportional to," and C and D refer to the circumference and diameter of a circle, respectively. The next step is to change this proportionality to an equation, which will make it possible to link the two quantities numerically. This merely entails inserting a proportionality constant, which in many cases is determined by measurement. (In some cases it can be chosen arbitrarily, if it involves only the definition of a new unit.) The ancients found that the ratio of the circumference to the diameter of any circle was 3.1416 (to keep only the first few decimal places). This number is designated by the Greek letter π. It is the constant of proportionality for the relationship $C \propto D$. To obtain an equation, we insert π into the proportion and change the $\propto$ to $=$. Thus, $C = \pi D$.

Direct proportion

Other kinds of proportionality occur as well. For example, the area of a circle is proportional to the *square* of its radius. That is, if the radius is doubled, the area becomes four times as large; and so on. In this case we can write $A \propto r^2$, where A stands for the area and r for the radius of the circle.

Sometimes two quantities are related in such a way that an increase in one leads to a proportional *decrease* in the other. This is called *inverse proportion*. For example, the time required to travel a given distance is inversely proportional to the speed of travel. The greater the speed, the less time it takes. We can write this inverse proportion as time $\propto$ 1/speed. The larger the denominator of a fraction, the lower the value of the fraction is as a whole. For example, $\frac{1}{4}$ is less than $\frac{1}{2}$. Thus, if the speed is doubled, the time is halved, which is what we want to express by this inverse proportionality relationship.

Inverse proportion

†When one quantity affects another, we often use the expression "is a function of" to indicate this dependence; for example, we say that the pressure in a tire is a function of the temperature.

Whatever kind of proportion is found to hold, it can be changed to an equality by insertion of the proper proportionality constant. Quantitative statements or predictions about the physical world can then be made with the equation.

A–2 Exponents

When we write 10^4, we mean that you multiply 10 by itself four times: $10^4 = 10 \times 10 \times 10 \times 10 = 10{,}000$. The superscript 4 is called an *exponent*, and 10 is said to be raised to the fourth power. Any number or symbol can be raised to a power; special names are used when the exponent is 2 (a^2 is "a squared") or 3 (a^3 is "a cubed"). For any other power, we say a^n is "a to the nth power." If the exponent is 1, it is usually dropped: $a^1 = a$, since no multiplication is involved.

The rules for multiplying numbers expressed as powers are as follows:

$$(a^n)(a^m) = a^{n+m}. \qquad \textbf{(A–1)}$$

That is, the exponents are added. To see why, consider the result of the multiplication of 3^3 by 3^4:

$$(3^3)(3^4) = (3)(3)(3) \times (3)(3)(3)(3) = (3)^7.$$

Here the sum of the exponents is $3 + 4 = 7$, so rule A–1 works. Notice that this rule works only if the base numbers (a in Eq. A–1) are the same. Thus we *cannot* use the rule of summing exponents for $(6^3)(5^2)$; these numbers would have to be written out. However, if the base numbers are different but the exponents are the same, we can write a second rule:

$$(a^n)(b^n) = (ab)^n. \qquad \textbf{(A–2)}$$

For example, $(5^3)(6^3) = (30)^3$ since

$$(5)(5)(5)(6)(6)(6) = (30)(30)(30).$$

The third rule involves a power raised to another power: $(a^3)^2$ means $(a^3)(a^3)$, which is equal to $a^{3+3} = a^6$. The general rule is then

$$(a^n)^m = a^{nm}. \qquad \textbf{(A–3)}$$

In this case, the exponents are multiplied.

Negative exponents are used for reciprocals. Thus,

$$\frac{1}{a} = a^{-1}, \qquad \frac{1}{a^3} = a^{-3},$$

and so on. The reason for using negative exponents is to allow us to use the multiplication rules given above. For example, $(a^5)(a^{-3})$ means

$$\frac{(a)(a)(a)(a)(a)}{(a)(a)(a)} = a^2.$$

Rule A–1 gives us the same result:

$$(a^5)(a^{-3}) = a^{5-3} = a^2.$$

What does an exponent of zero mean? That is, what is a^0? Any number raised to the zeroth power is defined as being equal to 1:

$$a^0 = 1.$$

This definition is used because it follows from the rules for adding exponents. For example,

$$a^3a^{-3} = a^{3-3} = a^0 = 1.$$

But *does* a^3a^{-3} actually equal 1? Yes, because

$$a^3a^{-3} = \frac{a^3}{a^3} = 1.$$

Fractional exponents are used to represent *roots*. For example, $a^{\frac{1}{2}}$ means the square root of a; that is, $a^{\frac{1}{2}} = \sqrt{a}$. Similarly, $a^{\frac{1}{3}}$ means the cube root of a, and so on. The fourth root of a means that if you multiply the fourth root of a by itself four times, you again get a:

$$\left(a^{\frac{1}{4}}\right)^4 = a.$$

This is consistent with rule A–3 since $\left(a^{\frac{1}{4}}\right)^4 = a^{\frac{4}{4}} = a^1 = a.$

A–3 Powers of 10, or Exponential Notation

Writing out very large and very small numbers such as the distance of Neptune from the Sun, 4,500,000,000 km, or the diameter of a typical atom, 0.00000001 cm, is inconvenient and prone to error. It also leaves in question (see Section 1–4) the number of significant figures. (How many of the zeros are significant in the number 4,500,000,000 km?) We therefore make use of the "powers of 10," or exponential notation. The distance from Neptune to the Sun is then expressed as 4.50×10^9 km (assuming that the value is significant to three digits), and the diameter of an atom 1.0×10^{-8} cm. This way of writing numbers is based on the use of exponents, where a^n signifies a multiplied by itself n times. For example, $10^4 = 10 \times 10 \times 10 \times 10 = 10{,}000.$ Thus, $4.50 \times 10^9 = 4.50 \times 1{,}000{,}000{,}000 = 4{,}500{,}000{,}000.$ Notice that the exponent (9 in this case) is just the number of places the decimal point is moved to the right to obtain the fully written-out number (4.500,000,000.)

When two numbers are multiplied (or divided), you first multiply (or divide) the simple parts and then the powers of 10. Thus, 2.0×10^3 multiplied by 5.5×10^4 equals $(2.0 \times 5.5) \times \left(10^3 \times 10^4\right) = 11 \times 10^7$, where we have used the rule for adding exponents (Appendix A–2). Similarly, 8.2×10^5 divided by 2.0×10^2 equals

$$\frac{8.2 \times 10^5}{2.0 \times 10^2} = \frac{8.2}{2.0} \times \frac{10^5}{10^2} = 4.1 \times 10^3.$$

For numbers less than 1, say 0.01, the exponent power of 10 is written with a negative sign: $0.01 = 1/100 = 1/10^2 = 1 \times 10^{-2}$. Similarly, $0.002 = 2 \times 10^{-3}$. The decimal point has again been moved the number of places expressed in the exponent. Thus, $0.020 \times 3600 = 72$; in exponential notation $\left(2.0 \times 10^{-2}\right) \times \left(3.6 \times 10^3\right) = 7.2 \times 10^1 = 72.$

Notice also that $10^1 \times 10^{-1} = 10 \times 0.1 = 1$, and by the law of exponents, $10^1 \times 10^{-1} = 10^0$. Therefore, $10^0 = 1$.

When writing a number in exponential notation, it is usual to make the simple number be between 1 and 10. Thus it is conventional to write 4.5×10^9 rather than 45×10^8, although they are the same number.† This notation also allows the number of *significant figures* to be clearly expressed. We write 4.50×10^9 if this value is accurate to three significant figures, but 4.5×10^9 if it is accurate to only two.

A–4 Algebra

Physical relationships between quantities can be represented as equations involving symbols (usually letters of the alphabet) that represent the quantities. The manipulation of such equations is the field of algebra, and it is used a great deal in physics. An equation involves an equals sign, which tells us that the quantities on either side of the equals sign have the same value. Examples of equations are

$$3 + 8 = 11$$
$$2x + 7 = 15$$
$$a^2b + c = 6.$$

The first equation involves only numbers, so is called an arithmetic equation. The other two equations are algebraic since they involve symbols. In the third equation, the quantity a^2b means the product of a times a times b: $a^2b = a \times a \times b.$

†Another convention used, particularly with computers, is that the simple number be between 0.1 and 1. Thus we could write 4,500,000,000 as 0.450×10^{10}.

Solving for an Unknown

Often we wish to solve for one (or more) symbols, and we treat it as an *unknown*. For example, in the equation $2x + 7 = 15$, x is the unknown; this equation is true, however, only when $x = 4$. Determining what value (or values) the unknown(s) can have to satisfy the equation(s) is called *solving the equation*. To solve an equation, the following rule can be used:

> *An equation will remain true if any operation performed on one side is also performed on the other side:* for example, (*a*) addition or subtraction of a number or symbol; (*b*) multiplication or division by a number or symbol; (*c*) raising each side of the equation to the same power, or taking the same root (such as square root).

EXAMPLE A–1 Solve for x in the equation

$$2x + 7 = 15.$$

APPROACH We perform the same operations on both sides of the equation to isolate x as the only variable on the left side of the equals sign.

SOLUTION We first subtract 7 from both sides:

$$2x + 7 - 7 = 15 - 7$$

or

$$2x = 8.$$

Then we divide both sides by 2 to get

$$\frac{2x}{2} = \frac{8}{2},$$

or, carrying out the divisions,

$$x = 4,$$

and this solves the equation.

EXAMPLE A–2 (*a*) Solve the equation

$$a^2b + c = 24$$

for the unknown a in terms of b and c. (*b*) Solve for a assuming that $b = 2$ and $c = 6$.

APPROACH We perform operations to isolate a as the only variable on the left side of the equals sign.

SOLUTION (*a*) We are trying to solve for a, so we first subtract c from both sides:

$$a^2b = 24 - c,$$

then divide by b:

$$a^2 = \frac{24 - c}{b},$$

and finally take square roots:

$$a = \sqrt{\frac{24 - c}{b}}.$$

(*b*) If we are given that $b = 2$ and $c = 6$, then

$$a = \sqrt{\frac{24 - 6}{2}} = 3.$$

NOTE Whenever we take a square root, the number can be either positive or negative. Thus $a = -3$ is also a solution. Why? Because $(-3)^2 = 9$, just as $(+3)^2 = 9$. So we actually get two solutions: $a = +3$ and $a = -3$.

To check a solution, we put it back into the original equation (this is really a check that we did all the manipulations correctly). In the equation

$$a^2b + c = 24,$$

we put in $a = 3, b = 2, c = 6$ and find

$$(3)^2(2) + (6) \stackrel{?}{=} 24$$
$$24 = 24,$$

which checks.

EXERCISE A Put $a = -3$ into the equation of Example A–2 and show that it works too.

Two or More Unknowns

If we have two or more unknowns, one equation is not sufficient to find them. In general, if there are n unknowns, n independent equations are needed. For example, if there are two unknowns, we need two equations. If the unknowns are called x and y, a typical procedure is to solve one equation for x in terms of y, and substitute this into the second equation.

EXAMPLE A–3 Solve the following pair of equations for x and y.

$$3x - 2y = 19$$
$$x + 4y = -3.$$

APPROACH We have two unknowns and two equations; we can start by solving the second equation for x in terms of y. Then we substitute this result for x into the first equation.

SOLUTION We subtract $4y$ from both sides of the second equation:

$$x = -3 - 4y.$$

We substitute this expression for x into the first equation, and simplify:

$$3(-3 - 4y) - 2y = 19$$
$$-9 - 12y - 2y = 19 \quad \text{(carried out the multiplication by 3)}$$
$$-14y = 28 \quad \text{(added 9 to both sides)}$$
$$y = -2. \quad \text{(divided both sides by } -14\text{)}$$

Now that we know $y = -2$, we substitute this into the expression for x:

$$x = -3 - 4y$$
$$= -3 - 4(-2) = -3 + 8 = 5.$$

Our solution is $x = 5, y = -2$. We check this solution by putting these values back into the original equations:

$$3x - 2y \stackrel{?}{=} 19$$
$$3(5) - 2(-2) \stackrel{?}{=} 19$$
$$15 + 4 \stackrel{?}{=} 19$$
$$19 = 19 \quad \text{(it checks)}$$

and

$$x + 4y \stackrel{?}{=} -3$$
$$5 + 4(-2) \stackrel{?}{=} -3$$
$$-3 = -3. \quad \text{(it checks)}$$

Other methods for solving two or more equations, such as the method of determinants, can be found in an algebra textbook.

The Quadratic Formula

We sometimes encounter equations that involve an unknown, say x, that appears not only to the first power, but squared as well. Such a *quadratic equation* can be written in the form

$$ax^2 + bx + c = 0.$$

The quantities a, b, and c are typically numbers or constants that are given.[†] The general solutions to such an equation are given by the *quadratic formula*:

Quadratic formula

$$x = \frac{-b \pm \sqrt{b^2 - 4ac}}{2a}. \qquad \textbf{(A–4)}$$

The $\pm$ sign indicates that there are two solutions for x: one where the plus sign is used, the other where the minus sign is used.

EXAMPLE A–4 Find the solutions for x in the equation

$$3x^2 - 5x = 2.$$

APPROACH Here x appears both to the first power and squared, so we use the quadratic equation.

SOLUTION First we write this equation in the standard form

$$ax^2 + bx + c = 0$$

by subtracting 2 from both sides:

$$3x^2 - 5x - 2 = 0.$$

In this case, a, b, and c in the standard formula take the values $a = 3, b = -5$, and $c = -2$. The two solutions for x are

$$x = \frac{+5 + \sqrt{25 - (4)(3)(-2)}}{(2)(3)} = \frac{5 + 7}{6} = 2$$

and

$$x = \frac{+5 - \sqrt{25 - (4)(3)(-2)}}{(2)(3)} = \frac{5 - 7}{6} = -\frac{1}{3}.$$

In this Example, the two solutions are $x = 2$ and $x = -\frac{1}{3}$. In physics problems, it sometimes happens that only one of the solutions corresponds to a real-life situation; in this case, the other solution is discarded. In other cases, both solutions may correspond to physical reality.

Notice, incidentally, that b^2 must be greater than $4ac$, so that $\sqrt{b^2 - 4ac}$ yields a real number. If $(b^2 - 4ac)$ is less than zero (negative), there is no real solution. The square root of a negative number is called *imaginary*.

A second-order equation—one in which the highest power of x is 2—has two solutions; a third-order equation—involving x^3—has three solutions; and so on.

A–5 The Binomial Expansion

Sometimes we end up with a quantity of the form $(1 + x)^n$. That is, the quantity $(1 + x)$ is raised to the nth power. This can be written as an infinite sum of terms, known as a *series expansion*, as follows:

$$(1 + x)^n = 1 + nx + \frac{n(n - 1)}{2!}x^2 + \cdots. \qquad \textbf{(A–5)}$$

This formula is useful for us mainly when x is very small compared to one $(x \ll 1)$. In this case, each successive term is much smaller than the preceding

[†] Or one or more of them could be variables, in which case additional equations are needed.

term. For example, if $x = 0.01$, and $n = 2$, say, then whereas the first term equals 1, the second term is $nx = (2)(0.01) = 0.02$, and the third term is $[(2)(1)/2](0.01)^2 = 0.0001$, and so on. Thus, when x is small, we can ignore all but the first two (or three) terms and can write

$$(1 + x)^n \approx 1 + nx. \tag{A–6}$$

This approximation often allows us to solve an equation easily that otherwise might be very difficult. Some examples are

$$(1 + x)^2 \approx 1 + 2x,$$

$$\frac{1}{1 + x} = (1 + x)^{-1} \approx 1 - x,$$

$$\sqrt{1 + x} = (1 + x)^{\frac{1}{2}} \approx 1 + \tfrac{1}{2}x,$$

$$\frac{1}{\sqrt{1 + x}} = (1 + x)^{-\frac{1}{2}} \approx 1 - \tfrac{1}{2}x,$$

where $x \ll 1$.

As a numerical example, let us evaluate $\sqrt{1.02}$ using the binomial expansion since $x = 0.02$ is much smaller than 1:

$$\sqrt{1.02} = (1.02)^{\frac{1}{2}} = (1 + 0.02)^{\frac{1}{2}} \approx 1 + \tfrac{1}{2}(0.02) = 1.01.$$

You can check with a calculator (and maybe not even more quickly) that $\sqrt{1.02} \approx 1.01$.

A–6 Plane Geometry

We review here a number of theorems involving angles and triangles that are useful in physics.

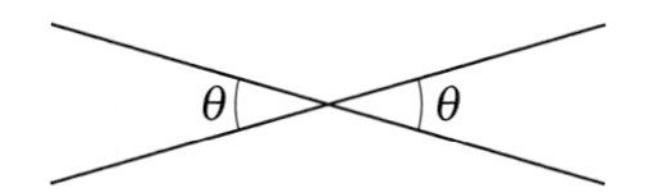

FIGURE A–1

1. *Equal angles.* Two angles are equal if any of the following conditions are true:
   (*a*) They are vertical angles (Fig. A–1); *or*
   (*b*) the left side of one is parallel to the left side of the other, and the right side of one is parallel to the right side of the other (the left and right sides are as seen from the vertex, where the two sides meet; Fig. A–2); *or*
   (*c*) the left side of one is perpendicular to the left side of the other, and the right sides are likewise perpendicular (Fig. A–3).

FIGURE A–2

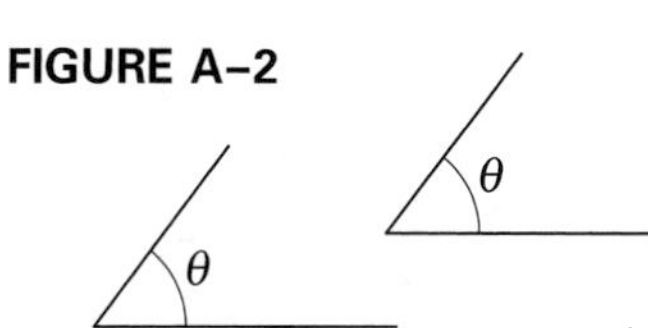

FIGURE A–3

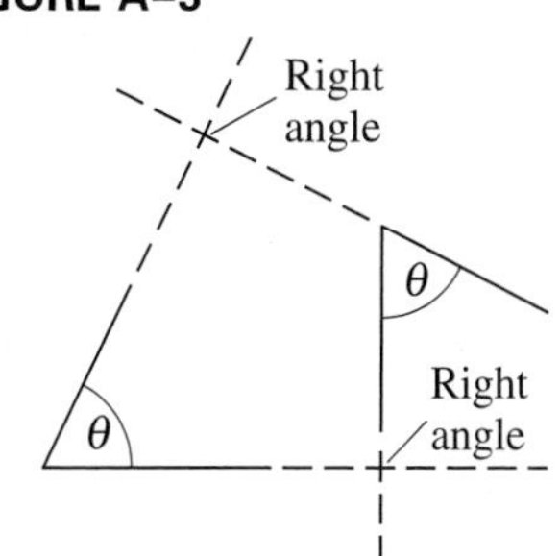

2. *The sum of the angles* in any plane triangle is 180°.

3. *Similar triangles.* Two triangles are said to be similar if all three of their angles are equal (in Fig. A–4, $\theta_1 = \phi_1$, $\theta_2 = \phi_2$, and $\theta_3 = \phi_3$). Similar triangles thus have the same basic shape but may be different sizes and have different orientations. Two useful theorems about similar triangles are:
   (*a*) Two triangles are similar if any two of their angles are equal. (This follows because the third angles must also be equal since the sum of the angles of a triangle is 180°.)
   (*b*) The ratios of corresponding sides of two similar triangles are equal. That is (Fig. A–4),

$$\frac{a_1}{b_1} = \frac{a_2}{b_2} = \frac{a_3}{b_3}.$$

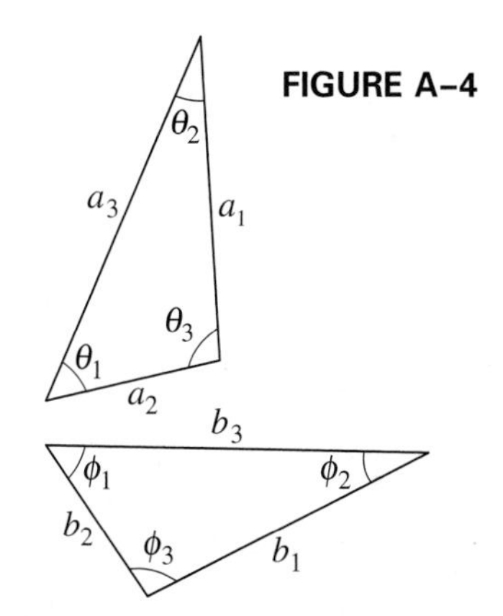

FIGURE A–4

4. *Congruent triangles.* Two triangles are congruent if one can be placed precisely on top of the other. That is, they are similar triangles and they have the same size. Two triangles are congruent if any of the following holds:
   (*a*) The three corresponding sides are equal.
   (*b*) Two sides and the enclosed angle are equal (“side-angle-side”).
   (*c*) Two angles and the enclosed side are equal (“angle-side-angle”).

FIGURE A–5

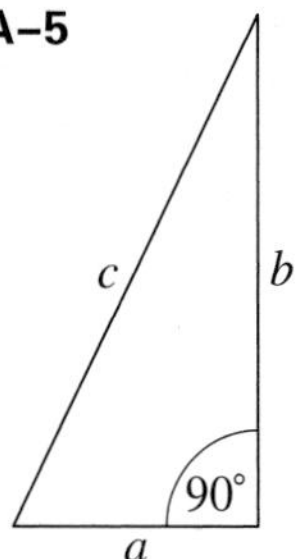

5. *Right triangles.* A right triangle has one angle that is 90° (a *right angle*); that is, the two sides that meet at the right angle are perpendicular (Fig. A–5). The two other (acute) angles in the right triangle add up to 90°.
6. *Pythagorean theorem.* In any right triangle, the square of the length of the hypotenuse (the side opposite the right angle) is equal to the sum of the squares of the lengths of the other two sides. In Fig. A–5,

$$c^2 = a^2 + b^2.$$

A–7 Trigonometric Functions and Identities

FIGURE A–6

Trigonometric functions for any angle θ are defined by constructing a right triangle about that angle as shown in Fig. A–6; opp and adj are the lengths of the sides opposite and adjacent to the angle θ, and hyp is the length of the hypotenuse:

$$\sin\theta = \frac{\text{opp}}{\text{hyp}} \qquad \csc\theta = \frac{1}{\sin\theta} = \frac{\text{hyp}}{\text{opp}}$$

$$\cos\theta = \frac{\text{adj}}{\text{hyp}} \qquad \sec\theta = \frac{1}{\cos\theta} = \frac{\text{hyp}}{\text{adj}}$$

$$\tan\theta = \frac{\text{opp}}{\text{adj}} = \frac{\sin\theta}{\cos\theta} \qquad \cot\theta = \frac{1}{\tan\theta} = \frac{\text{adj}}{\text{opp}}$$

$$\text{adj}^2 + \text{opp}^2 = \text{hyp}^2 \qquad \text{(Pythagorean theorem).}$$

FIGURE A–7

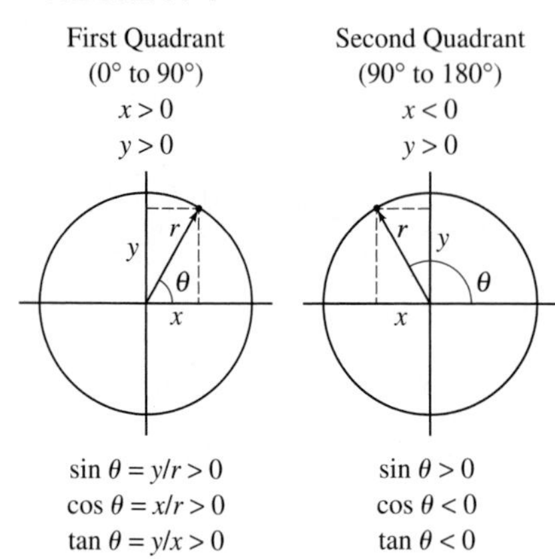

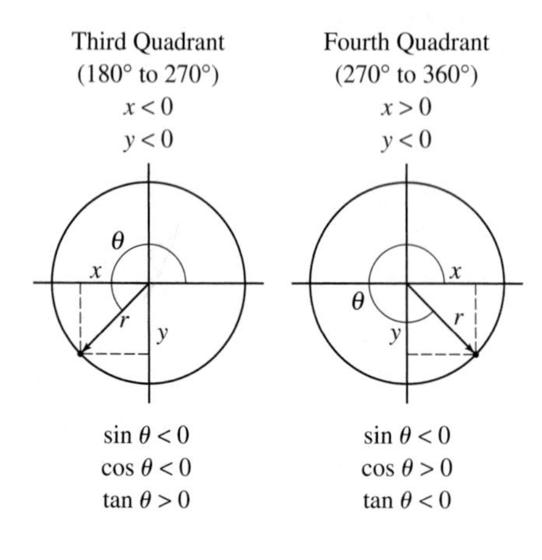

Figure A–7 shows the signs (+ or −) that cosine, sine, and tangent take on for angles θ in the four quadrants (0° to 360°). Note that angles are measured counterclockwise from the x axis as shown; negative angles are measured from *below* the x axis, clockwise: for example, $-30° = +330°$, and so on.

The following are some useful identities among the trigonometric functions:

$$\sin^2\theta + \cos^2\theta = 1$$

$$\sin 2\theta = 2\sin\theta\cos\theta$$

$$\cos 2\theta = \cos^2\theta - \sin^2\theta = 2\cos^2\theta - 1 = 1 - 2\sin^2\theta$$

$$\tan 2\theta = \frac{2\tan\theta}{1 - \tan^2\theta}$$

$$\sin(A \pm B) = \sin A\cos B \pm \cos A\sin B$$

$$\cos(A \pm B) = \cos A\cos B \mp \sin A\sin B$$

$$\tan(A \pm B) = \frac{\tan A \pm \tan B}{1 \mp \tan A\tan B}$$

$$\sin(180° - \theta) = \sin\theta$$

$$\cos(180° - \theta) = -\cos\theta$$

$$\sin(90° - \theta) = \cos\theta$$

$$\cos(90° - \theta) = \sin\theta$$

$$\sin\tfrac{1}{2}\theta = \sqrt{\frac{1 - \cos\theta}{2}}$$

$$\cos\tfrac{1}{2}\theta = \sqrt{\frac{1 + \cos\theta}{2}}$$

$$\tan\tfrac{1}{2}\theta = \sqrt{\frac{1 - \cos\theta}{1 + \cos\theta}}$$

$$\sin A \pm \sin B = 2\sin\left(\frac{A \pm B}{2}\right)\cos\left(\frac{A \mp B}{2}\right).$$

FIGURE A–8

For any triangle (see Fig. A–8):

$$\frac{\sin\alpha}{a} = \frac{\sin\beta}{b} = \frac{\sin\gamma}{c} \qquad \text{(law of sines)}$$

$$c^2 = a^2 + b^2 - 2ab\cos\gamma. \qquad \text{(law of cosines)}$$

Trigonometric Table: Numerical Values of Sin, Cos, Tan

Angle in Degrees	Angle in Radians	Sine	Cosine	Tangent	Angle in Degrees	Angle in Radians	Sine	Cosine	Tangent
0°	0.000	0.000	1.000	0.000					
1°	0.017	0.017	1.000	0.017	46°	0.803	0.719	0.695	1.036
2°	0.035	0.035	0.999	0.035	47°	0.820	0.731	0.682	1.072
3°	0.052	0.052	0.999	0.052	48°	0.838	0.743	0.669	1.111
4°	0.070	0.070	0.998	0.070	49°	0.855	0.755	0.656	1.150
5°	0.087	0.087	0.996	0.087	50°	0.873	0.766	0.643	1.192
6°	0.105	0.105	0.995	0.105	51°	0.890	0.777	0.629	1.235
7°	0.122	0.122	0.993	0.123	52°	0.908	0.788	0.616	1.280
8°	0.140	0.139	0.990	0.141	53°	0.925	0.799	0.602	1.327
9°	0.157	0.156	0.988	0.158	54°	0.942	0.809	0.588	1.376
10°	0.175	0.174	0.985	0.176	55°	0.960	0.819	0.574	1.428
11°	0.192	0.191	0.982	0.194	56°	0.977	0.829	0.559	1.483
12°	0.209	0.208	0.978	0.213	57°	0.995	0.839	0.545	1.540
13°	0.227	0.225	0.974	0.231	58°	1.012	0.848	0.530	1.600
14°	0.244	0.242	0.970	0.249	59°	1.030	0.857	0.515	1.664
15°	0.262	0.259	0.966	0.268	60°	1.047	0.866	0.500	1.732
16°	0.279	0.276	0.961	0.287	61°	1.065	0.875	0.485	1.804
17°	0.297	0.292	0.956	0.306	62°	1.082	0.883	0.469	1.881
18°	0.314	0.309	0.951	0.325	63°	1.100	0.891	0.454	1.963
19°	0.332	0.326	0.946	0.344	64°	1.117	0.899	0.438	2.050
20°	0.349	0.342	0.940	0.364	65°	1.134	0.906	0.423	2.145
21°	0.367	0.358	0.934	0.384	66°	1.152	0.914	0.407	2.246
22°	0.384	0.375	0.927	0.404	67°	1.169	0.921	0.391	2.356
23°	0.401	0.391	0.921	0.424	68°	1.187	0.927	0.375	2.475
24°	0.419	0.407	0.914	0.445	69°	1.204	0.934	0.358	2.605
25°	0.436	0.423	0.906	0.466	70°	1.222	0.940	0.342	2.747
26°	0.454	0.438	0.899	0.488	71°	1.239	0.946	0.326	2.904
27°	0.471	0.454	0.891	0.510	72°	1.257	0.951	0.309	3.078
28°	0.489	0.469	0.883	0.532	73°	1.274	0.956	0.292	3.271
29°	0.506	0.485	0.875	0.554	74°	1.292	0.961	0.276	3.487
30°	0.524	0.500	0.866	0.577	75°	1.309	0.966	0.259	3.732
31°	0.541	0.515	0.857	0.601	76°	1.326	0.970	0.242	4.011
32°	0.559	0.530	0.848	0.625	77°	1.344	0.974	0.225	4.331
33°	0.576	0.545	0.839	0.649	78°	1.361	0.978	0.208	4.705
34°	0.593	0.559	0.829	0.675	79°	1.379	0.982	0.191	5.145
35°	0.611	0.574	0.819	0.700	80°	1.396	0.985	0.174	5.671
36°	0.628	0.588	0.809	0.727	81°	1.414	0.988	0.156	6.314
37°	0.646	0.602	0.799	0.754	82°	1.431	0.990	0.139	7.115
38°	0.663	0.616	0.788	0.781	83°	1.449	0.993	0.122	8.144
39°	0.681	0.629	0.777	0.810	84°	1.466	0.995	0.105	9.514
40°	0.698	0.643	0.766	0.839	85°	1.484	0.996	0.087	11.43
41°	0.716	0.656	0.755	0.869	86°	1.501	0.998	0.070	14.301
42°	0.733	0.669	0.743	0.900	87°	1.518	0.999	0.052	19.081
43°	0.750	0.682	0.731	0.933	88°	1.536	0.999	0.035	28.636
44°	0.768	0.695	0.719	0.966	89°	1.553	1.000	0.017	57.290
45°	0.785	0.707	0.707	1.000	90°	1.571	1.000	0.000	∞

A–8 Logarithms

Logarithms are defined in the following way:

$$\text{if } y = A^x, \qquad \text{then } x = \log_A y.$$

Common logs

That is, the logarithm of a number y to the base A is that number which, as the exponent of A, gives back the number y. For *common logarithms*, the base is 10, so

$$\text{if } y = 10^x, \qquad \text{then } x = \log y.$$

Natural logs

The subscript 10 on $\log_{10}$ is usually omitted when dealing with common logs. Another base sometimes used is the exponential base $e = 2.718\cdots$, a natural number.† Such logarithms are called *natural logarithms* and are written ln. Thus,

$$\text{if } y = e^x, \qquad \text{then } x = \ln y.$$

For any number y, the two types of logarithm are related by

$$\ln y = 2.3026 \log y.$$

Some simple rules for logarithms are as follows:

$$\log(ab) = \log a + \log b. \tag{A–7}$$

This is true because if $a = 10^n$ and $b = 10^m$, then $ab = 10^{n+m}$. From the definition of logarithm, $\log a = n$, $\log b = m$, and $\log(ab) = n + m$; hence, $\log(ab) = n + m = \log a + \log b$. In a similar way, we can show that

$$\log\left(\frac{a}{b}\right) = \log a - \log b \tag{A–8}$$

and

$$\log a^n = n \log a. \tag{A–9}$$

These three rules apply not only to common logs but to natural or any other kind of logarithm.

Logs were once used as a technique for simplifying certain types of calculation. Because of the advent of electronic calculators and computers, they are not often used any more for this purpose. However, logs do appear in certain physical equations, so it is helpful to know how to deal with them. If you do not have a calculator that calculates logs, you can easily use a *log table*, such as the small one shown here (Table A–1). The number N is given to two digits (some tables give N to three or more digits); the first digit is in the vertical column to the left, the second digit is in the horizontal row across the top. For example, the Table tells us that $\log 1.0 = 0.000$, $\log 1.1 = 0.041$, and $\log 4.1 = 0.613$. Table A–1 does not include the decimal point—it is understood. The Table gives logs for numbers between 1.0 and 9.9; for larger or smaller numbers, we use rule A–7:

$$\log(ab) = \log a + \log b.$$

For example,

$$\log(380) = \log(3.8 \times 10^2) = \log(3.8) + \log(10^2).$$

From the Table, $\log 3.8 = 0.580$; and from rule A–9,

$$\log(10^2) = 2\log(10) = 2,$$

since $\log(10) = 1$. [This follows from the definition of the logarithm: if

†The exponential base e can be written as an infinite series:

$$e = 1 + \frac{1}{1} + \frac{1}{1\cdot 2} + \frac{1}{1\cdot 2\cdot 3} + \frac{1}{1\cdot 2\cdot 3\cdot 4} + \cdots.$$

TABLE A–1 Short Table of Common Logarithms

N	0.0	0.1	0.2	0.3	0.4	0.5	0.6	0.7	0.8	0.9
1	000	041	079	114	146	176	204	230	255	279
2	301	322	342	362	380	398	415	431	447	462
3	477	491	505	519	531	544	556	568	580	591
4	602	613	623	633	643	653	663	672	681	690
5	699	708	716	724	732	740	748	756	763	771
6	778	785	792	799	806	813	820	826	833	839
7	845	851	857	863	869	875	881	886	892	898
8	903	908	914	919	924	929	935	940	944	949
9	954	959	964	968	973	978	982	987	991	996

$10 = 10^1$, then $1 = \log(10)$.] Thus,

$$\begin{aligned} \log(380) &= \log(3.8) + \log\left(10^2\right) \\ &= 0.580 + 2 \\ &= 2.580. \end{aligned}$$

Similarly,

$$\begin{aligned} \log(0.081) &= \log(8.1) + \log\left(10^{-2}\right) \\ &= 0.908 - 2 = -1.092. \end{aligned}$$

Sometimes we need to do the reverse process: find the number N whose log is, say, 2.670. This is called "taking the antilogarithm." To do so, we separate our number 2.670 into two parts, making the separation at the decimal point:

Antilogs

$$\begin{aligned} \log N &= 2.670 = 2 + 0.670 \\ &= \log 10^2 + 0.670. \end{aligned}$$

We now look at Table A–1 to see what number has its log equal to 0.670; none does, so we must *interpolate*: we see that $\log 4.6 = 0.663$ and $\log 4.7 = 0.672$. So the number we want is between 4.6 and 4.7, and closer to the latter by $\frac{7}{9}$. Approximately we can say that $\log 4.68 = 0.670$. Thus

Interpolation

$$\begin{aligned} \log N &= 2 + 0.670 \\ &= \log\left(10^2\right) + \log(4.68) = \log\left(4.68 \times 10^2\right), \end{aligned}$$

so $N = 4.68 \times 10^2 = 468$.

If the given logarithm is negative, say, -2.180, we proceed as follows:

$$\begin{aligned} \log N &= -2.180 = -3 + 0.820 \\ &= \log 10^{-3} + \log 6.6 = \log 6.6 \times 10^{-3}, \end{aligned}$$

so $N = 6.6 \times 10^{-3}$. Notice that we added to our given logarithm the next largest integer (3 in this case) so that we have an integer, plus a decimal number between 0 and 1.0 whose antilogarithm can be looked up in the Table.

APPENDIX B

Selected Isotopes

(1) Atomic Number Z	(2) Element	(3) Symbol	(4) Mass Number A	(5) Atomic Mass[†]	(6) % Abundance (or Radioactive Decay[‡] Mode)	(7) Half-life (if radioactive)
0	(Neutron)	n	1	1.008665	β^-	10.24 min
1	Hydrogen	H	1	1.007825	99.9885%	
	Deuterium	d or D	2	2.014102	0.0115%	
	Tritium	t or T	3	3.016049	β^-	12.33 yr
2	Helium	He	3	3.016029	0.000137%	
			4	4.002603	99.999863%	
3	Lithium	Li	6	6.015122	7.59%	
			7	7.016004	92.41%	
4	Beryllium	Be	7	7.016929	EC, γ	53.29 days
			9	9.012182	100%	
5	Boron	B	10	10.012937	19.9%	
			11	11.009306	80.1%	
6	Carbon	C	11	11.011434	β^+, EC	20.39 min
			12	12.000000	98.93%	
			13	13.003355	1.07%	
			14	14.003242	β^-	5730 yr
7	Nitrogen	N	13	13.005739	β^+, EC	9.965 min
			14	14.003074	99.632%	
			15	15.000109	0.368%	
8	Oxygen	O	15	15.003065	β^+, EC	122.24 s
			16	15.994915	99.757%	
			18	17.999160	0.205%	
9	Fluorine	F	19	18.998403	100%	
10	Neon	Ne	20	19.992440	90.48%	
			22	21.991386	9.25%	
11	Sodium	Na	22	21.994437	β^+, EC, γ	2.6019 yr
			23	22.989770	100%	
			24	23.990963	β^-, γ	14.951 h
12	Magnesium	Mg	24	23.985042	78.99%	
13	Aluminum	Al	27	26.981538	100%	
14	Silicon	Si	28	27.976927	92.2297%	
			31	30.975363	β^-, γ	157.3 min

[†] The masses given in column (5) are those for the neutral atom, including the Z electrons.
[‡] Chapter 30; EC = electron capture.

(1) Atomic Number Z	(2) Element	(3) Symbol	(4) Mass Number A	(5) Atomic Mass	(6) % Abundance (or Radioactive Decay Mode)	(7) Half-life (if radioactive)
15	Phosphorus	P	31	30.973762	100%	
			32	31.973907	β^-	14.262 days
16	Sulfur	S	32	31.972071	94.9%	
			35	34.969032	β^-	87.38 days
17	Chlorine	Cl	35	34.968853	75.78%	
			37	36.965903	24.22%	
18	Argon	Ar	40	39.962383	99.600%	
19	Potassium	K	39	38.963707	93.258%	
			40	39.963999	0.0117%	
					β^-, EC, γ, β^+	1.277×10^9 yr
20	Calcium	Ca	40	39.962591	96.94%	
21	Scandium	Sc	45	44.955910	100%	
22	Titanium	Ti	48	47.947947	73.72%	
23	Vanadium	V	51	50.943964	99.750%	
24	Chromium	Cr	52	51.940512	83.789%	
25	Manganese	Mn	55	54.940363	100%	
26	Iron	Fe	56	55.934942	91.75%	
27	Cobalt	Co	59	58.933200	100%	
			60	59.933822	β^-, γ	5.2708 yr
28	Nickel	Ni	58	57.935348	68.077%	
			60	59.930791	26.223%	
29	Copper	Cu	63	62.929601	69.17%	
			65	64.927794	30.83%	
30	Zinc	Zn	64	63.929147	48.6%	
			66	65.926037	27.9%	
31	Gallium	Ga	69	68.925581	60.108%	
32	Germanium	Ge	72	71.922076	27.5%	
			74	73.921178	36.3%	
33	Arsenic	As	75	74.921596	100%	
34	Selenium	Se	80	79.916522	49.6%	
35	Bromine	Br	79	78.918338	50.69%	
36	Krypton	Kr	84	83.911507	57.00%	
37	Rubidium	Rb	85	84.911789	72.17%	
38	Strontium	Sr	86	85.909262	9.86%	
			88	87.905614	82.58%	
			90	89.907738	β^-	28.79 yr
39	Yttrium	Y	89	88.905848	100%	
40	Zirconium	Zr	90	89.904704	51.4%	
41	Niobium	Nb	93	92.906378	100%	
42	Molybdenum	Mo	98	97.905408	24.1%	
43	Technetium	Tc	98	97.907216	β^-, γ	4.2×10^6 yr
44	Ruthenium	Ru	102	101.904350	31.55%	
45	Rhodium	Rh	103	102.905504	100%	
46	Palladium	Pd	106	105.903483	27.33%	
47	Silver	Ag	107	106.905093	51.839%	
			109	108.904756	48.161%	

(1) Atomic Number Z	(2) Element	(3) Symbol	(4) Mass Number A	(5) Atomic Mass	(6) % Abundance (or Radioactive Decay Mode)	(7) Half-life (if radioactive)
48	Cadmium	Cd	114	113.903358	28.7%	
49	Indium	In	115	114.903878	95.71%; β^-	4.41×10^{14} yr
50	Tin	Sn	120	119.902197	32.58%	
51	Antimony	Sb	121	120.903818	57.21%	
52	Tellurium	Te	130	129.906223	34.1%; $\beta^-\beta^-$	$> 5.6 \times 10^{22}$ yr
53	Iodine	I	127	126.904468	100%	
			131	130.906124	β^-, γ	8.0207 days
54	Xenon	Xe	132	131.904155	26.89%	
			136	135.907220	8.87%; $\beta^-\beta^-$	$> 3.6 \times 10^{20}$ yr
55	Cesium	Cs	133	132.905447	100%	
56	Barium	Ba	137	136.905821	11.232%	
			138	137.905241	71.70%	
57	Lanthanum	La	139	138.906348	99.910%	
58	Cerium	Ce	140	139.905434	88.45%	
59	Praseodymium	Pr	141	140.907648	100%	
60	Neodymium	Nd	142	141.907719	27.2%	
61	Promethium	Pm	145	144.912744	EC, α	17.7 yr
62	Samarium	Sm	152	151.919728	26.75%	
63	Europium	Eu	153	152.921226	52.19%	
64	Gadolinium	Gd	158	157.924101	24.84%	
65	Terbium	Tb	159	158.925343	100%	
66	Dysprosium	Dy	164	163.929171	28.2%	
67	Holmium	Ho	165	164.930319	100%	
68	Erbium	Er	166	165.930290	33.6%	
69	Thulium	Tm	169	168.934211	100%	
70	Ytterbium	Yb	174	173.938858	31.8%	
71	Lutetium	Lu	175	174.940768	97.41%	
72	Hafnium	Hf	180	179.946549	35.08%	
73	Tantalum	Ta	181	180.947996	99.988%	
74	Tungsten (wolfram)	W	184	183.950933	30.64%; α	$> 4 \times 10^{18}$ yr
75	Rhenium	Re	187	186.955751	62.60%; β^-	4.35×10^{10} yr
76	Osmium	Os	191	190.960928	β^-, γ	15.4 days
			192	191.961479	40.78%	
77	Iridium	Ir	191	190.960591	37.3%	
			193	192.962924	62.7%	
78	Platinum	Pt	195	194.964774	33.832%	
79	Gold	Au	197	196.966552	100%	
80	Mercury	Hg	199	198.968262	16.87%	
			202	201.970626	29.9%	
81	Thallium	Tl	205	204.974412	70.476%	
82	Lead	Pb	206	205.974449	24.1%	
			207	206.975881	22.1%	
			208	207.976636	52.4%	
			210	209.984173	β^-, γ, α	22.3 yr
			211	210.988731	β^-, γ	36.1 min
			212	211.991887	β^-, γ	10.64 h
			214	213.999798	β^-, γ	26.8 min

(1) Atomic Number Z	(2) Element	(3) Symbol	(4) Mass Number A	(5) Atomic Mass	(6) % Abundance (or Radioactive Decay Mode)	(7) Half-life (if radioactive)
83	Bismuth	Bi	209	208.980383	100%	
			211	210.987258	α, γ, β^-	2.14 min
84	Polonium	Po	210	209.982416	α, γ, EC	138.376 days
			214	213.995186	α, γ	164.3 μs
85	Astatine	At	218	218.008681	α, β^-	1.5 s
86	Radon	Rn	222	222.017570	α, γ	3.8235 days
87	Francium	Fr	223	223.019731	β^-, γ, α	22.00 min
88	Radium	Ra	226	226.025403	α, γ	1600 yr
89	Actinium	Ac	227	227.027747	β^-, γ, α	21.773 yr
90	Thorium	Th	228	228.028731	α, γ	1.9116 yr
			232	232.038050	100%; α, γ	1.405×10^{10} yr
91	Protactinium	Pa	231	231.035879	α, γ	3.276×10^4 yr
92	Uranium	U	232	232.037146	α, γ	68.9 yr
			233	233.039628	α, γ	1.592×10^5 yr
			235	235.043923	0.720%; α, γ	7.038×10^8 yr
			236	236.045562	α, γ	2.342×10^7 yr
			238	238.050783	99.274%; α, γ	4.468×10^9 yr
			239	239.054288	β^-, γ	23.45 min
93	Neptunium	Np	237	237.048167	α, γ	2.144×10^6 yr
			239	239.052931	β^-, γ	2.3565 days
94	Plutonium	Pu	239	239.052157	α, γ	24,110 yr
			244	244.064198	α	8.00×10^7 yr
95	Americium	Am	243	243.061373	α, γ	7370 yr
96	Curium	Cm	247	247.070347	α, γ	1.56×10^7 yr
97	Berkelium	Bk	247	247.070299	α, γ	1380 yr
98	Californium	Cf	251	251.079580	α, γ	898 yr
99	Einsteinium	Es	252	252.082970	α, EC, γ	471.7 days
100	Fermium	Fm	257	257.095099	α, γ	100.5 days
101	Mendelevium	Md	258	258.098425	α, γ	51.5 days
102	Nobelium	No	259	259.10102	α, EC	58 min
103	Lawrencium	Lr	262	262.1097	α, EC, fission	3.6 h
104	Rutherfordium	Rf	263	263.11831	fission	10 min
105	Dubnium	Db	262	262.11415	α, fission, EC	34 s
106	Seaborgium	Sg	266	266.1219	α, fission	21 s
107	Bohrium	Bh	264	264.1247	α	0.44 s
108	Hassium	Hs	269	269.1341	α	9 s
109	Meitnerium	Mt	268	268.1388	α	0.07 s
110	Darmstadtium	Ds	271	271.14608	α	0.06 ms
111		Uuu	272	272.1535	α	1.5 ms
112		Uub	277	277	α	0.24 ms

APPENDIX C

Rotating Frames of Reference; Inertial Forces; Coriolis Effect

Inertial and Noninertial Reference Frames

In Chapters 5 and 8 we examined the motion of objects, including circular and rotational motion, from the outside, as observers fixed on the Earth. Sometimes it is convenient to place ourselves (in theory, if not physically) into a reference frame that is rotating. Let us examine the motion of objects from the point of view, or frame of reference, of persons seated on a rotating platform such as a merry-go-round. It looks to them as if the rest of the world is going around *them*. But let us focus on what they observe when they place a tennis ball on the floor of the rotating platform, which we assume is frictionless. If they put the ball down gently, without giving it any push, they will observe that it accelerates from rest and moves outward as shown in Fig. C–1a. According to Newton's first law, an object initially at rest should stay at rest if no force

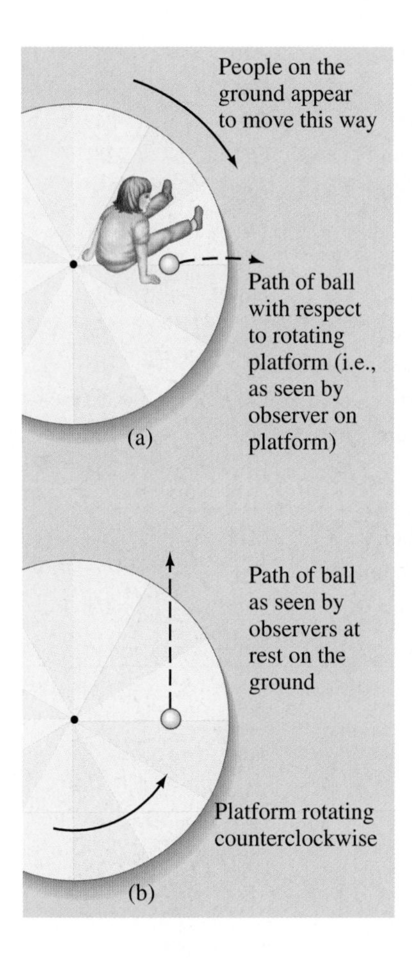

FIGURE C–1 Path of a ball released on a rotating merry-go-round as seen (a) in the reference frame of the merry-go-round, and (b) in a reference frame fixed on the ground.

acts on it. But, according to the observers on the rotating platform, the ball starts moving even though there is no force applied to it. To observers on the ground, this is all very clear: the ball has an initial velocity when it is released (because the platform is moving), and it simply continues moving in a straight-line path as shown in Fig. C–1b, in accordance with Newton's first law.

But what shall we do about the frame of reference of the observers on the rotating platform? Clearly, Newton's first law, the law of inertia, does not hold in this rotating frame of reference. For this reason, such a frame is called a **noninertial reference frame**. An **inertial reference frame** (as discussed in Chapter 4) is one in which the law of inertia—Newton's first law—does hold, and so do Newton's second and third laws. In a noninertial reference frame, such as our rotating platform, Newton's second law also does not hold. For instance in the situation described above, there is no net force on the ball; yet, with respect to the rotating platform, the ball accelerates.

Fictitious (Inertial) Forces

Because Newton's laws do not hold when observations are made with respect to a rotating frame of reference, calculation of motion can be complicated. However, we can still apply Newton's laws in such a reference frame if we make use of a trick. The ball on the rotating platform of Fig. C–1a flies outward when released (as if a force were acting on it—though as we saw above, no force actually does act on it); so the trick we use is to write down the equation $\Sigma F = ma$ as if a force equal to mv^2/r (or $m\omega^2 r$) were acting radially outward on the object in addition to any other forces that may be acting. This extra force, which might be designated as "centrifugal force" since it *seems* to act outward, is called a **fictitious force** or **pseudoforce**. It is a pseudoforce ("pseudo" means "false") because there is no object that exerts this force. Furthermore, when viewed from an inertial reference frame, the effect doesn't exist at all. We have made up this pseudoforce so that we can make calculations in a noninertial frame using Newton's second law, $\Sigma F = ma$. Thus the observer in the noninertial frame of Fig. C–1a uses Newton's second law for the ball's outward motion by assuming that a force equal to mv^2/r acts on it. Such pseudoforces are also called **inertial forces** since they arise only because the reference frame is not an inertial one.

Fictitious force (pseudoforce)

Inertial force

We can examine the motion of a particle in a centrifuge (Section 5–5) from the frame of reference of the rotating test tube. In this frame of reference, the particles move in a more-or-less straight path down the tube. (From the reference frame of the Earth, the particles go round and round.) The acceleration of a particle with respect to the rotating tube can then be calculated using $F = ma$ if we include a pseudoforce, "F," equal to $m\omega^2 r = m(v^2/r)$ acting down the tube, in addition to the drag force F_D exerted by the fluid on the particle (Fig. C–2) up the tube.

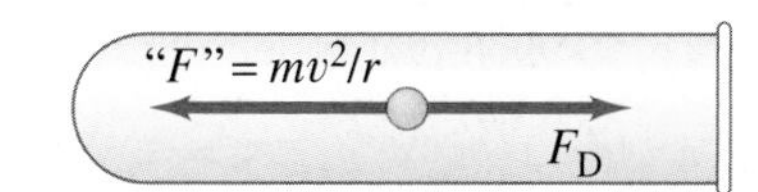

FIGURE C–2 The forces on a particle in a test tube rotating in a centrifuge, seen in the reference frame of the test tube.

In Section 5–3 we discussed the forces on a person in a car going around a curve (Fig. 5–11) from the point of view of an inertial frame. The car, on the other hand, is not an inertial frame. Passengers in such a car could interpret this being pressed outward as the effect of a "centrifugal" force. But they need to recognize that it is a pseudoforce because there is no identifiable object exerting it. It is an effect of being in a noninertial frame of reference.

The Earth itself is rotating on its axis. Thus, strictly speaking, Newton's laws are not valid on the Earth. However, the effect of the Earth's rotation is usually so small that it can be ignored, although it does influence the movement of large air masses and ocean currents. Because of the Earth's rotation, the material of the Earth is concentrated slightly more at the equator. The Earth is thus not a perfect sphere but is slightly fatter at the equator than at the poles.

Coriolis Effect

In a reference frame that rotates at a constant angular speed ω (relative to an inertial frame), there exists another pseudoforce known as the *Coriolis force*. It appears to act on a body in a rotating reference frame only if the body is moving relative to that reference frame, and it acts to deflect the body sideways. It, too, is an effect of the reference frame being noninertial and hence is referred to as an *inertial force*. To see how the Coriolis force arises, consider two people, A and B, at rest on a platform rotating with angular speed ω, as shown in Fig. C–3a. They are situated at distances r_A and r_B, respectively, from the axis of rotation (at O). The woman at A throws a ball with a horizontal velocity $\vec{\mathbf{v}}$ (in her reference frame) radially outward toward the man at B on the outer edge of the platform. In Fig. C–3a, we view the situation from an inertial reference frame. The ball initially has not only the velocity $\vec{\mathbf{v}}$ radially outward, but also a tangential velocity $\vec{\mathbf{v}}_A$ due to the rotation of the platform. Now Eq. 8–4 tells us that $v_A = r_A\omega$, where r_A is the woman's radial distance from the axis of rotation at O. If the man at B had this same velocity v_A, the ball would reach him perfectly. But his speed is greater than v_A (Fig. C–3a) since he is farther from the axis of rotation. His speed is $v_B = r_B\omega$, which is greater than v_A because $r_B > r_A$. Thus, when the ball reaches the outer edge of the platform, it passes a point that the man at B has already passed because his speed in that direction is greater than the ball's. So the ball passes behind him.

Figure C–3b shows the situation as seen from the rotating platform as frame of reference. Both A and B are at rest, and the ball is thrown with velocity $\vec{\mathbf{v}}$ toward B, but the ball deflects to the right as shown and passes behind B as previously described. This is not a centrifugal-force effect, for the latter acts radially outward. Instead, this effect acts sideways, perpendicular to $\vec{\mathbf{v}}$, and is called a **Coriolis acceleration**; it is said to be due to the Coriolis force, which is a fictitious inertial force. Its explanation as seen from an inertial system was given above: it is an effect of being in a rotating system, wherein points that are farther from the rotation axis have higher linear speeds. On the other hand, when viewed from the rotating system, we can describe the motion using Newton's second law, $\Sigma\vec{\mathbf{F}} = m\vec{\mathbf{a}}$, if we add a "pseudoforce" term corresponding to this Coriolis effect.

Let us determine the magnitude of the Coriolis acceleration for the simple case described above. (We assume v is large and distances are short, so we can ignore

FIGURE C–3 The origin of the Coriolis effect. Looking down on a rotating platform, (a) as seen from a nonrotating inertial system, and (b) as seen from the rotating platform as frame of reference.

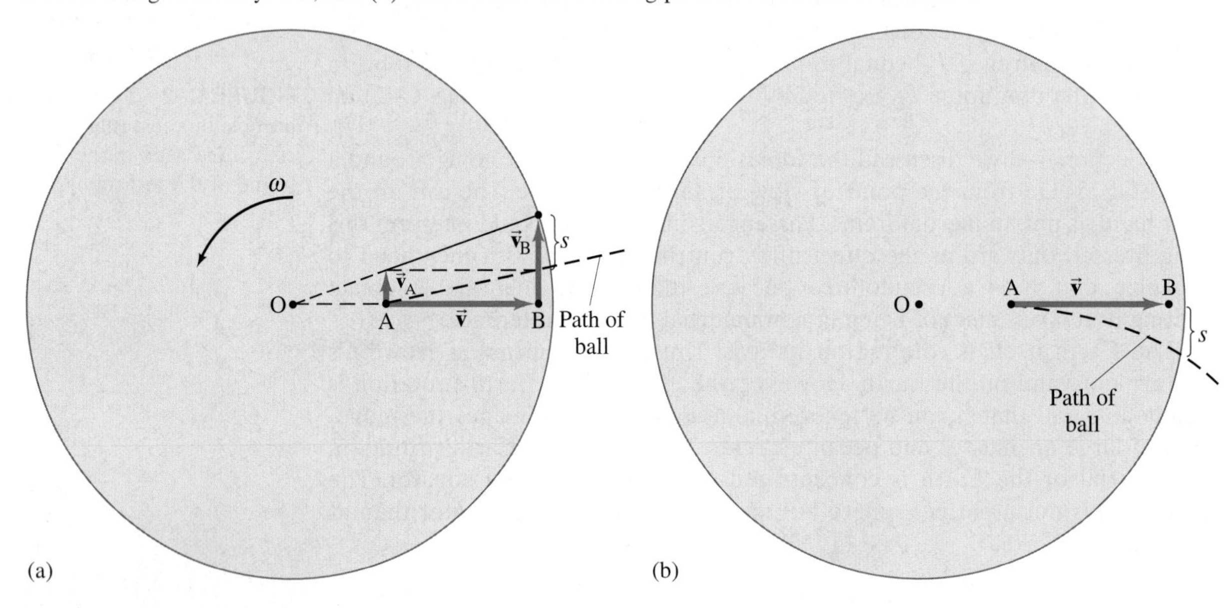

gravity.) We do the calculation from the inertial reference frame (Fig. C–3a). The ball moves radially outward a distance $r_B - r_A$ at speed v in a time t given by

$$r_B - r_A = vt.$$

During this time, the ball moves to the side a distance s_A given by

$$s_A = v_A t.$$

The man at B, in this time t, moves a distance

$$s_B = v_B t.$$

The ball therefore passes behind him a distance s (Fig. C–3a) given by

$$s = s_B - s_A = (v_B - v_A)t.$$

We saw earlier that $v_A = r_A\omega$ and $v_B = r_B\omega$, so

$$s = (r_B - r_A)\omega t.$$

We substitute $r_B - r_A = vt$ (see above) and get

$$s = \omega v t^2. \quad \textbf{(C–1)}$$

This same s equals the sideways displacement as seen from the noninertial rotating system (Fig. C–3b).

We see immediately that Eq. C–1 corresponds to motion at constant acceleration. For as we saw in Chapter 2 (see Eq. 2–11b), $y = \frac{1}{2}at^2$ for a constant acceleration (with zero initial velocity in the y direction). Thus, if we write Eq. C–1 in the form $s = \frac{1}{2}a_{Cor}t^2$, we see that the Coriolis acceleration a_{Cor} is

$$a_{Cor} = 2\omega v. \quad \textbf{(C–2)}$$

This relation is valid for any velocity in the plane of rotation—that is, in the plane perpendicular to the axis of rotation (in Fig. C–3, the axis through point O perpendicular to the page).

Because the Earth rotates, the Coriolis effect has some interesting manifestations on the Earth. It affects the movement of air masses and thus has an influence on weather. In the absence of the Coriolis effect, air would rush directly into a region of low pressure, as shown in Fig. C–4a. But because of the Coriolis effect, the winds are deflected to the right in the Northern Hemisphere (Fig. C–4b), since the Earth rotates from west to east. So there tends to be a counterclockwise wind pattern around a low-pressure area. The reverse is true in the Southern Hemisphere. Thus cyclones rotate counterclockwise in the Northern Hemisphere and clockwise in the Southern Hemisphere. The same effect explains the easterly trade winds near the equator: any winds heading south toward the equator will be deflected toward the west (that is, as if coming from the east).

The Coriolis effect also acts on a falling body. A body released from the top of a high tower will not hit the ground directly below the release point, but will be deflected slightly to the east. Viewed from an inertial frame, this is because the top of the tower revolves with a slightly higher speed than does the bottom of the tower.

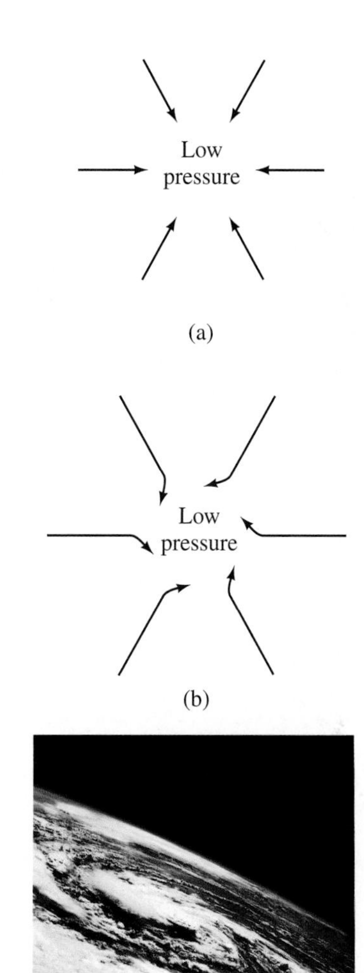

FIGURE C–4 (a) Winds (moving air masses) would flow directly toward a low-pressure area if the Earth did not rotate; (b) and (c): because of the Earth's rotation, the winds are deflected to the right in the Northern Hemisphere (as in Fig. C–3) as if a fictitious (Coriolis) force were acting.

APPENDIX D

Molar Specific Heats for Gases, and the Equipartition of Energy

Molar Specific Heats for Gases

The values of the specific heats for gases depend on how the thermodynamic process is carried out. Two important processes are those in which either the volume or the pressure is kept constant, and Table D–1 shows how different they can be.

The difference in specific heats for gases is nicely explained in terms of the first law of thermodynamics and kinetic theory. For gases we usually use **molar specific heats**, C_V and C_P, which are defined as the heat required to raise 1 mol of a gas by 1 C° at constant volume and at constant pressure, respectively. In analogy to Eq. 14–2, the heat Q needed to raise the temperature of n moles of gas by ΔT is

Molar specific heats

$$Q = nC_V\,\Delta T \qquad \text{[volume constant]} \qquad \textbf{(D–1a)}$$

$$Q = nC_P\,\Delta T. \qquad \text{[pressure constant]} \qquad \textbf{(D–1b)}$$

It is clear from the definition of molar specific heat (compare Eqs. 14–2 and D–1) that

$$C_V = Mc_V \quad \text{and} \quad C_P = Mc_P,$$

where M is the molecular mass of the gas ($M = m/n$ in grams/mol). The values for molar specific heats are included in Table D–1. These values are nearly the same for different gases that have the same number of atoms per molecule.

Now we use kinetic theory of gases to see, first, why the specific heats of gases are higher for constant-pressure processes than for constant-volume processes.

TABLE D–1 Specific Heats of Gases at 15°C

	Specific Heats (kcal/kg · K)		Molar Specific Heats (cal/mol · K)		
Gas	c_V	c_P	C_V	C_P	$C_P - C_V$ (cal/mol · K)
Monatomic					
He	0.75	1.15	2.98	4.97	1.99
Ne	0.148	0.246	2.98	4.97	1.99
Diatomic					
N_2	0.177	0.248	4.96	6.95	1.99
O_2	0.155	0.218	5.03	7.03	2.00
Triatomic					
CO_2	0.153	0.199	6.80	8.83	2.03
H_2O (100°C)	0.350	0.482	6.20	8.20	2.00
Polyatomic					
C_2H_6	0.343	0.412	10.30	12.35	2.05

Imagine that an ideal gas is slowly heated via these two processes—first at constant volume, and then at constant pressure. In both processes, we let the temperature increase by the same amount, ΔT. In the constant-volume process, no work is done since $\Delta V = 0$. Thus, according to the first law of thermodynamics, the heat added (denoted by Q_V) all goes into increasing the internal energy of the gas:

$$Q_V = \Delta U.$$

In the constant-pressure process, work *is* done. Hence the heat added, Q_P, must not only increase the internal energy but also is used to do work $W = P\,\Delta V$. Thus, for the same ΔT, more heat must be added in the process at constant pressure than at constant volume. For the process at constant pressure, the first law of thermodynamics gives

$$Q_P = \Delta U + P\,\Delta V.$$

Since ΔU is the same in the two processes (we chose ΔT to be the same), we can combine the two above equations:

$$Q_P - Q_V = P\,\Delta V.$$

From the ideal gas law, $V = nRT/P$, so for a process at constant pressure $\Delta V = nR\,\Delta T/P$. Putting this into the above equation and using Eqs. D–1, we find

$$nC_P\,\Delta T - nC_V\,\Delta T = P\left(\frac{nR\,\Delta T}{P}\right)$$

or, after cancellations,

$$C_P - C_V = R. \qquad \textbf{(D–2)}$$

Since the gas constant $R = 8.315\,\text{J/mol}\cdot\text{K} = 1.99\,\text{cal/mol}\cdot\text{K}$, we predict that C_P will be larger than C_V by about $1.99\,\text{cal/mol}\cdot\text{K}$. Indeed, this is very close to what is obtained experimentally, as the last column in Table D–1 shows.

Now we calculate the molar specific heat of a monatomic gas using the kinetic theory. For a process carried out at constant volume, no work is done, so the first law of thermodynamics tells us that

$$\Delta U = Q_V.$$

For an ideal monatomic gas, the internal energy, U, is the total kinetic energy of all the molecules,

$$U = N\left(\tfrac{1}{2}m\overline{v^2}\right) = \tfrac{3}{2}nRT$$

as we saw in Section 14–2. Then, using Eq. D–1a, we write $\Delta U = Q_V$ as

$$\Delta U = \tfrac{3}{2}nR\,\Delta T = nC_V\,\Delta T \qquad \textbf{(D–3)}$$

or

$$C_V = \tfrac{3}{2}R. \qquad \textbf{(D–4)}$$

Since $R = 8.315\,\text{J/mol}\cdot\text{K} = 1.99\,\text{cal/mol}\cdot\text{K}$, kinetic theory predicts that $C_V = 2.98\,\text{cal/mol}\cdot\text{K}$ for an ideal monatomic gas. This is very close to the experimental values for monatomic gases such as helium and neon (Table D–1). From Eq. D–2, C_P is predicted to be about $4.97\,\text{cal/mol}\cdot\text{K}$, also in agreement with experiment (Table D–1).

Equipartition of Energy

The measured molar specific heats for more complex gases (Table D–1), such as diatomic (two-atom) and triatomic (three-atom) gases, increase with the increased number of atoms per molecule. We can explain this by assuming that the internal energy includes not only translational kinetic energy but other forms of energy as well. For example, in a diatomic gas (Fig. D–1), the two atoms can rotate about two different axes (but rotation about a third axis passing through the two atoms would give rise to neglible energy, since the moment of inertia is so small). The molecules can have rotational as well as translational kinetic energy.

FIGURE D–1 A diatomic molecule can rotate about two different axes.

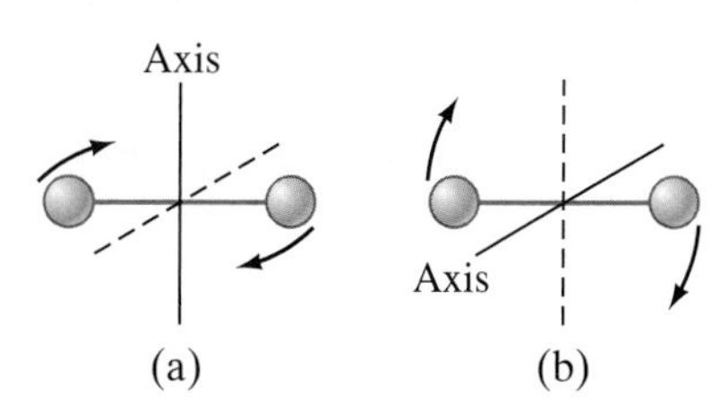

Degrees of freedom

It is useful to introduce the idea of **degrees of freedom**, by which we mean the number of independent ways molecules can possess energy. For example, a monatomic gas has three degrees of freedom, because an atom can have velocity along the x, y, and z axes. These are considered to be three independent motions because a change in any one of the components would not affect the others. A diatomic molecule has the same three degrees of freedom associated with translational kinetic energy plus two more degrees of freedom associated with rotational kinetic energy (Fig. D–1), for a total of five degrees of freedom.

Equipartition of energy

Table D–1 indicates that the C_V for diatomic gases is about $\frac{5}{3}$ times as great as for a monatomic gas—that is, in the same ratio as their degrees of freedom. This led nineteenth-century physicists to the **principle of equipartition of energy**. This principle states that energy is shared equally among the active degrees of freedom, and each active degree of freedom of a molecule has on the average an energy equal to $\frac{1}{2}kT$. Thus, the average energy for a molecule of a monatomic gas would be $\frac{3}{2}kT$ (which we already knew) and of a diatomic gas $\frac{5}{2}kT$. Hence the internal energy of a diatomic gas would be $U = N\left(\frac{5}{2}kT\right) = \frac{5}{2}nRT$, where n is the number of moles. Using the same argument we did for monatomic gases, we see that for diatomic gases the molar specific heat at constant volume would be $\frac{5}{2}R = 4.97\ \text{cal/mol}\cdot\text{K}$, in accordance with measured values. More complex molecules have even more degrees of freedom and thus greater molar specific heats.

However, measurements showed that for diatomic gases at very low temperatures, C_V has a value of only $\frac{3}{2}R$, as if there were only three degrees of freedom. And at very high temperatures, C_V was about $\frac{7}{2}R$, as if there were seven degrees of freedom. The explanation is that at low temperatures, nearly all molecules have only translational kinetic energy, so, no energy goes into rotational energy and only three degrees of freedom are "active." At very high temperatures, all five degrees of freedom are active plus two additional ones. We interpret the two new degrees of freedom as being associated with the two atoms vibrating, as if they were connected by a spring (Fig. D–2). One degree of freedom comes from the kinetic energy of the vibrational motion, and the second from the potential energy of vibrational motion $\left(\frac{1}{2}kx^2\right)$. At room temperature, these two degrees of freedom are apparently not active. Why fewer degrees of freedom are "active" at lower temperatures was eventually explained by Einstein using the quantum theory.

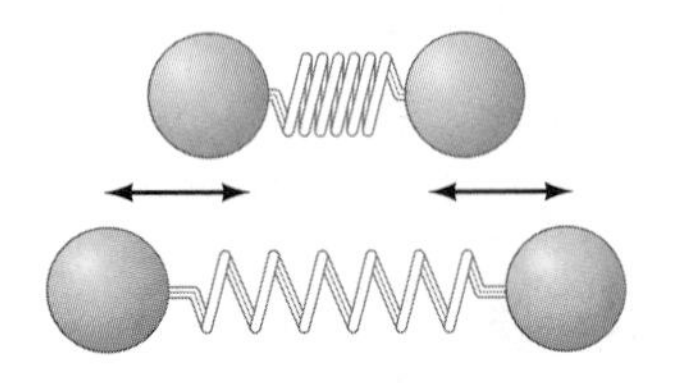

FIGURE D–2 A diatomic molecule can vibrate, as if the two atoms were connected by a spring. Of course they are not, but rather they exert forces on each other that are electrical in nature—of a form that resembles a spring force.

Solids

The principle of equipartition of energy can be applied to solids as well. The molar specific heat of any solid at high temperature is close to $3R$ ($6.0\ \text{cal/mol}\cdot\text{K}$), Fig. D–3. This is called the *Dulong and Petit value* after the scientists who first measured it in 1819. (Note that Table 14–1 gave the specific heats per kilogram, not per mole.) At high temperatures, each atom apparently has six degrees of freedom, although some are not active at low temperatures. Each atom in a crystalline solid can vibrate about its equilibrium position as if it were connected by springs to each of its neighbors (Fig. D–4). Thus it can have three degrees of freedom for kinetic energy and three more associated with potential energy of vibration in each of the x, y, and z directions, which is in accord with measured values.

FIGURE D–3 Molar specific heats of solids as a function of temperature.

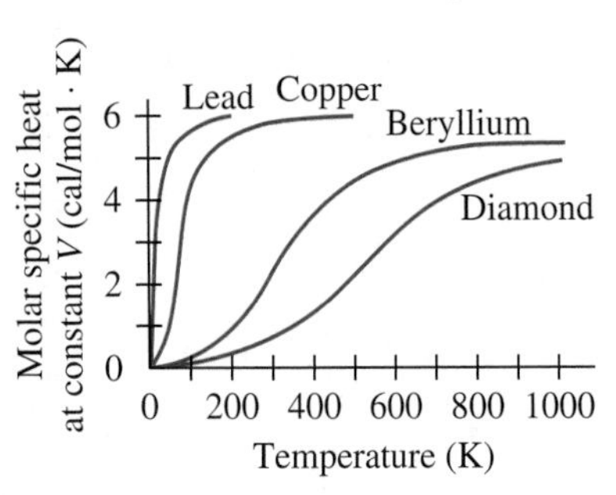

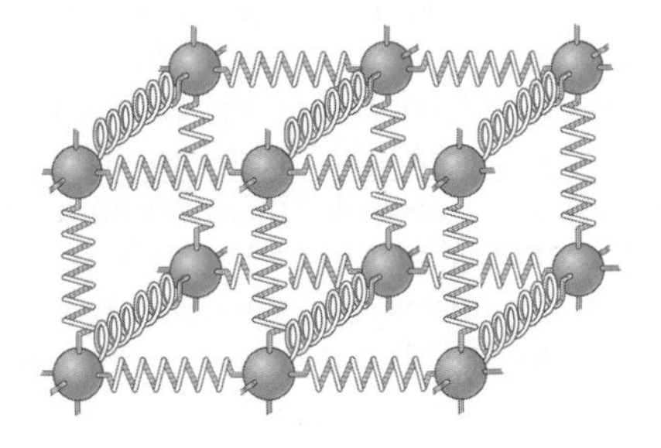

FIGURE D–4 The atoms in a crystalline solid can vibrate about their equilibrium positions as if they were connected to their neighbors by springs. (The forces between atoms are actually electrical in nature.)

APPENDIX E

Galilean and Lorentz Transformations

We now examine in detail the mathematics of relating quantities in one inertial reference frame to the equivalent quantities in another. In particular, we will see how positions and velocities *transform* (that is, change) as we go from one frame of reference to another.

We begin with the classical, or Galilean, viewpoint. Consider two reference frames S and S′ which are each characterized by a set of coordinate axes, Fig. E–1. The axes x and y (z is not shown) refer to S, and x' and y' refer to S′. The x' and x axes overlap one another, and we assume that frame S′ moves to the right (in the x direction) at speed v with respect to S. For simplicity let us assume the origins O and O′ of the two reference frames are superimposed at time $t = 0$.

Now consider an event that occurs at some point P (Fig. E–1) represented by the coordinates x', y', z' in reference frame S′ at the time t'. What will be the coordinates of P in S? Since S and S′ overlap precisely initially, after a time t, S′ will have moved a distance vt'. Therefore, at time t', $x = x' + vt'$. The y and z coordinates, on the other hand, are not altered by motion along the x axis; thus $y = y'$ and $z = z'$. Finally, since time is assumed to be absolute in Galilean–Newtonian physics, clocks in the two frames will agree with each other; so $t = t'$. We summarize these in the following **Galilean transformation** equations:

$$
\begin{aligned}
x &= x' + vt' \\
y &= y' \\
z &= z' \\
t &= t'.
\end{aligned}
\tag{E–1}
$$

Galilean transformations

These equations give the coordinates of an event in the S frame when those in the S′ frame are known. If those in the S system are known, then the S′ coordinates are obtained from

$$x' = x - vt, \qquad y' = y, \qquad z' = z, \qquad t' = t.$$

These four equations are the "inverse" transformation and are very easily obtained from Eqs. E–1. Notice that the effect is merely to exchange primed and unprimed quantities and replace v by $-v$. This makes sense because from the S′ frame, S moves to the left (negative x direction) with speed v.

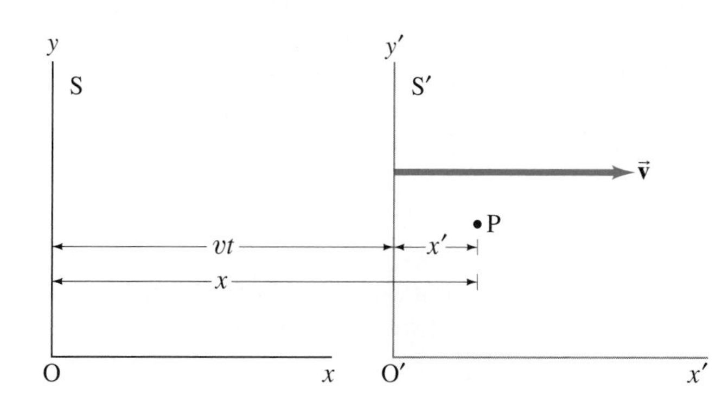

FIGURE E–1 Inertial reference frame S′ moves to the right at speed v with respect to inertial frame S.

Now suppose that the point P in Fig. E–1 represents an object that is moving. Let the components of its velocity vector in S′ be u'_x, u'_y, and u'_z (we use u to distinguish it from the relative velocity of the two frames, v). Now $u'_x = \Delta x'/\Delta t'$, $u'_y = \Delta y'/\Delta t'$, and $u'_z = \Delta z'/\Delta t'$, where all quantities are as measured in the S′ frame. For example, if at time t'_1 the particle is at x'_1 and a short time later, t'_2, it is at x'_2, then

$$u'_x = \frac{x'_2 - x'_1}{t'_2 - t'_1} = \frac{\Delta x'}{\Delta t'}.$$

Now the velocity of P as seen from S will have components u_x, u_y, and u_z. We can show how these are related to the velocity components in S′ by using Eqs. E–1. For example,

$$\begin{aligned} u_x = \frac{\Delta x}{\Delta t} = \frac{x_2 - x_1}{t_2 - t_1} &= \frac{(x'_2 + vt'_2) - (x'_1 + vt'_1)}{t'_2 - t'_1} \\ &= \frac{(x'_2 - x'_1) + v(t'_2 - t'_1)}{t'_2 - t'_1} \\ &= \frac{\Delta x'}{\Delta t'} + v = u'_x + v. \end{aligned}$$

For the other components, $u'_y = u_y$ and $u'_z = u_z$, so we have

Galilean velocity transformations

$$\begin{aligned} u_x &= u'_x + v, \\ u_y &= u'_y, \\ u_z &= u'_z. \end{aligned} \qquad \textbf{(E–2)}$$

These are known as the **Galilean velocity transformation** equations. We see that the y and z components of velocity are unchanged, but the x components differ by v. This is just what we have used before when dealing with relative velocity. For example, if S′ is a train and S the Earth, and the train moves with speed v with respect to Earth, a person walking toward the front of the train with speed u'_x will have a speed with respect to the Earth of $u_x = u'_x + v$.

Relativity theory

The Galilean transformations, Eqs. E–1 and E–2, are valid only when the velocities involved are not relativistic (Chapter 26)—that is, much less than the speed of light, c. We can see, for example, that the first of Eqs. E–2 will not work for the speed of light, c, which is the same in all inertial reference frames (a basic postulate in the theory of relativity). That is, light traveling in S′ with speed $u'_x = c$ will have speed $c + v$ in S, according to Eq. E–2, whereas the theory of relativity insists it must be c in S. Clearly, then, a new set of transformation equations is needed to deal with relativistic velocities.

We will derive the required equations in a simple way, again looking at Fig. E–1. We assume the transformation is linear and of the form

$$x = \gamma(x' + vt'), \qquad y = y', \qquad z = z'.$$

That is, we modify the first of Eqs. E–1 by multiplying by a factor γ which is yet to be determined. But we assume the y and z equations are unchanged because we expect no length contraction in these directions. We won't assume a form for t, but will derive it. The inverse equations must have the same form with v replaced by $-v$. (The principle of relativity demands it, since S′ moving to the right with respect to S is equivalent to S moving to the left with respect to S′.) Therefore

$$x' = \gamma(x - vt).$$

Now if a light pulse leaves the common origin of S and S′ at time $t = t' = 0$, after a time t it will have traveled along the x axis a distance

$x = ct$ (in S), or $x' = ct'$ (in S′). Therefore, from the equations for x and x' above,

$$ct = \gamma(ct' + vt') = \gamma(c + v)t',$$
$$ct' = \gamma(ct - vt) = \gamma(c - v)t.$$

We substitute t' from the second equation into the first and find $ct = \gamma(c + v)\gamma(c - v)(t/c) = \gamma^2(c^2 - v^2)t/c$. We cancel out the t on each side and solve for γ to find

$$\gamma = \frac{1}{\sqrt{1 - v^2/c^2}}.$$

Now that we have found γ, we need only find the relation between t and t'. To do so, we combine $x' = \gamma(x - vt)$ with $x = \gamma(x' + vt')$:

$$x' = \gamma(x - vt) = \gamma[\gamma(x' + vt') - vt].$$

We solve for t and find $t = \gamma(t' + vx'/c^2)$. In summary,

$$\begin{aligned} x &= \frac{1}{\sqrt{1 - v^2/c^2}}(x' + vt'), \\ y &= y', \\ z &= z', \\ t &= \frac{1}{\sqrt{1 - v^2/c^2}}\left(t' + \frac{vx'}{c^2}\right). \end{aligned} \qquad \textbf{(E–3)}$$

Lorentz transformations

These are called the **Lorentz transformation** equations. They were first proposed, in a slightly different form, by Lorentz in 1904 to explain the null result of the Michelson–Morley experiment and to make Maxwell's equations take the same form in all inertial systems. A year later, Einstein derived them independently based on his theory of relativity. Notice that not only is the x equation modified as compared to the Galilean transformation but so is the t equation. Indeed, we see directly in this last equation, as well as in the first, how the space and time coordinates mix.

The relativistically correct velocity equations are readily obtained. For example, using Eqs. E–3 (we let $\gamma = 1/\sqrt{1 - v^2/c^2}$),

$$\begin{aligned} u_x = \frac{\Delta x}{\Delta t} &= \frac{\gamma(\Delta x' + v\,\Delta t')}{\gamma(\Delta t' + v\,\Delta x'/c^2)} = \frac{(\Delta x'/\Delta t') + v}{1 + (v/c^2)(\Delta x'/\Delta t')} \\ &= \frac{u'_x + v}{1 + vu'_x/c^2}. \end{aligned}$$

The others are obtained in the same way, and we collect them here:

$$\begin{aligned} u_x &= \frac{u'_x + v}{1 + vu'_x/c^2}, \\ u_y &= \frac{u'_y\sqrt{1 - v^2/c^2}}{1 + vu'_x/c^2}, \\ u_z &= \frac{u'_z\sqrt{1 - v^2/c^2}}{1 + vu'_x/c^2}. \end{aligned} \qquad \textbf{(E–4)}$$

Relativistic velocity transformations

The first of these equations is Eq. 26–9, which we used in Section 26–11 where we discussed how velocities do not add in our commonsense (Galilean) way, because of the denominator $(1 + vu'_x/c^2)$. We can now also see that the y and z components of velocity are also altered and that they depend on the x' component of velocity.

EXAMPLE E–1 **Length contraction.** Derive the length contraction formula, Eq. 26–2, from the Lorentz transformation equations.

SOLUTION Let an object of length L_0 be at rest on the x axis in S. The coordinates of its two end points are x_1 and x_2, so that $x_2 - x_1 = L_0$. At any instant in S′, the end points will be at x_1' and x_2' as given by the Lorentz transformation equations. The length measured in S′ is $L = x_2' - x_1'$. An observer in S′ measures this length by measuring x_2' and x_1' at the same time (in the S′ frame), so $t_2' = t_1'$. Then, from the first of Eqs. E–3,

$$L_0 = x_2 - x_1 = \frac{1}{\sqrt{1 - v^2/c^2}}\left(x_2' + vt_2' - x_1' - vt_1'\right).$$

Since $t_2' = t_1'$, we have

$$L_0 = \frac{1}{\sqrt{1 - v^2/c^2}}\left(x_2' - x_1'\right) = \frac{L}{\sqrt{1 - v^2/c^2}},$$

or

$$L = L_0\sqrt{1 - \frac{v^2}{c^2}},$$

which is Eq. 26–2.

EXAMPLE E–2 **Time dilation.** Derive the time dilation formula, Eq. 26–1, from the Lorentz transformation equations.

SOLUTION The time Δt_0 between two events that occur at the same place $(x_2' = x_1')$ in S′ is measured to be $\Delta t_0 = t_2' - t_1'$. Since $x_2' = x_1'$, then from the last of Eqs. E–3, the time Δt between the events as measured in S is

$$\Delta t = t_2 - t_1 = \frac{1}{\sqrt{1 - v^2/c^2}}\left(t_2' + \frac{vx_2'}{c^2} - t_1' - \frac{vx_1'}{c^2}\right)$$
$$= \frac{1}{\sqrt{1 - v^2/c^2}}\left(t_2' - t_1'\right)$$
$$= \frac{\Delta t_0}{\sqrt{1 - v^2/c^2}},$$

which is Eq. 26–1. Notice that we chose S′ to be the frame in which the two events occur at the same place, so that $x_1' = x_2'$, and the terms containing x_1' and x_2' cancel out.

Answers to Odd-Numbered Problems

CHAPTER 1

1. (*a*) 1.4×10^{10} years;
(*b*) 4.4×10^{17} s.

3. (*a*) 1.156×10^{0};
(*b*) 2.18×10^{1};
(*c*) 6.8×10^{-3};
(*d*) 2.7635×10^{1};
(*e*) 2.19×10^{-1};
(*f*) 4.44×10^{2}.

5. 1%.

7. (*a*) 4%;
(*b*) 0.4%;
(*c*) 0.07%.

9. 1.7 m.

11. 9%.

13. (*a*) 1 megavolt;
(*b*) 2 micrometers;
(*c*) 6 kilodays;
(*d*) 18 hectobucks;
(*e*) 8 nanopieces.

15. (*a*) 1.5×10^{11} m;
(*b*) 150 gigameters.

17. 3.8 s.

19. 3.76 m.

21. 7.3%.

23. (*a*) 3.80×10^{13} m^2;
(*b*) 13.4.

25. $\approx 7 \times 10^5$ books.

27. ≈ 11 hr.

29. 8×10^4 cm^3.

31. 4×10^8 kg/yr.

33. (*a*) Cannot be correct;
(*b*) can be correct;
(*c*) can be correct.

35. 50,000 chips.

37. 2×10^{-4} m.

39. (*a*) 10^{12} protons or neutrons;
(*b*) 10^{10} protons or neutrons;
(*c*) 10^{29} protons or neutrons;
(*d*) 10^{68} protons or neutrons.

41. 1500 gumballs.

43. ≈ 3 ft.

45. ≈ 3500 km.

47. 150 m long, 25 m wide, 15 m high; 6×10^4 m^3.

49. 210 yd, 190 m.

51. 2.21×10^{19} m^3, 49.3 Moons.

53. (*a*) 3%, 3%;
(*b*) 0.7%, 0.2%.

CHAPTER 2

1. 72.3 km/h.

3. 61 m.

5. -2.5 cm/s.

7. (*a*) 2.6×10^2 km;
(*b*) 77 km/h.

9. (*a*) 4.3 m/s;
(*b*) 0 m/s.

11. 2.7 min.

13. 6.8 h, 8.7×10^2 km/h.

15. 6.73 m/s.

17. (*a*) 7.41 m/s^2;
(*b*) 9.60×10^4 km/h^2.

19. -5.5 m/s^2, -0.56 *g*'s.

21. 2.0 m/s^2, 114 m.

23. 1.8×10^2 m.

25. 63.0 m.

27. -36 *g*'s.

31. 3.1 s.

33. 51.8 m.

35. (*a*) 8.8 s;
(*b*) 86 m/s.

37. 15 m/s, 11 m.

39. 5.61 s.

43. 4.1×10^{-2} s.

45. 46 m.

47. (*a*) 5.20 s;
(*b*) 38.9 m/s;
(*c*) 84.7 m.

49. (*a*) 48 s;
(*b*) 90 s to 108 s;
(*c*) 0 s to 38 s, 65 s to 83 s, 90 s to 108 s;
(*d*) 65 s to 83 s.

51. (*a*) 0 s to 18 s;
(*b*) 27 s;
(*c*) 38 s;
(*d*) both directions.

53. (*a*) 4 m/s^2;
(*b*) 3 m/s^2;
(*c*) 0.35 m/s^2;
(*d*) 1.6 m/s^2.

55.

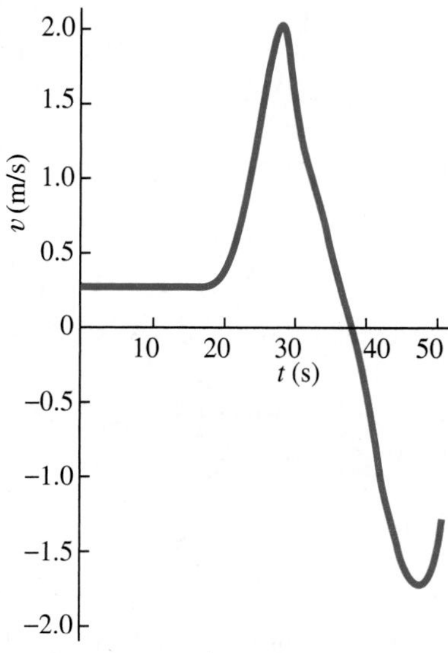

57. (*a*) -150 m/s^2;
(*b*) loosen.

59. 1.3 m.

61. (*b*) 14 m;
(*c*) 39.4 m.

63. 31 m/s.

65. (*a*) 8.8 min;
(*b*) 7.5 min.

67. 4.9 m/s to 5.7 m/s, 6.0 m/s to 6.9 m/s, smaller range of initial velocities.

69. 29.0 m.

71. 5.1×10^{-2} m/s^2.

73. 3.3 min; 5.2 km; 23.3 s, 0.61 km.

75. (*a*) 88 m/s;
(*b*) 27 s;
(*c*) 1590 m;
(*d*) 36 s;
(*e*) -177 m/s;
(*f*) 54 s.

77.
(*a*)

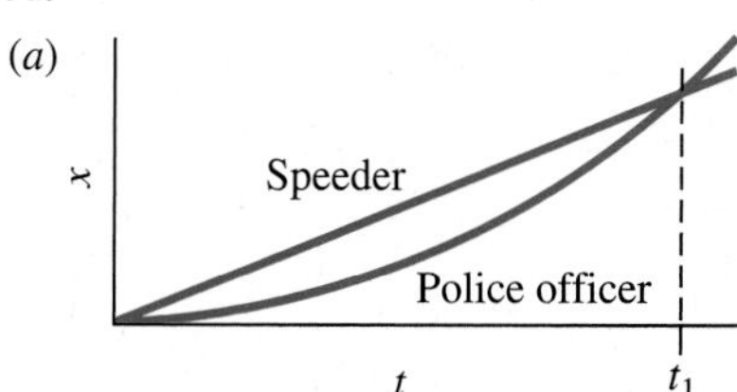

(*b*) 23 s;
(*c*) 3.0 m/s^2;
(*d*) 67 m/s.

79. 18 m/s.

81. 0.44 m/min, 2.9 burgers/min.

83. 12 m/s.

85. (*a*) Near the midpoint of the time interval;
(*b*) A;
(*c*) at the times when the two graphs cross; at the first crossing, bicycle B is passing bicycle A; at the second crossing, bicycle A is passing bicycle B;
(*d*) A;
(*e*) they have the same average velocity.

CHAPTER 3

1. 282 km, 12° south of west.

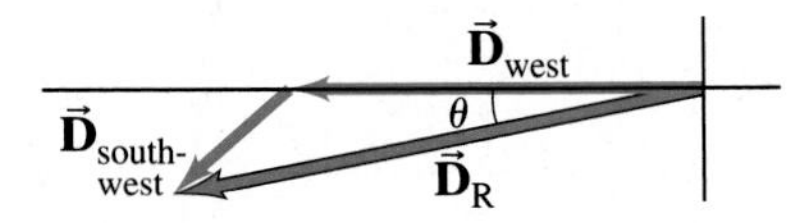

3. $\vec{V}_2 - \vec{V}_1$.

5. 58 m, 48°. $\vec{V}_R = \vec{V}_1 + \vec{V}_2 + \vec{V}_3$

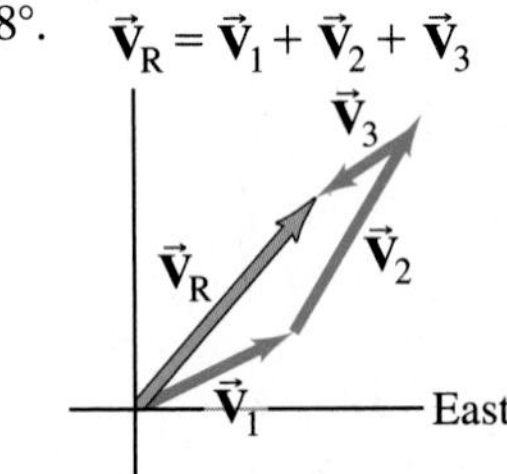

7. (*a*)

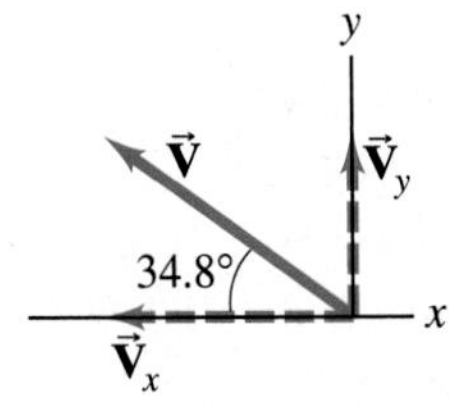

(*b*) −11.7 units, 8.16 units;
(*c*) 14.3 units, 34.8° above the $-x$ axis.

9. (*a*) 550 km/h, 487 km/h;
(*b*) 1650 km, 1460 km.

11. 64.6, 53.1°.

13. (*a*) 62.6, 329°;
(*b*) 77.5, 71.9°;
(*c*) 77.5, 251.9°.

15. −2450 m, 3870 m, 2450 m; 5190 m.

17. 4.0 m.

19. 13° and 77°.

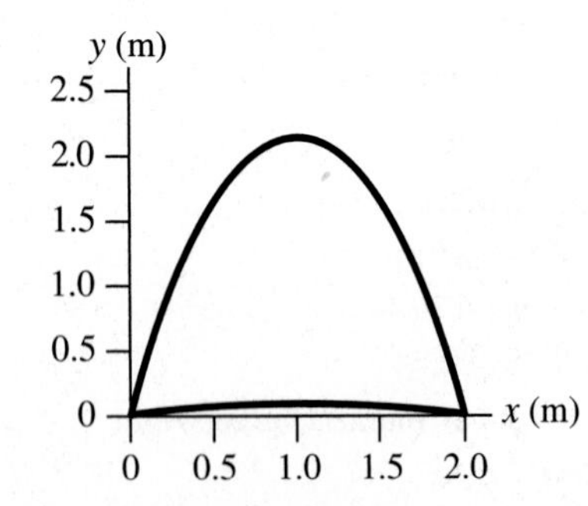

21. 7.92 m/s.

23. 12.9 m.

25. 6 times farther.

27. 5.71 s.

29. The football will not clear the bar. It is 0.76 m too low when it reaches the goal post.

31. (*a*) 10.4 s;
(*b*) 541 m;
(*c*) 51.9 m/s, −63.1 m/s;
(*d*) 81.7 m/s;
(*e*) 50.6° below the horizon;
(*f*) 78.1 m.

33. 76°.

35. (*a*) 481 m;
(*b*) 8.37 m/s down;
(*c*) 97.4 m/s.

37. 1.8 m/s, 19° to the river bank.

39. (*a*) 2.59 m/s, 62° from the shore;
(*b*) 3.60 m downstream, 6.90 m across the river.

41. (*a*) 543 km/h, 7.61° east of south;
(*b*) 17 km.

43. 1.41 m/s.

45. (*a*) 1.24 m/s;
(*b*) 2.28 m/s.

47. (*a*) 67 m;
(*b*) 170 s.

49. 42.2° north of east.

51. 114 km/h.

53. 6.2°.

55. 4.7 m/s² left (opposite to the truck's motion), 2.8 m/s² down.

57. $v_T/\tan\theta$.

59. 180 s, 4.8 km; 21.2 s, 0.56 km.

61. 1.9 m/s².

63. 1.9 m/s, 2.7 s.

65. 49.6°.

67. 63 m/s, 66° above the horizontal.

69. 10.8 m/s to 11.0 m/s.

71. (*a*) 36 m/s;
(*b*) 20 m/s.

73. 7.0 m/s, 97°.

75. 39 m.

CHAPTER 4

1. 75.0 N.

3. 1.15×10^3 N.

5. (*a*) 196 N, 196 N;
(*b*) 294 N, 98.0 N.

7. 68.4 N.

9. 780 N, backward.

11. 2.00 g's, 9.51×10^3 N.

13. 5.08×10^4 N, 4.43×10^4 N.

15. 2.5 m/s², down.

17. (*a*) 7.4 m/s², down;
(*b*) 1.29×10^3 N.

19. (*a*) 47.0 N;
(*b*) 17.0 N;
(*c*) 0 N.

21.

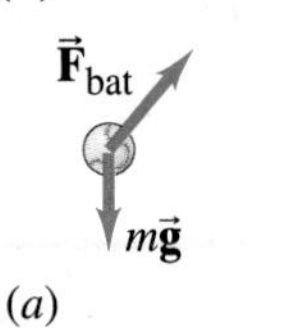

23. 1.41×10^3 N.

25. (*a*) 63 N, 31 N;
(*b*) 73 N, 36 N.

27. 6.9×10^3 N, 8.9×10^3 N.

29. (*a*) 320 N;
(*b*) 1.5 m/s².

31. (*a*)

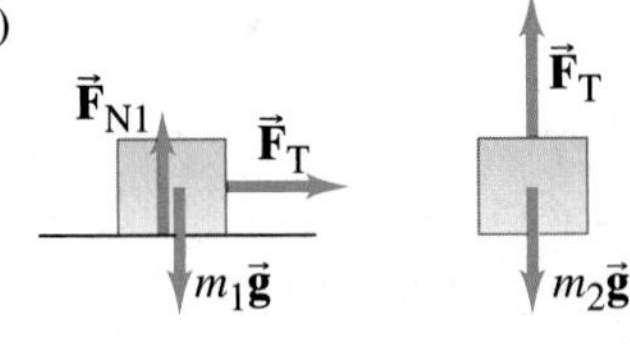

(*b*) $a = g\dfrac{m_2}{m_1 + m_2}$,

$F_T = m_1 a = g\dfrac{m_1 m_2}{m_1 + m_2}$.

33. (*a*)

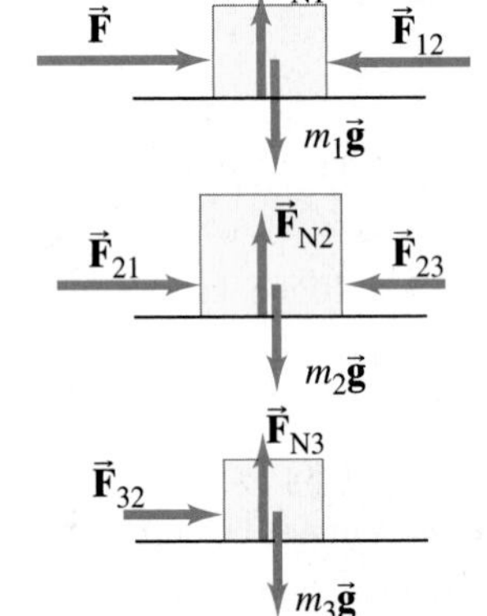

(*b*) $a = \dfrac{F}{m_1 + m_2 + m_3}$;

(*c*) $F_{1\,net} = \dfrac{m_1 F}{m_1 + m_2 + m_3}$,

$F_{2\,net} = \dfrac{m_2 F}{m_1 + m_2 + m_3}$,

$F_{3\,net} = \dfrac{m_3 F}{m_1 + m_2 + m_3}$;

(*d*) $F_{12} = F_{21} = \dfrac{(m_2 + m_3)F}{m_1 + m_2 + m_3}$,

$F_{23} = F_{32} = \dfrac{m_3 F}{m_1 + m_2 + m_3}$;

(*e*) 2.67 m/s²; 32.0 N, 64.0 N, 32.0 N.

35. $1.74\,m/s^2$, 22.6 N, 20.9 N.
37. (*a*) 0.98;
(*b*) 0.91.
39. $7.8\,m/s^2$.
41. 73 N, 0.59.
43. (*a*)

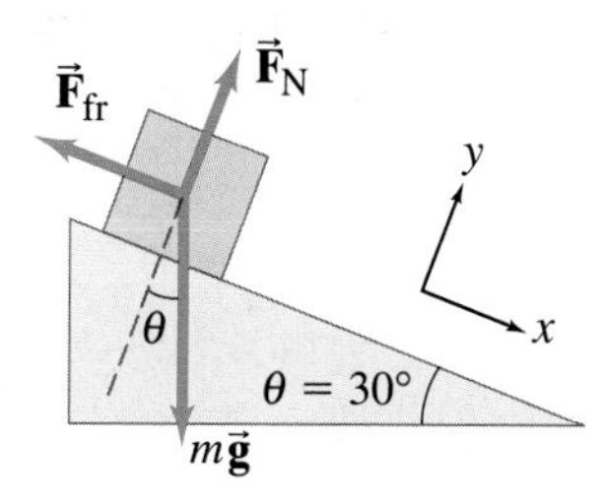

(*b*) No change;
(*c*) friction force direction would be reversed.
45. 40 N.
47. 4.1 m.
49. $-7.4\,m/s^2$.
51. 0.40.
53. (*a*) 1.2 m;
(*b*) 1.6 s.
55. 101 N, 0.719.
57. (*a*) 0.58;
(*b*) 5.7 m/s;
(*c*) 15 m/s.
59. 0.36.
61. 5.3×10^2 N.
63. (*a*) $g\dfrac{(m_1 \sin\theta - m_2)}{(m_1 + m_2)}$;
(*b*) $m_1 \sin\theta > m_2$ (down the plane), $m_1 \sin\theta < m_2$ (up the plane).
65. 1.3×10^2 N.
67. 1.3 m.
69. 1.54×10^3 N.
71. (*a*) 16 m/s;
(*b*) 13 m/s.
73. Yes, 3.8 m/s.
75. 82 m/s.
77. 5.9°.
79. 940 N, 79° above the horizontal.
81. (*a*) 9.43×10^4 N;
(*b*) 1.33×10^4 N;
(*c*) 1.33×10^4 N.
83. 12 m/s.
85. (*a*) 45 N (10 lb);
(*b*) 37 N (8.4 lb);
(*c*) not when pulled vertically.
87. (*a*) $4.1\,m/s^2$, $3.2\,m/s^2$;
(*b*) $4.1\,m/s^2$, $3.2\,m/s^2$;
(*c*) $3.5\,m/s^2$.
89. 5.3×10^2 N, 2.6×10^2 N.

CHAPTER 5

1. (*a*) $1.42\,m/s^2$;
(*b*) 35.5 N.
3. $5.97 \times 10^{-3}\,m/s^2$, 3.56×10^{22} N, the Sun.
5. 0.9 *g*'s.
7. (*a*) 3.73 N;
(*b*) 9.61 N.
9. 25 m/s, yes.
11. 30.4 m/s, 0.403 rev/s.
13. 8.5 m/s.
15. 11 rpm.
17. 3.38×10^4 rpm.
21. 0.22.
23. $4\pi^2 f^2(m_1 r_1 + m_2 r_2)$, $4\pi^2 m_2 r_2 f^2$.
25. 3.5×10^3 N, 5.0×10^2 N.
27. (*a*) 1.27 m/s;
(*b*) 3.05 m/s.
29. (*a*) 21.0 kg, 21.0 kg;
(*b*) 206 N, 252 N.
31. $4.4\,m/s^2$.
33. 3.9 kg, 0.1 kg.
35. 2.02×10^7 m.
37. $4.38 \times 10^7\,m/s^2$.
39. 3.2×10^{-8} N toward center of square.
41. 6.4×10^{23} kg.
43. 6.32×10^3 m/s.
45. 10 s/rev.
47. 7.90×10^3 m/s.
49. 2.0×10^4 s, 7.1×10^4 s.
51. (*a*) 21 N, toward the Moon;
(*b*) 2.0×10^2 N, away from Moon.
53. (*a*) 5.4×10^2 N;
(*b*) 5.4×10^2 N;
(*c*) 7.2×10^2 N;
(*d*) 3.6×10^2 N;
(*e*) 0 N.
55. (*b*) $5.4 \times 10^3\,kg/m^3$.
57. 1.62×10^{11} m.
59. 2690×10^6 km, yes, Pluto.
61. (*a*) 1.90×10^{27} kg;
(*b*) 1.90×10^{27} kg, 1.89×10^{27} kg, 1.90×10^{27} kg, yes.
63. 671×10^3 km, 1070×10^3 km, 1880×10^3 km.
65. 9.0 d.
67. 2.64×10^6 m.
69. 0.344%.
71. $2.6\,m/s^2$ upward.
73. (*a*) 2.2×10^3 m;
(*b*) 5.4×10^3 N;
(*c*) 3.8×10^3 N.
75. (*a*) $\theta = \tan^{-1} m_M R^2_{Earth}/M_{Earth} D^2_M$;
(*b*) 5×10^{13} kg;
(*c*) $(8 \times 10^{-4})^\circ$.
77. 5.07×10^3 s.
79. 26.9 m/s.
81. 5.2×10^{39} kg, 2.6×10^9 solar masses.
83. (*a*) 3.86×10^3 m/s;
(*b*) 4.36×10^4 s.
85. (*a*) ≈12 h;
(*b*) 1.8×10^3 m.
87. $5 \times 10^{-5}\,N\cdot m^2/kg^2$.
89. 3.8×10^{-10} N, upward.
91. $1.6 \times 10^{-4}\,m/s^2$.
93. $v_{min} = v_0\sqrt{\dfrac{(1 - \mu_s Rg/v_0^2)}{(1 + \mu_s v_0^2/Rg)}}$,
$v_{max} = v_0\sqrt{\dfrac{(1 + Rg\mu_s/v_0^2)}{(1 - \mu_s v_0^2/Rg)}}$.

CHAPTER 6

1. 7.27×10^3 J.
3. (*a*) 9.2×10^2 J;
(*b*) 5.2×10^3 J.
5. 4.9×10^2 J.
9. (*a*) 1.10*Mg*;
(*b*) 1.10*Mgh*.
11. 5.0×10^3 J.
13. 8.4×10^{-2} J.
15. 484 m/s.
17. -1.64×10^{-18} J.
19. 44 m/s.
21. 2.25.
23. 1.1 N.
25. (*a*) 3.24×10^3 N;
(*b*) 9.83×10^3 J;
(*c*) 7.13×10^4 J;
(*d*) -6.14×10^4 J;
(*e*) 8.31 m/s.
27. 82 J.
29. 8.1×10^4 N/m.
31. (*a*) 9.2×10^5 J;
(*b*) 9.2×10^5 J;
(*c*) yes.
33. 1.4 m, no unless length <0.7 m.
35. 5.14 m/s.
37. (*a*) 9.2 m/s;
(*b*) −0.31 m.

39. (*a*) 8.3 m/s;
    (*b*) 3.64 m.
41. $\frac{1}{2}mv^2 + \frac{1}{2}kx^2 = \frac{1}{2}kx_0^2$.
43. 26 m/s, 12 m/s, 20 m/s.
45. 12 Mg/h.
47. 5.3×10^6 J.
49. (*a*) 21 m/s;
    (*b*) 2.4×10^2 m.
51. (*a*) 25%;
    (*b*) 5.4 m/s;
    (*c*) heat, sound, non-elastic deformation.
53. 23 m/s.
55. 0.40.
57. (*a*) 1.1×10^3 km/h;
    (*b*) 2×10^3 N.
59. 5.5×10^2 N.
61. (*b*) 0.10 hp.
63. 2.2×10^4 W, 3.0×10^1 hp.
65. 480 W.
67. 1.0×10^3 W.
69. 18°.
71. 9.0×10^2 W.
73. 1.5×10^3 J.
75. (*a*) 2.5*r*;
    (*b*) 11*mg*;
    (*c*) 5*mg*;
    (*d*) *mg*.
77. (*a*) $\sqrt{2gL}$;
    (*b*) $\sqrt{1.2\,gL}$.
79. (*a*) 2.5×10^5 J;
    (*b*) 23 m/s;
    (*c*) −1.56 m.
81. (*a*) 4.0×10^1 m/s;
    (*b*) 3.0×10^5 W.
83. (*a*) 1.4×10^3 m;
    (*b*) 1.6×10^2 m/s.
85. 4.2×10^4 N.
87. 3.9×10^2 W.
89. 2*k*.
91. 4.6 s.
93. (*a*) 1×10^2 m/s;
    (*b*) 4×10^7 W.

CHAPTER 7

1. 0.24 kg·m/s.
3. 4.40×10^3 N toward the pitcher.
5. 6.0×10^7 N upward.
7. 12.6 m/s.
9. 8×10^2 N, $F_{wind} > F_{fr} \approx 7 \times 10^2$ N.
11. 4.2×10^3 m/s.
13. (*a*) 6.9×10^3 m/s away from Earth, 4.7×10^3 m/s away from Earth;
    (*b*) 5.9×10^8 J.
15. (*a*) 2.0 kg·m/s;
    (*b*) 5.8×10^2 N.
17. 2.1 kg·m/s to the left.
19. (*a*) 3.8×10^2 kg·m/s;
    (*b*) -3.8×10^2 kg·m/s;
    (*c*) 3.8×10^2 kg·m/s;
    (*d*) 5.1×10^2 N.
21. 69 m.
23. 1.00 m/s west, 2.00 m/s east.
25. 0.88 m/s and 2.23 m/s, both in direction of tennis ball's initial motion.
27. (*a*) 3.62 m/s, 4.42 m/s;
    (*b*) -4.0×10^2 kg·m/s, 4.0×10^2 kg·m/s.
29. 0.35 m, 1.4 m.
31. $v_2 = \sqrt{2}\,v_1$.
33. (*a*) $-M/(m + M)$;
    (*b*) −0.96.
35. 23 m/s.
37. (*b*) $e = \sqrt{h'/h}$.
39. (*a*) 1.7 m/s for both;
    (*b*) −2.1 m/s, 7.4 m/s;
    (*c*) 0, 4.3 m/s, reasonable;
    (*d*) 2.8 m/s, 0, not reasonable;
    (*e*) −4.0 m/s, 10.3 m/s, not reasonable.
41. 60° relative to eagle A, 6.7 m/s.
43. 141°.
45. 39.9 u.
47. 6.5×10^{-11} m.
49. (1.04 m, −1.04 m) relative to raft's center.
51. (1.2*l*, 0.9*l*) relative to back left corner.
53. 17% of the whole body mass.
55. 21.7 cm horizontal, 7.6 cm vertical.
57. (*a*) 4.66×10^6 m from center of Earth.
59. 24.8 cm.
61. $vm/(m + M)$ upward, the balloon stops.
63. $v'_x = \frac{3}{2}v_0$, $v'_y = -v_0$.
65. (*a*) 0.194 m/s;
    (*b*) 8.8×10^2 N.
67. $m_B = \frac{5}{3}m$.
69. 4.00 m.
71. 3.8×10^2 m/s.
73. (*a*) 2.5×10^{-13} m/s;
    (*b*) 1.7×10^{-17};
    (*c*) 0.19 J.
75. (*a*)

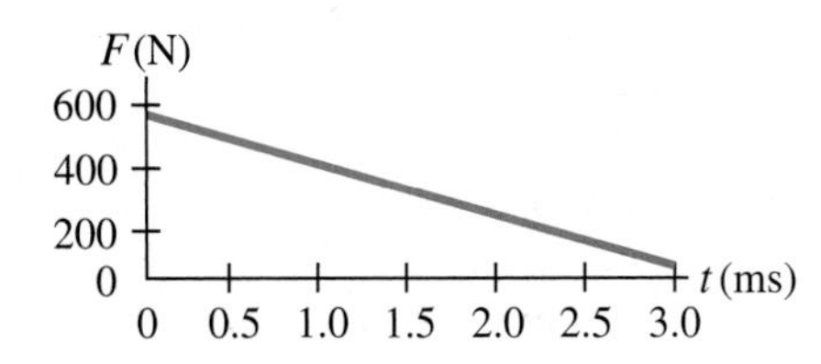

    (*b*) 0.93 N·s;
    (*c*) 4.2×10^{-3} kg.
77. 6.7×10^3 m/s.
79. (*a*) −4.4 m/s, 4.0 m/s;
    (*b*) 2.0 m.
81. −29.6 km/s.

CHAPTER 8

1. (*a*) 0.52 rad, $\pi/6$ rad;
   (*b*) 0.99 rad, $19\pi/60$ rad;
   (*c*) 1.57 rad, $\pi/2$ rad;
   (*d*) 6.28 rad, 2π rad;
   (*e*) 7.33 rad, $7\pi/3$ rad.
3. 5.3×10^3 m.
5. 7.4×10^{-2} m.
7. (*a*) 2.6×10^2 rad/s;
   (*b*) 46 m/s, 1.2×10^4 m/s^2.
9. (*a*) 1.99×10^{-7} rad/s;
   (*b*) 7.27×10^{-5} rad/s.
11. 3.6×10^4 rpm.
13. $\omega_1/\omega_2 = R_2/R_1$.
15. 2.8×10^4 rev.
17. (*a*) 4.0×10^1 rev/min^2;
    (*b*) 4.0×10^1 rpm.
19. (*a*) −0.42 rad/s^2;
    (*b*) 210 s.
21. (*a*) −4.1 rad/s^2;
    (*b*) 7.6 s.
23. (*a*) 41 m·N;
    (*b*) 29 m·N.
25. $mg(L_2 - L_1)$, clockwise.
27. 1.81 kg·m^2.
29. (*a*) 0.94 kg·m^2;
    (*b*) 2.4×10^{-2} m·N.
31. (*a*) 6.1 kg·m^2;
    (*b*) 0.61 kg·m^2;
    (*c*) vertical axis.
33. 20 N.
35. 62 m·N.
37. 993 rev, 10.9 s.
39. (*a*) 92 rad/s^2;
    (*b*) 7.9×10^2 N.

41. $a = \frac{(m_2 - m_1)}{(m_1 + m_2 + I/r^2)} g < a_{I=0}$,
$a_{I=0} = \frac{(m_2 - m_1)}{(m_1 + m_2)} g$.

43. 1.40×10^4 J.

45. 56 J.

47. 1.42×10^4 J.

49. 3.22 m/s.

51. 2.64 kg·m²/s.

53. (a) His rotational inertia increases;
(b) 1.6.

55. 0.77 kg·m², by pulling her arms in toward the center of her body.

57. (a) 14 kg·m²/s;
(b) −2.7 m·N.

59. $\omega/2$.

61. (a) 1.2 rad/s;
(b) 1.8×10^3 J, 1.1×10^3 J.

63. 5×10^{-2} rad/s, 2×10^4 KE_i.

65. (2.7×10^{-16})%.

67. −0.30 rad/s.

69. 8.21×10^{-6}.

71. 53 m·N.

73. (a) $\omega_R/\omega_F = N_F/N_R$.
(b) 4.0;
(c) 1.5.

75. (b) 2.2×10^3 rad/s; (c) 25 min.

77. (a) 4.3 m;
(b) 5.2 s.

79. $Mg\sqrt{2Rh - h^2}/(R - h)$.

81. 2.8 m·N, from his arm muscles.

83. (a) 7.8 kg·m²/s;
(b) 3.9 m·N;
(c) 2.9 rad/s.

85. $2.7(R - r)$.

CHAPTER 9

1. 430 N, 112° clockwise from $\vec{F}_A$.

3. 6.52 kg.

5. 1.1×10^3 N.

7. 5.8×10^3 N, 8.1×10^3 N.

9. (a) 2.3 m from adult;
(b) 2.5 m from adult.

11. 2.6×10^3 N, 3.1×10^3 N.

13. 0.32 m.

15. 6.1×10^3 N, 5.9×10^3 N.

17. 34.6 N.

19. 9.05×10^{-1} m.

21. (a) 4.25×10^2 N;
(b) 4.25×10^2 N, 3.28×10^2 N.

23. (a) 0.78 N;
(b) 0.98 N.

25. 55.2 N, 63.7 N.

27. 0.50.

29. 1.0×10^2 N.

31. 9.9×10^2 N.

33. 2.7×10^3 N.

35. $2.4w$.

37. (b) Yes, by 1/24 of a brick length;
(c) $D = \sum_{i=1}^{n} \frac{L}{2i}$;
(d) 35 bricks.

39. (a) 2.0×10^5 N/m²;
(b) 4.1×10^{-6}.

41. (a) 1.4×10^5 N/m²;
(b) 6.9×10^{-7};
(c) 6.5×10^{-6} m.

43. 9.6×10^6 N/m².

45. (-2×10^{-2})%.

47. (a) 1.1×10^2 m·N, clockwise;
(b) the wall.

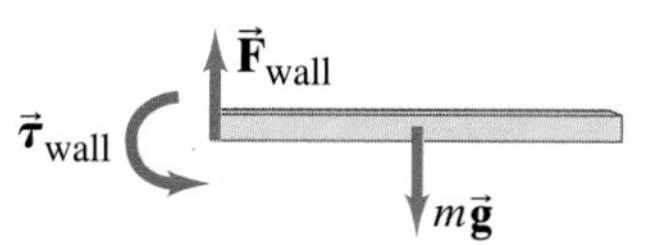

49. (a) 393 N;
(b) thicker.

51. (a) 4.4×10^{-5} m²;
(b) 2.7×10^{-3} m.

53. 1.2×10^{-2} m.

55. 12 m.

57. 2.94×10^{-1} kg, 2.29×10^{-1} kg, 6.56×10^{-2} kg.

59. (a) $Mg\sqrt{h/(2R - h)}$;
(b) $Mg\sqrt{h(2R - h)}/(R - h)$.

61. (a)

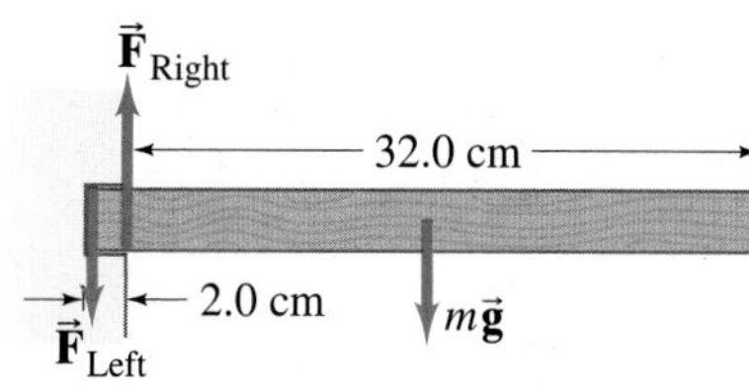

(b) $F_{Left} = 3.7 \times 10^2$ N,
$F_{Right} = 4.2 \times 10^2$ N,
$mg = 49$ N;
(c) 8.3 m·N.

63. 29°.

65. 7.7×10^{-6} m.

67. (a) $0.29mg$;
(b) $0.58mg$;
(c) horizontal at lowest point, 60° above the horizontal at points of attachment.

69. (a) $\mu_s < l/2h$;
(b) $\mu_s > l/2h$.

71. (a) $F_{Left} = 3.3 \times 10^2$ N up,
$F_{Right} = 2.3 \times 10^2$ N down;
(b) 0.65 m;
(c) 1.2 m.

73. $F_{Left} = 1.0 \times 10^2$ N,
$F_{Right} = 1.9 \times 10^2$ N.

75. Average force per area = 4.5×10^5 N/m².

77. (a) 3.5×10^8 N/m²;
(b) the bone will break;
(c) 8.2×10^6 N/m², the bone will not break.

79. 2.34 m.

CHAPTER 10

1. 3×10^{11} kg.

3. 5.8×10^2 kg.

5. 0.8477.

7. (a) 7×10^7 N/m²;
(b) 2×10^5 N/m².

9. (a) 4.7×10^5 N;
(b) 4.7×10^5 N.

11. 2.2×10^3 kg.

13. 13 m.

15. 1.60×10^4 m.

17. (a) 9.6×10^5 N/m²;
(b) 98 m.

19. (a) 1.41×10^5 Pa;
(b) 9.8×10^4 Pa.

21. 1.06×10^3 kg/m³, 3% higher.

23. 0.199.

25. 920 kg.

27. Iron or steel.

29. (a) 7.4×10^5 N;
(b) 1.0×10^4 N.

31. (a) 1.03×10^3 kg/m³;
(b) $\rho_{liquid} = \rho_{object}(m_{object} - m_{apparent})/m_{object}$.

33. 0.105.

35. 0.90 m/s.

39. 4.4×10^5 s (5.1 days).

41. 5.6×10^{-3} m³/s.

43. 1.9×10^5 N.

45. 9.7×10^4 Pa (≈0.96 atm).

47. (b) 0.24 m/s.

49. (a) $2\sqrt{h_1(h_2 - h_1)}$;
(b) $h_1' = h_2 - h_1$.

51. new time = 0.13 (previous time).

53. 9.9×10^2 Pa.

55. 0.9 Pa/cm.

57. (*a*) $Re = 2500$, so turbulent;
(*b*) $Re = 5000$, so turbulent.

59. 3.6×10^{-2} N/m.

61. No, 8.3×10^{-6} kg is the maximum mass that could be supported.

63. (*a*) 0.75 m;
(*b*) 0.65 m;
(*c*) 0.24 m.

65. 150 N to 220 N.

67. 0.047 atm.

69. 0.6 atm.

71. 0.142 m.

73. 1.3×10^2 N.

75. 1.1 m.

77. 0.33 kg.

79. 1.1 W.

81. 4.6 m.

83. (*a*) 9.1 m/s;
(*b*) 0.26 L/s;
(*c*) 0.91 m/s.

85. 4.0×10^{-4} m^3/s.

87. 4.2×10^{-3} Pa·s.

CHAPTER 11

1. 0.72 m.

3. 1.5 Hz.

5. 3.8 Hz.

7. (*a*) 0.16 N/m;
(*b*) 2.8 Hz.

9. (*a*) 2.5 m/s;
(*b*) ± 1.6 m/s;
(*c*) 1.8 J;
(*d*) $x = (0.13\text{ m})\cos(6.0\pi t)$.

11. $\pm \frac{1}{2} x_0$.

13. (*a*) 6.0×10^{-2} m;
(*b*) 0.58 m/s.

15. (*a*) 4.2×10^2 N/m;
(*b*) 3.3 kg.

17. $\pm 0.707A$.

19. (*a*) $y = (0.18\text{ m})\cos(2\pi t/0.65\text{ s})$;
(*b*) 0.16 s;
(*c*) 1.7 m/s;
(*d*) 17 m/s^2, at the release point.

21. (*a*) 0.38 m;
(*b*) 1.03 Hz;
(*c*) 0.967 s;
(*d*) 0.92 J;
(*e*) 5.1×10^{-2} J, 0.86 J.
(*f*)

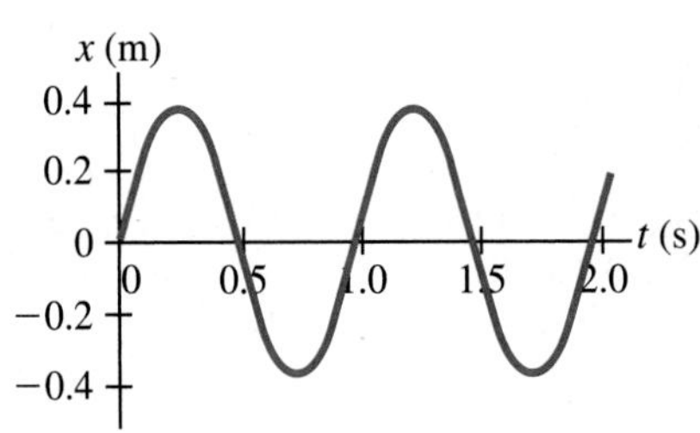

23. (*a*) 0.490 s, 2.04 Hz;
(*b*) 0.231 m;
(*c*) 37.9 m/s^2;
(*d*) $y = (0.231\text{ m})\sin(4.08\pi t)$;
(*e*) 3.31 J.

27. 114 N/m, 19.4 m.

29. 0.99 m.

31. (*a*) 1.8 s;
(*b*) the pendulum will not oscillate.

33. Shorten the pendulum by 0.7 mm.

35. (*a*) $-11°$;
(*b*) 15°;
(*c*) 15°.

37. 1.31 m.

39. (*a*) 1.4×10^3 m/s;
(*b*) 4.1×10^3 m/s;
(*c*) 5.1×10^3 m/s.

41. 0.35 s.

43. 2.1×10^3 m.

45. 0.99 m.

47. (*a*) 4.6×10^9 W/m^2;
(*b*) 2.3×10^{10} W.

49. 1.73.

51. (a)

(b)

(*c*) All the energy is kinetic energy.

53. 441 Hz.

55. 9.7×10^{-2} m.

57. 290 Hz, 580 Hz, 870 Hz.

59. (*a*) 1.3 kg;
(*b*) 0.32 kg;
(*c*) 5.2×10^{-2} kg.

61. 1.1 m/s.

63. 25°.

65. 44°.

67. 10 min.

69. (*a*) 3.2×10^{-2} m;
(*b*) 1.5 m.

71. (*a*) 1.8×10^4 N/m;
(*b*) 0.71 s.

73. 220 Hz.

75. (*a*) $1.22f$;
(*b*) $0.71f$.

77. (*a*) G: 784 Hz, 1180 Hz; A: 880 Hz, 1320 Hz;
(*b*) 1.26;
(*c*) 1.12;
(*d*) 0.794.

79. (*a*) 3.0 m/s;
(*b*) 5.0×10^3 m/s^2.

81. $\lambda = 4L/(2n - 1), n = 1, 2, 3, \cdots$.

83. Horizontal period is longer by a factor of $\sqrt{1 + l_0 k/mg}$.

85. 6.44 m from the origin of the first pulse.

87. 0.40 s.

CHAPTER 12

1. 3.4×10^2 m.

3. (*a*) 17 cm to 17 m;
(*b*) 3.4×10^{-5} m.

5. 55 m.

7. (*a*) 8%;
(*b*) 4%.

9. 63 dB.

11. 114 dB.

13. (*a*) 9×10^{-6} W;
(*b*) 1×10^7 people.

15. (*a*) 122 dB, 114 dB;
(*b*) no.

17. 1.3.

19. 4.

21. 25 dB.

23. (*a*) 10^9;
(*b*) 10^{12}.

25. (*a*) 76.6 Hz, 230 Hz, 383 Hz, 536 Hz;
(*b*) 153 Hz, 306 Hz, 459 Hz, 613 Hz.

27. 8.6 mm to 8.6 m.

29. (*a*) 0.18 m;
(*b*) 1.1 m;
(*c*) 440 Hz, 0.78 m.

31. -2.6%.

33. (*a*) 0.583 m;
(*b*) 862 Hz.

35. (*a*) 55 Hz;
(*b*) 2.0×10^2 m/s.

37. (*a*) 248 overtones;
(*b*) 249 overtones.

39. ± 0.50 Hz.

41. 28.5 kHz.

43. 3.0 Hz.

45. $f_A = 438$ Hz or 444 Hz,
$f_C = 437$ Hz or 445 Hz,
$f_{beat} = 1$ Hz or 7 Hz.

47. (a) 130.5 Hz, 133.5 Hz;
(b) increase by 2.3%, decrease by 2.2%.

49. (a) 1690 Hz;
(b) 1410 Hz.

51. (a) 2091 Hz and 2087 Hz;
(b) 3550 Hz and 2870 Hz;
(c) 16,000 Hz and 3750 Hz.

53. 4.32×10^4 Hz.

55. 2 Hz.

57. 0.171 m/s.

59. (a) 110 m/s;
(b) 260 m/s.

61. (a) 120;
(b) 0.48°.

65. 0.3 s.

67. (a) 57 Hz, 69 Hz, 86 Hz, 110 Hz, 170 Hz.

69. 88 dB.

71. 15 W.

73. 50 dB.

75. (a) 2.8×10^2 m/s, 48 N;
(b) 0.195 m;
(c) 880 Hz, 1320 Hz.

77. 7.4×10^2 N.

79. 504 Hz.

81. 17 m/s.

83. 2.84 m.

85. 2.29×10^3 Hz.

87. 11.5 m.

89. 34 Hz, 43 Hz, 61 Hz.

91. 10^6.

93. 17 km/h.

CHAPTER 13

1. 3.3×10^{22} atoms.

3. (a) 20°C;
(b) 3300°F.

5. (a) 5°F;
(b) −26°C.

7. 4.3×10^{-3} m.

9. 8×10^{-2} m.

11. 981 kg/m³.

13. 5.12 mL.

15. (a) −140°C;
(b) 180°C.

17. (b) 5.7×10^{-3} (0.57% increase).

21. (a) 6.1 cm;
(b) $\delta L = \dfrac{V_{0\,\text{bulb}}}{\pi r_0^2}(\beta_{Hg} - \beta_{glass})\,\Delta T$.

23. 3.5×10^7 N/m².

25. (a) 27°C;
(b) 4.3×10^3 N.

27. −459.67°F.

29. 1.07 m³.

31. 1.43 kg/m³.

33. (a) 14.8 m³;
(b) 1.83×10^5 Pa.

35. 2.40×10^8 Pa.

37. 37°C.

39. 3.43 atm.

41. 2.69×10^{25} molecules/m³.

43. (a) 7×10^{22} moles;
(b) 4×10^{46} molecules.

45. 19 molecules/breath.

47. 6×10^3 m/s.

49. 899°C.

51. 25.9°C.

55. 3.9×10^2 m/s.

57. 3.34×10^{-9} m.

61. (a) solid or vapor;
(b) 5.11 atm $\leq P \leq$ 73 atm, −56.6°C $\leq T \leq$ 31°C.

63. 14°C.

65. 91°C.

67. 1.1×10^3 Pa.

69. 3.1 kg.

71. 0.28 s, $v_{diffuse} = 5.4 \times 10^{-5}$ m/s,
$v_{rms} = 3.1 \times 10^2$ m/s,
$v_{diffuse}/v_{rms} = 1.7 \times 10^{-7}$.

73. (a) low;
(b) (1.7×10^{-2})%.

75. 0.21.

77. 260 m/s, 4×10^{-22} atm.

79. 11 L, not advisable.

81. 1.65, 1.29.

83. 1.1×10^{44} molecules.

85. 15 hours.

87. 0.66×10^3 kg/m³, −3.5%.

89. 1.6×10^{-3} cm.

91. (a) 2.20×10^3 L;
(b) 92 min;
(c) 30 min.

93. 6.8 balls/s.

95. (a) 1.7×10^3 Pa;
(b) 7.0×10^2 Pa.

97. 6% decrease.

99. 3.0 kg.

CHAPTER 14

1. 1.0×10^7 J.

3. (a) 1.0×10^7 J;
(b) 2.9 kWh;
(c) $0.29 per day, no.

5. 220 kg/h.

7. 100 kcal.

9. 2.0×10^3 J/kg·C°.

11. 40.1°C.

13. (1.9×10^2)°C.

15. 425 s.

17. 2.3×10^3 J/kg·C°.

19. 0.32 C°.

21. 5.0×10^6 J.

23. 1.3 kg.

25. 9.90×10^{-3} kg.

27. 4.7×10^3 kcal.

29. 1.12×10^4 J/kg.

31. 1.7 g.

33. 83 W.

35. (a) 95 W;
(b) 33 W.

37. 23 bulbs.

39. (1.6×10^2)°C.

41. 10 C°.

43. (b) $\dfrac{Q}{t} = A \dfrac{(T_1 - T_2)}{\sum_{i=1}^{n} l_i/k_i}$.

45. 6.4 Calories.

47. 4×10^{15} J.

49. (a) 3.2×10^{26} W;
(b) 1.1×10^3 W/m².

51. 0.80 C°.

53. (a) 46 W;
(b) 7.3×10^3 W.

55. 20 W, only about 9% of the required heat loss rate.

57. (a) 44 C°;
(b) none of the bullet will melt.

59. 4.1 g/h.

61. (a) 1.2×10^{18} J;
(b) $Q_{Sun} = 1.3 \times 10^4\, Q_{interior}$.

63. A mixture of liquid water and steam at 100°C, with the mass of liquid water equal to twice the mass of the steam.

65. (a) 3.1×10^7 J;
(b) 3.3×10^3 s.

CHAPTER 15

1. (a) 0 J;
(b) 3.40×10^3 J.

3.

P
B A
P_0
$P_0/2$
C
0 0.5 1.0 V (L)

5.

P (atm)
5.0
4.0
3.0
2.0
1.0
A
C
B
0 1.0 2.0 3.0 4.0 5.0 V (L)

7. (a) 0 J;
(b) 1850 J;
(c) rise.

9. -4.0×10^2 K.

11. (a, c)

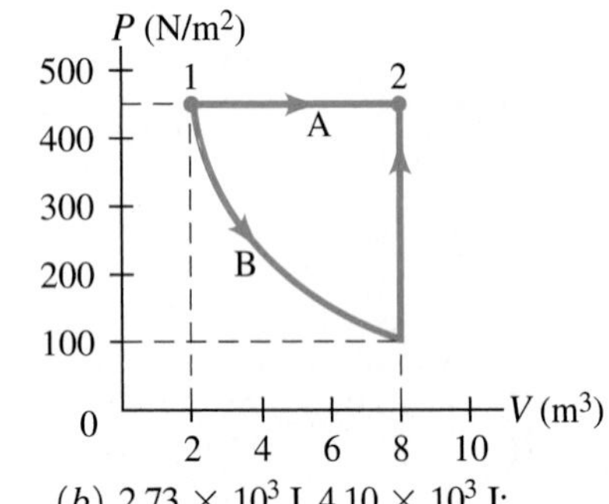

(b) 2.73×10^3 J, 4.10×10^3 J;
(d) 4.10×10^3 J.

13. (a) 25 J;
(b) 63 J;
(c) −95 J;
(d) −120 J;
(e) −15 J.

15. 162 W.

17. 0.28.

19. 0.23.

21. 1.6×10^{13} J/h.

23. 440°C.

25. 9.0×10^2 MW (MJ/s).

27. 250°C.

29. 5.7.

31. −21°C.

33. 76 L.

35. -1.5×10^3 J/K.

37. -1.22×10^6 J/K.

39. 0.15 J/K.

41. $4.35 \times 10^{-2} \frac{\text{J/K}}{\text{s}}$.

43. 1.1 J/K.

45. (a) 1/9;
(b) 1/18.

47. (a) 5/16;
(b) 1/64.

49. (a) 1.32×10^6 kWh;
(b) 7.09×10^4 kW.

51. Yes, the proposed engine operates at a higher than ideal efficiency.

53. (a) 4.0×10^4 J/s;
(b) 1.6×10^5 J/s;
(c) 220 s.

55. (a) 0.077.

57. (a) 45°C;
(b) 0.58 J/K.

59. 0.24.

61. (a) $P_A(V_C - V_A)$;
(b) $P_C(V_C - V_A)$;
(c) $\frac{1}{2}(P_C + P_A)(V_C - V_A)$.

63. (a) 5.3C°;
(b) 77 J/kg·K.

65. 200 J.

67. 180 W.

CHAPTER 16

1. 13 N.

3. 2.7×10^{-3} N.

5. 5.5×10^3 N.

7. 4.88 cm.

9. -5.4×10^7 C.

11. (a) $q_1 = q_2 = 0.5Q_T$;
(b) $q_1 = 0, q_2 = Q_T$.

13. top charge: 83.7 N, 90°; lower left charge: 83.7 N, 210°; lower right charge: 83.7 N, 330°.

15. 2.96×10^7 N, toward center of square.

17. $\vec{F}_1 = 0.30$ N at 265°, $\vec{F}_2 = 0.26$ N at 139°, $\vec{F}_3 = 0.26$ N at 30°.

19. $0.40Q_0$, $0.37l$ from $-Q_0$ toward $-3Q_0$.

21. (a) 69.9×10^{-6} C, 22.1×10^{-6} C;
(b) 104.4×10^{-6} C, -14.4×10^{-6} C.

23. 3.78×10^{-16} N, west.

25. 9.5×10^5 N/C, up.

27. 1.32×10^{14} m/s², the direction of the acceleration is opposite to the direction of the field.

29.

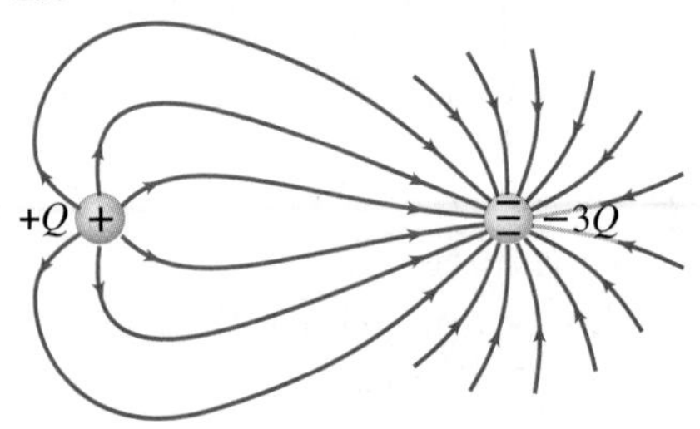

31. 6.54×10^{-10} N/C, south.

33. 4.70×10^6 N/C at 45°.

35. $\frac{4kQxa}{(x^2 - a^2)^2}$, to the left.

37. (a) $\frac{\sqrt{3}\,kQ}{l^2}$, 240°;
(b) $\frac{kQ}{l^2}$, 330°.

39. 1/4.

41. (a) 7.49×10^6 m/s.

43. 1.28×10^{-8} C.

45. (a) -1.1×10^5 N·m²/C;
(b) 0.

47. 1.15×10^{-9} C.

49. (a) 0;
(b) 0;
(c) 3.27×10^3 N/C;
(d) 8.74×10^2 N/C;
(e) no difference.

51. (a) 4.6×10^{-10} N;
(b) 7.1×10^{-10} N;
(c) 6×10^{-5} N.

53. $1/(3.5 \times 10^9)$.

55. 6.8×10^5 C, negative.

57. 1.0×10^7 electron charges.

59. 2.1×10^{-10} m.

61. (a) 0.115 m;
(b) 2.14×10^{-8} s.

63. $\frac{1.08 \times 10^7}{[3.00 - \cos(12.5t)]^2}$ N/C (upward).

65. 5×10^{-9} C.

67. 7.8×10^{-7} C, positive.

69. -7.0×10^8 C, 0 C.

71. $x = \frac{d}{\sqrt{2} - 1} \approx 2.41d$, no.

73. -7.66×10^{-6} C, unstable.

CHAPTER 17

1. 4.2×10^{-4} J.
3. 3.7×10^{-15} J, 2.3×10^{4} eV.
5. 3.8×10^{4} V/m.
7. 3.0×10^{-2} m.
9. 7×10^{-5} m.
11. (*a*) 1.6×10^{7} m/s;
(*b*) 3.4×10^{7} m/s.
13. 1.63×10^{7} m/s.
15. 2.1×10^{-9} C.
17.

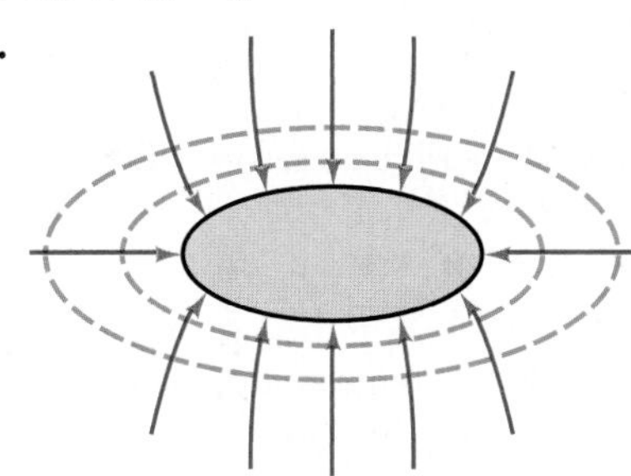

19. $\frac{\sqrt{2}kQ}{2L}(\sqrt{2}+1)$.
21. 4.8×10^{3} m/s.
23. 6.9×10^{-18} J.
25. 4.2×10^{6} V.
27. (*a*) 27 V;
(*b*) 2.2×10^{-18} J, 14 eV;
(*c*) -2.2×10^{-18} J, -14 eV;
(*d*) 2.2×10^{-18} J, 14 eV.
29. (*a*) 3.6×10^{-2} V;
(*b*) 2.5×10^{-2} V;
(*c*) -2.5×10^{-2} V.
31. 2.9×10^{-6} F.
33. 7.9×10^{-13} F.
35. 5.0×10^{7} m^2.
37. 2.63×10^{-8} C.
39. 4.5×10^{4} V/m.
41. $C_{2.50}$: 712 V, 1.78×10^{-3} C;
$C_{6.80}$: 712 V, 4.84×10^{-3} C.
43. 1.5×10^{-10} F.
45. 4.82×10^{-9} F, 0.283 m^2.
47. 9.6×10^{-5} F.
49. (*a*) 7×10^{-12} F;
(*b*) 7×10^{-11} C;
(*c*) 200 V/m;
(*d*) 3×10^{-10} J;
(*e*) capacitance, charge, energy.
51. (*a*) quadrupled;
(*b*) doubled.
53. 2.9×10^{5} V/m.
55. (*a*) 6.3 KeV;
(*b*) 42.8:1.
57. 1.0×10^{-7} J/m^3.
59. 620 V.
61. 1.5×10^{-7} C.
63. (*a*) 11 cm from − charge, on opposite side of − charge from + charge;
(*b*) 0.7 cm from − charge, on same side of − charge as + charge; 5.2 cm from − charge, on opposite side of − charge from + charge.
65. 6.5°.
67. 9×10^{-16} m, no.
69. (*a*) 23 J;
(*b*) 3.4×10^{5} W.
71. 1.03×10^{6} m/s.
73. 2.5×10^{-10} C.
75. (*a*) 4.2×10^{-11} C;
(*b*) 4.2×10^{-11} C;
(*c*) 18 V;
(*d*) 1.3×10^{-10} J.
77. (*a*) 2.7×10^{3} m/s;
(*b*) 2.2×10^{3} m/s.

CHAPTER 18

1. 8.13×10^{18} electrons/s.
3. 5.5×10^{-11} A.
5. 950 V.
7. (*a*) 25 A;
(*b*) 7.5×10^{4} C.
9. 2.8×10^{-3} V.
11. (*a*) 20 Ω;
(*b*) 430 J.
13. 3.3×10^{-2} Ω.
15. yes, tungsten diameter = 4.6 mm.
17. 22 C°.
19. 1800°C.
21. (*a*) 3.8×10^{-4} Ω;
(*b*) 1.5×10^{-3} Ω;
(*c*) 6.0×10^{-3} Ω.
23. 58.3°C.
25. $R_{\text{carbon}} = 2090$ Ω;
$R_{\text{Nichrome}} = 2610$ Ω.
27. 0.96 W.
29. (*a*) 190 Ω, 0.63 A;
(*b*) 33 Ω, 3.7 A.
31. (*a*) 850 W;
(*b*) 17 Ω;
(*c*) 12 Ω.
33. 0.14 kWh, 20 cents/month.
35. (*a*) 6.7 Ω, 1.4 W;
(*b*) 4.
37. 18 bulbs.
39. 7500 W.
41. (*a*) 10 A;
(*b*) 1.2 Ω.
43. 0.39 A, 0.55 A.
45. 390 V.
47. (*a*) 4500 W;
(*b*) 13 A.
49. 5.1×10^{-10} m/s.
51. 2.6 A/m^2, north.
53. 35 m/s.
55. 3×10^{-14} W.
57. 6.2 A.
59. 2.9×10^{-4} m.
61. $1200 per hour per meter.
63. 1/4.
65. 3.8×10^{-3} m.
67. (*a*) 1500 W;
(*b*) 12 A.
69. 2:1.
71. (*a*) 26 Ω;
(*b*) 26 s;
(*c*) 0.17 cents.
73. 2.58×10^{-4} m, 38.8 m.
75. 1.4×10^{12} protons.
77. 1.79×10^{-4} m.
79. (*a*) $I_A = 0.33$ A, $I_B = 3.3$ A;
(*b*) $R_A = 360$ Ω, $R_B = 3.6$ Ω;
(*c*) $Q_A = 1.2 \times 10^{3}$ C,
$Q_B = 1.2 \times 10^{4}$ C;
(*d*) $E_A = E_B = 1.4 \times 10^{5}$ J;
(*e*) B.
81. 1.34×10^{-4} Ω.
83. 2200°C.

CHAPTER 19

1. (*a*) 8.41 V;
(*b*) 8.49 V.
3. 0.048 Ω, 0.11 Ω.
5. 960 Ω, 60 Ω.
7. 9.3 V.
9. (*a*) 2820 Ω;
(*b*) 300 Ω.
11. 720 Ω (all in series), 80 Ω (all in parallel), 360 Ω (two in parallel, in series with third), 160 Ω (two in series, in parallel with third).
13. (*a*) 14 V;
(*b*) 28 Ω, 6.9 W.
15. 27 Ω.
17. (*a*) 840 Ω;
(*b*) $V_{470} = 6.7$ V;
$V_{680} = V_{820} = 5.3$ V.
19. (*a*) V_1, V_2 increase;
V_3, V_4 decrease;
(*b*) I_1, I_2 increase;
I_3, I_4 decrease;
(*c*) increases;
(*d*) before: $I_1 = 0.117$ A, $I_2 = 0$,
$I_3 = I_4 = 0.059$ A;
after: $I_1 = 0.132$ A,
$I_2 = I_3 = I_4 = 0.044$ A; yes.

21. (a) V_{left} decreases, V_{middle} increases, V_{right} goes to 0;
(b) I_{left} decreases, I_{middle} increases, I_{right} goes to 0;
(c) increases;
(d) 14.1 V;
(e) 14.3 V.

23. 0.41 A.

25. (a) −25.7 V;
(b) $V_{80} = 77.4$ V, $V_{45} = 43.3$ V.

27. $I_1 = 0.68$ A, left; $I_2 = 0.40$ A, left.

29. $I_1 = 0.13$ A, right; $I_2 = 0.31$ A, left; $I_3 = 0.18$ A, up.

31. 2 Ω: 0.26 A, 6 Ω: 0.028 A, 8 Ω: 0.29 A, 10 Ω: 0.26 A, 12 Ω: 0.29 A.

33. 1.30 A.

35. (a) 28.2 μF;
(b) 0.78 μF.

37. 3.71 μF.

39. 7300 pF, yes.

41. $C_1 + \frac{C_2 C_3}{C_2 + C_3}$.

43. $Q_1 = 48.0\ \mu C$, $Q_3 = 24.0\ \mu C$; $V_1 = 3.00$ V, $V_2 = 1.50$ V, $V_3 = 1.50$ V; $V = 3.00$ V.

45. (a) $V_{0.40} = 5.4$ V, $V_{0.60} = 3.6$ V;
(b) $Q_{0.40} = Q_{0.60} = 2.2 \times 10^{-6}$ C;
(c) $V_{0.40} = V_{0.60} = 9.0$ V, $Q_{0.40} = 3.6 \times 10^{-6}$ C, $Q_{0.60} = 5.4 \times 10^{-6}$ C.

47. in parallel, 500 pF.

49. 1.0×10^6 Ω.

51. 9.3×10^{-2} s.

53. 7.5×10^6 Ω.

55. (a) 5.0×10^{-5} Ω in parallel;
(b) 5.0×10^6 A in series.

57. 1000 Ω in series, 100 Ω/V.

59. 5.52×10^{-3} A.

61. 10 V.

63. 10.4 V, 2.6 Ω.

65. (b) 290 Ω, 140 Ω.

67. 7×10^{-3} A.

69. 1.1×10^{-5} Ω.

71. (a) $R_x = R_2 R_3 / R_1$;
(b) 65.7 Ω.

73. $\frac{1}{4}C, \frac{2}{5}C, \frac{3}{5}C, \frac{3}{4}C, C, \frac{4}{3}C, \frac{5}{2}C, 4C$.

75. 50.1 V, 1.25 Ω.

77. 52.3 V, −28.3 V (two answers because current direction through 4.0-kΩ resistor is unknown.)

79. (a) 6.7×10^{-5} A, upward;
(b) 16 V.

81. (a) 3.3 Ω;
(b) 2.2 V.

83. 100 Ω.

85. (a) 7.6 Ω;
(b) 0.33 A;
(c) 0.33 A;
(d) 0.95 W.

87. 7.2 Ω.

CHAPTER 20

1. (a) 7.6 N/m;
(b) 5.3 N/m.

3. 1.95 A.

5. 0.264 T.

7. (a) south pole;
(b) 4.1 A;
(c) 6.4×10^{-2} N.

9. 1.3 T.

11. (a) left;
(b) left;
(c) upward;
(d) inward;
(e) no force;
(f) downward.

13. clockwise circular motion of radius 2.77×10^{-5} m.

15. 1.6 T, east.

17. (a) 2.7×10^{-2} m;
(b) 3.8×10^{-7} s.

23. 6.20×10^{-7} m.

25. (a) 45°;
(b) 3.5×10^{-3} m.

27. 69 A.

29. 13 A, upward.

31. 2.5 A.

33. 1.1×10^{-4} T up.

35. 4.1×10^{-5} T, 11° below horizontal.

37. (a) $(2.0 \times 10^{-5}\ \mathrm{T/A})(I - 15\ \mathrm{A})$;
(b) $(2.0 \times 10^{-5}\ \mathrm{T/A})(I + 15\ \mathrm{A})$.

39. near wire: 4.5×10^{-2} N, attract; far wire: 2.2×10^{-2} N, repel.

41. 2.6×10^{-6} N, toward straight wire.

43. 4.1×10^{-5} T.

45. M: 5.8×10^{-4} N/m, 90°; N: 3.4×10^{-4} N/m, 300°; P: 3.4×10^{-4} N/m, 240°.

47. $\frac{\mu_0 I}{2\pi}\left(\frac{1}{x} - \frac{1}{d - x}\right)$, y direction.

49. 94.3 A.

51. short and fat.

55. 61.6 μA.

57. 0.88.

59. (a) 4.01×10^{-5} m·N;
(b) north.

61. 70 u, 72 u, 73 u, 74 u.

63. 2.5 m.

65. 41 T.

67. 3.0 T, upward.

69. 0.25 N, northerly, 68° above horizontal.

71. 1.12×10^{-6} m/s, west.

73. 1.6 A, down.

75. (a) $\frac{IlB}{m} t$;
(b) $\left(\frac{IlB}{m} - \mu_k g\right)t$;
(c) east.

77. (c) 48 MeV.

79. They will miss second tube, 9.1°.

81. 1×10^9 A.

83. (a) 2.1×10^{-3} T;
(b) out of plane formed by velocity and electric field directions;
(c) 5.8×10^7 Hz.

85. 1.3×10^4 turns.

87. 5.3×10^{-5} m, 3.3×10^{-4} m.

CHAPTER 21

1. −420 V.

3. to the left.

5. 8.5×10^{-2} V.

7. (a) 8.8×10^{-3} Wb;
(b) 55°;
(c) 5.1×10^{-3} Wb.

9. (a) clockwise;
(b) counterclockwise;
(c) clockwise;
(d) no induced current.

11. (a) 6.1×10^{-2} V;
(b) clockwise.

13. (a) clockwise;
(b) 4.3×10^{-2} V;
(c) 1.7×10^{-2} A.

15. 0.548 N.

17. (a) 0.17 V;
(b) 6.1×10^{-3} A;
(c) 6.4×10^{-4} N.

19. 5.86 C.

21. 28.2 V.

23. 2.08 rev/s.

25. (a) 99.0 A;
(b) 1.3×10^{-2} m².

27. 100 V.

29. 13 A.

31. step down, 0.375, 2.67.

33. 50, 4.8 V.

35. (a) step up;
(b) 2.8.

37. (*a*) 48 kV (rms);
(*b*) 0.056.

39. 7.7 V.

41. 0.14 H.

43. (*a*) 1.7×10^{-2} H,
(*b*) 81 turns.

45. $\mu_0 N_1 N_2 A/l$.

47. 29 J.

49. 5.1×10^{15} J.

51. (*a*) 2.3;
(*b*) 4.6;
(*c*) 6.9.

53. (*a*) 368 Ω;
(*b*) 2.21×10^{-2} Ω.

55. 9.90 Hz.

57.

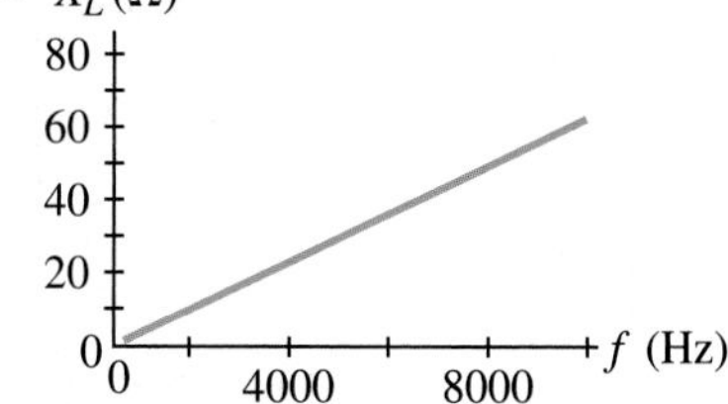

59. 4.97×10^{-2} H.

61. (*a*) 30 kΩ;
(*b*) 31 kΩ.

63. 1700 Ω.

65. 8.78 kΩ, −7.62°, 8.26×10^{-2} A.

67. (*a*) 6.65×10^{-2} A;
(*b*) 4.19°;
(*c*) 119.7 V, 8.77 V.

69. 3.63×10^5 Hz.

71. (*a*) 1.32×10^{-7} F;
(*b*) 34.1 A.

73. 7.05×10^{-3} J.

75. 200 kV.

77. (*a*) 41 kV;
(*b*) 3.1×10^7 W;
(*c*) 8.8×10^5 W;
(*d*) 3.0×10^7 W.

81. Put a 120-mH inductor in series with the device.

83. 102 V.

85. 0.10 H.

87. 7.5×10^{-2} H, 14 Ω.

89. (*b*) 2.5×10^{-6} H, 2.9×10^{-2} Ω.

CHAPTER 22

1. 7.9×10^{14} V/m/s.

3. 5.25 V/m.

5. 1.88×10^{10} Hz.

7. 3.11×10^{-7} m, ultraviolet.

9. 499 s (8.31 min).

11. 4.0×10^{16} m.

13. (*a*) 261 s;
(*b*) 1260 s.

15. 2.1×10^6 rev/s.

17. 9040 wavelengths, 3.54×10^{-15} s.

19. 1.21×10^7 s (≈ 140 days).

21. 0.95 W/m^2, 19 V/m.

23. 3.80×10^{26} W.

25. (*a*) 280 J;
(*b*) 2.6×10^9 V/m.

27. (*a*) 2.78 m to 3.41 m;
(*b*) 176 m to 561 m.

29. AM is longer, by a factor of 100.

31. 330 pF.

33. 2.6×10^{-9} H to 3.9×10^{-9} H.

35. 1.5 V/m.

37. 499 s (8.31 min).

39. (*a*) 1.28 s;
(*b*) 260 s (4.3 min).

41. (*a*) 0.07 V/m;
(*b*) 8 km.

43. 469 V/m.

45. (*a*) 150 m;
(*b*) 75 m.

47. (*a*) 1.8×10^{-10} J;
(*b*) 8.7×10^{-6} V/m, 2.9×10^{-14} T.

49. (*a*) parallel;
(*b*) 8.9 pF to 11 pF;
(*c*) 1.1 mH.

53. (*a*) 4.0×10^{-7} J;
(*b*) 1.2×10^{-2} V/m;
(*c*) 1.2×10^{-2} V.

CHAPTER 23

1. 5.0 m.

3. 5°.

5. 5.9×10^{-6} m^2.

7. 36.0 cm.

9. 2.09 cm behind the front surface of the ball, virtual, upright.

11. concave, 5.66 cm.

13. (*b*) −6.7 cm; (*c*) 1.0 mm.

15. (*a*) at the center of curvature;
(*b*) real;
(*c*) inverted;
(*d*) −1.

21. (*a*) convex;
(*b*) 22 cm behind the mirror;
(*c*) −98 cm;
(*d*) −196 cm.

23. (*a*) 1.97×10^8 m/s;
(*b*) 1.99×10^8 m/s;
(*c*) 2.21×10^8 m/s.

25. 1.49.

27. 64.0°.

29. 46.8°.

31. 4.6 m.

33. 81.1°.

37. 1.35.

39. $n \geq 1.5$.

43. (*a*)

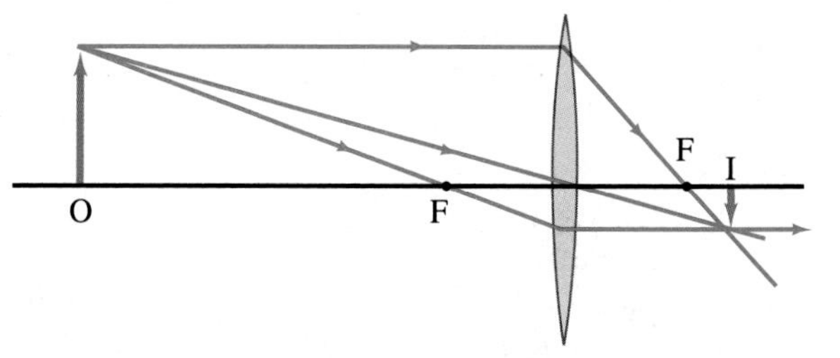

(*b*) 390 mm.

45. converging, 41.1 cm, real.

47. (*a*) −72 cm;
(*b*) 4.0.

49. (*a*) 80.6 mm;
(*b*) 82.2 mm;
(*c*) 87.0 mm;
(*d*) one at 24 cm.

51. (*a*) 3.0 cm away from lens;
(*b*) 0.5 cm toward lens.

53. (*a*) 75.0 mm;
(*b*) 25.0 mm.

55. (*a*) 15.2 cm, −2.54 mm, real and inverted;
(*b*) −12.1 cm, 2.02 mm, virtual and upright.

57. 49.2 cm, 16.8 cm.

59. 7.41 cm behind diverging lens.

61. (*a*) 7.14 cm to right of lens B;
(*b*) −0.357.
(*c*)

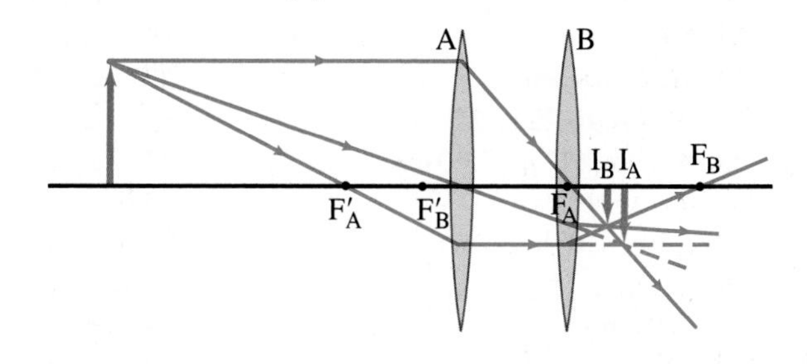

63. (*a*) 16 cm to the left of the converging lens;
(*b*) 1.8 cm to the right of the diverging lens.
67. 1.54.
69. −36.1 cm.
71. 5.6 m.
73. 5.16 m.
75. (*a*) $r = \infty$;
(*b*) $d_o = -d_i$;
(*c*) +1;
(*d*) yes.
77. The object is virtual and is closer to the lens than the focal point.
79. 49°.
81. (*a*)

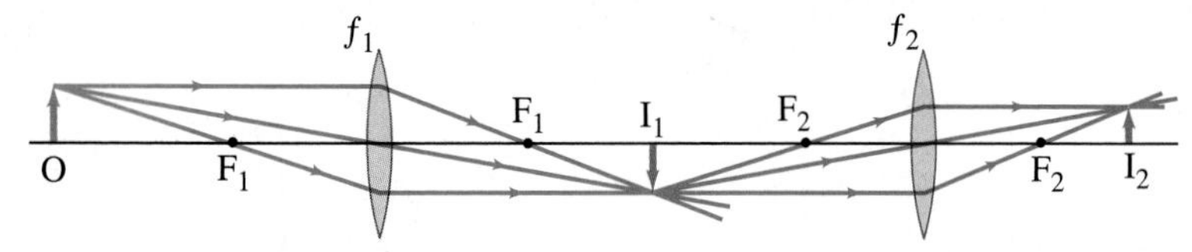

(*b*) 21 cm to the right of the second lens, 0.65.
83. 0.106 m, 2.7 m.
85. real and upright, real and upright.
87. (*a*) 0.26 mm in diameter;
(*b*) 0.47 mm in diameter;
(*c*) 1.3 mm in diameter;
(*d*) 0.56, 2.7.
89. 20.0 cm.
91. 7.5 cm, 10 cm.
93. 20.0 cm, converging.

CHAPTER 24

1. 4.9×10^{-7} m.
3. 6.2×10^{-7} m, 4.8×10^{14} Hz.
5. 1.4×10^{-4} m.
7. 610 nm.
9. reverse of the usual double-slit pattern.
11. 2.7×10^{-3} m.
13. 533 nm.
15. 0.22°.
17. 1.51°.
19. 2.51 m.
21. 3.6×10^{-6} m.
23. 9.53×10^{-7} m.
25. 2.6 cm.
27. 4.43°.
29. 1640 lines/cm.
31. 2.20×10^{-6} m, 53.3°.
33. 534 nm, 612 nm, 760 nm.
35. 7140 lines/cm.
37. 5.79×10^{5} lines/m.
39. 643 nm.
41. 169 nm.
43. 8.5×10^{-6} m.
45. 113 nm, 225 nm.
47. 471 nm.
49. 699 nm.
51. 2.50×10^{-4} m.
53. 0.089.
55. 61.2°.
57. (*a*) 35.3°;
(*b*) 63.4°.
59. 44.5%.
61. $0.031 I_0$.
63. (*a*) constructive;
(*b*) destructive.
65. 360 nm.
67. 31°.
69. 590 nm.
71. 600 nm to 700 nm of second-order spectrum overlaps 400 nm to 467 nm of third-order spectrum.
73. 0.7 m.
75. 810 nm.
77. 320 nm, 160 nm.
79. 36.9°.
81. (*a*) 0.086;
(*b*) none.
83. 10 mm.
85. 240 nm.
87. (*a*) 6.0°;
(*b*) 8.7°.

CHAPTER 25

1. 2.5 mm to 39 mm.
3. $1/62.5\text{ s} \approx 1/60\text{ s}$.
5. 17 mm.
7. 50 mm.
9. 0.16 s.
11. Glasses would be better.
13. −8.3 D, −7.1 D.
15. (*a*) nearsighted; (*b*) 20.6 cm.
17. 0.2 cm.
19. (*a*) 2.0 cm; (*b*) 1.9 cm.
21. 2.1.
23. (*a*) 17 cm; (*b*) 10 cm.
25. (*a*) 3.63; (*b*) 12.0 mm; (*c*) 6.88 cm.
27. (*a*) −64 cm; (*b*) 3.3×.
29. −27, 79 cm.
31. 22 cm.
33. −110×.
35. −94×.
37. $f_e = 0.73$ cm, $f_o = 124$ cm.
39. 440×.
41. 610×.
43. (*a*) 0.85 cm; (*b*) 230×.
45. (*a*) 14.4 cm; (*b*) 137×.
47. (*a*) 16.0 cm; (*b*) 14.3 cm; (*c*) 1.7 cm; (*d*) 0.71 mm.
49. 9.1 cm, 6.1×10^{-6} rad.
51. 1.7×10^{11} m.
53. yes.
55. 0.245 nm.
57. (*a*) 1;
(*b*) 1 to 2.7.
59. 16.
61. 100 mm, 200 mm.
63. 2.9×, 4.1×, the person with the normal eye.
65. (*a*) −2.3×;
(*b*) +4.5 D.
67. 34 cm.
69. −19×.
71. (*a*) 48 cm, −11×, 4.0-cm lens;
(*b*) 180 cm.
73. 1.7 m.

CHAPTER 26

1. 42.6 m.
3. (*a*) $1 - 2.2 \times 10^{-9}$;
(*b*) 0.9998;
(*c*) 0.980;
(*d*) 0.31;
(*e*) 0.20;
(*f*) 0.0447.
5. 2.32×10^{8} m/s.
7. 26 years.
9. 7×10^{-8} %.
11. (*a*) 2.7 years;
(*b*) 9.2 years.
13. (*a*) 6.39 m, 1.25 m;
(*b*) 15.0 s;
(*c*) $0.660c$;
(*d*) 15.0 s.
15. 8.1×10^{-19} kg·m/s.
17. $0.38c$.
19. (*a*) 310%;
(*b*) 140%.
21. 4×10^{-28} kg.
23. 938 MeV/c^2.
25. 9.0×10^{13} J, 3.7×10^{10} kg.
27. $0.60c$.
29. 944 MeV, 1630 MeV/c.
31. $0.437c$.
33. $0.30c$.
35. (*a*) $0.866c$;
(*b*) $0.745c$.
37. (*a*) 5.5×10^{19} J;
(*b*) 3.3%.
39. 237.0483 u.

41.

(a)

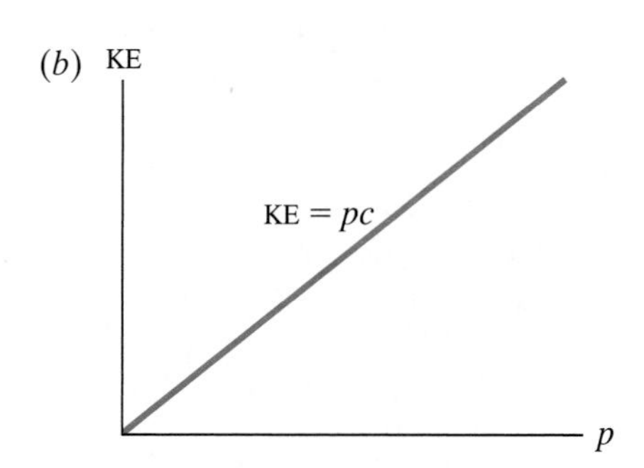

(b)

43. $0.80c$.

45. (a) $0.98c$;
(b) $0.42c$.

47. $0.92c$.

49. (a) $0.73c$;
(b) 5.9 years.

51. (a) $c - v = 0.77$ m/s;
(b) 0.21 m.

53. 1.02 MeV (or 1.64×10^{-13} J).

55. 234 MeV.

57. (a) 4×10^9 kg/s;
(b) 4×10^7 years;
(c) 1×10^{13} years.

59. 28.3 MeV (or 4.53×10^{-12} J).

61. (a) 2470 MeV/c;
(b) 0;
(c) 1.40×10^4 MeV/c.

63. 2.5×10^7 kg.

65. yes in the barn's frame of reference, but no in the boy's frame of reference.

67. $0.96c$.

69. c.

71. 6.8×10^{21} J, 68× larger.

73. (a) 1.17 s;
(b) 2.68 s.

CHAPTER 27

1. 6.2×10^4 C/kg.

3. 5 electrons.

5. (a) 1.06×10^{-5} m, infrared;
(b) 8.29×10^{-7} m, infrared;
(c) 7.25×10^{-4} m, microwave;
(d) 1.06×10^{-3} m, microwave.

7. 5.4×10^{-20} J, 0.34 eV.

9. 9.35×10^{-6} m.

11. 2.7×10^{-19} J to 5.0×10^{-19} J, 1.7 eV to 3.1 eV.

13. 2.4×10^{-13} Hz, 1.2×10^{-5} m.

15. 6.6×10^{-23} kg·m/s (or 0.12 MeV/c).

17. 6.5×10^{14} Hz.

19. copper and iron.

21. 0.63 eV.

23. 2.96 eV.

25. 1.9 eV, 43 kcal/mol.

27.

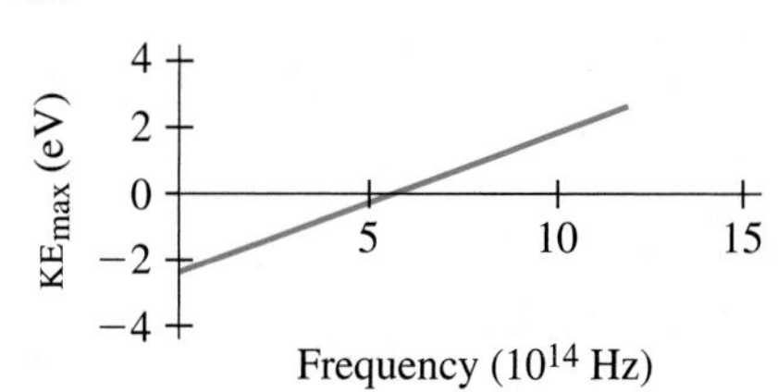

(a) 6.7×10^{-34} J·s;
(b) 5.5×10^{14} Hz;
(c) 2.30 eV.

29. (a) 2.42×10^{-12} m;
(b) 1.32×10^{-15} m.

31. (a) 55 eV;
(b) 0.105 nm.

33. 6.61×10^{-16} m.

35. 0.51 MeV, 170 MeV/c.

37. 2.9×10^{-32} m.

39. 26 V.

41. (a) 1.3×10^{-24} kg·m/s;
(b) 1.5×10^6 m/s;
(c) 6.0 V.

45. 4.7×10^{-38} m/s, 6.3×10^{38} times smaller.

47. 2.48×10^{-11} m.

49. 3.40 eV.

51. 122 eV.

55. 5.26×10^{-8} m.

57.

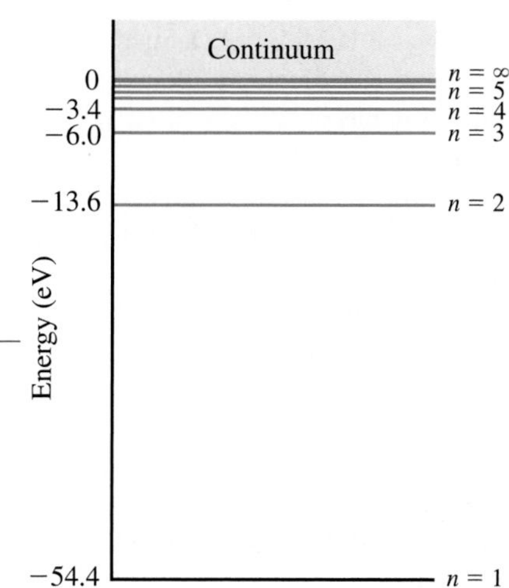

59. −27.2 eV, −13.6 eV.

61. yes, $7.3 \times 10^{-3}c$, 0.999973.

63. (a)

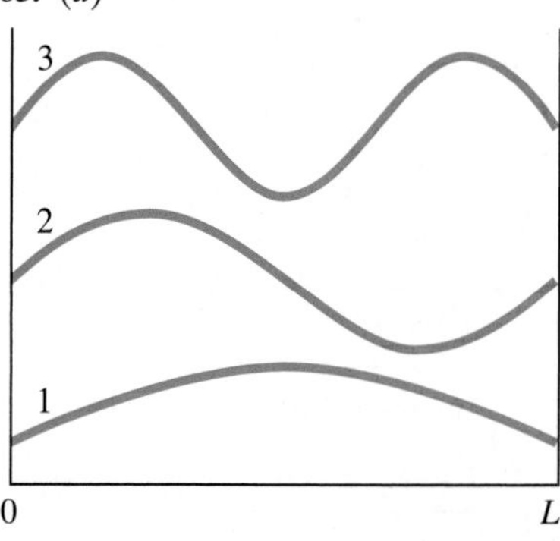

(c) 2.4×10^{-17} J, 150 eV;
(d) 1.6×10^{-66} J, 4.7×10^{-33} m/s;
(e) 1.3×10^{-10} m.

65. 3.27×10^{15} Hz.

67. 4.7×10^{26} photons/s.

69. 5.3×10^{18} photons/s.

71. 0.39 MeV each.

73. 4.7×10^{-14} m.

75. (4.4×10^{-40}) : 1, yes.

77. 0.64 eV.

79. 6.7 N.

81. 1.0 eV, no current would flow.

83. (a)

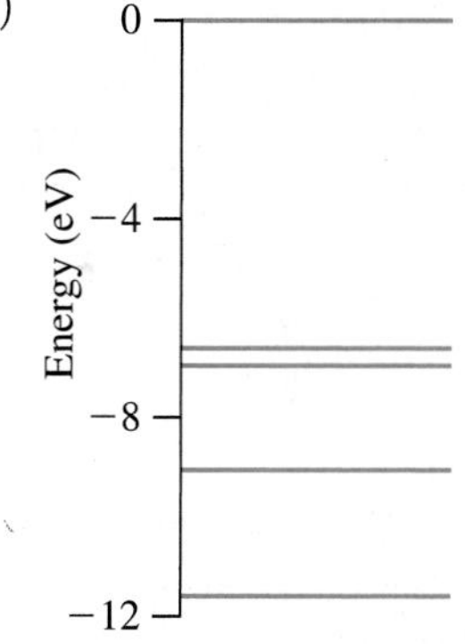

(b) ground state, 0.4 eV, 2.2 eV, 2.5 eV, 2.6 eV, 4.7 eV, 5.1 eV.

85. 1.8×10^{11} C/kg.

87. $E_n = -\dfrac{3 \times 10^{165}\text{ J}}{n^2}$,
$r_n = n^2(5 \times 10^{-129})$ m, no.

CHAPTER 28

1. 3.6×10^{-7} m.

3. 5.3×10^{-11} m.

5. 10^{-7} eV.

7. 3.00×10^{-10} eV/c^2.

9. $\Delta x_{electron} = 1.4 \times 10^{-3}$ m,
$\Delta x_{baseball} = 9.1 \times 10^{-33}$ m,
$\Delta x_{electron} \approx (1.5 \times 10^{29})\Delta x_{baseball}$.

13. 0, 1, 2, 3, 4, 5.

15. 14 electrons.

17. (*a*)

n	l	m_l	m_s
1	0	0	$\frac{1}{2}$
1	0	0	$-\frac{1}{2}$
2	0	0	$\frac{1}{2}$
2	0	0	$-\frac{1}{2}$
2	1	1	$\frac{1}{2}$
2	1	1	$-\frac{1}{2}$

Other combinations of quantum numbers are possible for the last two electrons.

(*b*)

n	l	m_l	m_s
1	0	0	$\frac{1}{2}$
1	0	0	$-\frac{1}{2}$
2	0	0	$\frac{1}{2}$
2	0	0	$-\frac{1}{2}$
2	1	1	$\frac{1}{2}$
2	1	1	$-\frac{1}{2}$
2	1	0	$\frac{1}{2}$
2	1	0	$-\frac{1}{2}$
2	1	−1	$\frac{1}{2}$
2	1	−1	$-\frac{1}{2}$
3	0	0	$\frac{1}{2}$
3	0	0	$-\frac{1}{2}$

19. $n \geq 5$; $m_l = 4, 3, 2, 1, 0, -1, -2, -3, -4$; $m_s = +\frac{1}{2}, -\frac{1}{2}$.
21. $n \geq 4$; $l \geq 3$; $m_s = +\frac{1}{2}, -\frac{1}{2}$.
23. (*a*) $1s^2 2s^2 2p^6 3s^2 3p^6 3d^7 4s^2$;
(*b*) $1s^2 2s^2 2p^6 3s^2 3p^6 3d^{10} 4s^2 4p^6$;
(*c*) $1s^2 2s^2 2p^6 3s^2 3p^6 3d^{10} 4s^2 4p^6 5s^2$.
25. (*a*) $n = 6$;
(*b*) −0.378 eV;
(*c*) $L = 0, l = 0$;
(*d*) $m_l = 0$.
29. (*a*) $5f, 5p, 4f, 4p, 3p, 2p$;
(*b*) four wavelengths.
31. 41 kV.
33. 0.18 nm.
35. 0.061 nm, partial shielding of the nucleus by the $n = 2$ shell.
37. 1.9×10^{-2} J, 6.1×10^{16} photons.
39. 5.64×10^{-4} rad;
(*a*) 170 m;
(*b*) 2.2×10^5 m.
41. 5.3×10^{-11} m, same as Bohr radius.
43. $L_{\text{min}} = 0$,
$L_{\text{max}} = 4.72 \times 10^{-34}\ \text{kg}\cdot\text{m}^2/\text{s}$.
45. (*a*) 3.1×10^{-34} m;
(*b*) $1.8 \times 10^{-32}\ \text{kg}\cdot\text{m}^2/\text{s}$.
47. 3600 C°/min.
49. $2870\ \text{m}^{-1/2}$.
53. $L_{\text{Bohr}} = 2\hbar$, $L_{\text{QM}} = 0$ or $\sqrt{2}\,\hbar$.
55. 6.3×10^{-14} s.
57. $\Delta p_{\text{proton}} : \Delta p_{\text{electron}} = 43:1$.
59. 1.8×10^{-35} m/s, yes, 10^{34} s.
61. copper.

CHAPTER 29

1. 5.1 eV.
3. 4.6 eV.
7. 1.10×10^{-10} m.
9. (*a*) 1.5×10^{-2} eV, 8.2×10^{-5} m;
(*b*) 3.0×10^{-2} eV, 4.1×10^{-5} m;
(*c*) 4.6×10^{-2} eV, 2.7×10^{-5} m.
11. (*a*) 720 N/m;
(*b*) 2.0×10^{-6} m.
13. 2.82×10^{-10} m.
17. 1.1×10^{-6} m.
19. 1.1×10^6 electrons.
21. 5×10^6.
23. 1.9 eV.
25. 13 mA.
27. (*a*) 2.4 mA;
(*b*) 4.8 mA.
29. (*a*) 8.1 mA;
(*b*) 5.7 mA.
31. 13 eV.
33. (*a*) −5.3 eV;
(*b*) 5.1 eV.
35. $1.94 \times 10^{-46}\ \text{kg}\cdot\text{m}^2$.
37. 5.50 eV.
39. (*a*) 9.4×10^{-2} eV;
(*b*) 6.3×10^{-10} m.
41. 6.47×10^{-4} eV.
43. (*a*) $146\ \text{V} \leq V_{\text{supply}} \leq 362\ \text{V}$;
(*b*) $3.34\ \text{k}\Omega \leq R_{\text{load}} < \infty$.

CHAPTER 30

1. 0.149 u.
3. $3726\ \text{MeV}/c^2$.
5. (*b*) 180 m; (*c*) 2.58×10^{-10} m.
7. 6×10^{26}, no, nucleon density is essentially the same for all nuclei.
9. 340 MeV.
11. 2.224 MeV.
13. 7.799 MeV.
15. ${}^{23}_{11}\text{Na}$: 8.11 MeV/ nucleon,
${}^{24}_{11}\text{Na}$: 8.06 MeV/ nucelon.
17. (*b*) stable.
19. 0.783 MeV.
21. (*a*) β^+;
(*b*) ${}^{22}_{11}\text{Na} \rightarrow {}^{22}_{10}\text{Ne} + e^+ + \nu$, 1.819 MeV.
23. (*a*) ${}^{234}_{90}\text{Th}$;
(*b*) 234.04367 u.
25. 0.0855 MeV.
27. (*a*) ${}^{32}_{16}\text{S}$;
(*b*) 31.97152 u.
29. 0.862 MeV.
31. 0.9602 MeV, 0.9602 MeV, 0 MeV.
33. 4.90 MeV.
37. (*a*) $1.5 \times 10^{-10}\ \text{yr}^{-1}$;
(*b*) 2.3 h.
39. 1/16 (or 0.0625).
41. 1.31×10^{-20} nuclei.
43. (*a*) 3.14×10^{12} decays/s;
(*b*) 3.13×10^{12} decays/s;
(*c*) 5.51×10^5 decays/s.
45. 0.77 g.
47. 1.68×10^{-13} kg.
49. 2.6 min.
51. 2.58×10^{-3} g.
53. $T_{\frac{1}{2}}({}^{218}_{84}\text{Po}) : T_{\frac{1}{2}}({}^{214}_{84}\text{Po}) = (1.2 \times 10^6):1$.
55. ${}^{232}_{90}\text{Th} \rightarrow {}^{228}_{88}\text{Ra} \rightarrow {}^{228}_{89}\text{Ac} \rightarrow {}^{228}_{90}\text{Th} \rightarrow {}^{224}_{88}\text{Ra} \rightarrow {}^{220}_{86}\text{Rn}$,
${}^{235}_{92}\text{U} \rightarrow {}^{231}_{90}\text{Th} \rightarrow {}^{231}_{91}\text{Pa} \rightarrow {}^{227}_{89}\text{Ac} \rightarrow {}^{227}_{90}\text{Th} \rightarrow {}^{223}_{88}\text{Ra}$.
57. $N_{\text{D}} = N_0(1 - e^{-\lambda t})$.
59. 2.3×10^4 yr.
61. 41 yr.
63. 6.64 half-lives.
65. (*b*) 98%.
67. (*a*) ${}^{191}_{77}\text{Ir}$;
(*b*)

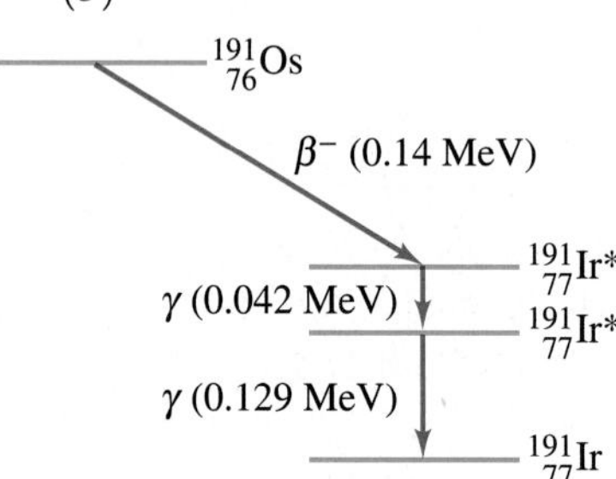

to the higher excited state.
69. 550 MeV, 2.5×10^{12} J.
71. (*a*) 2.4×10^5 yr;
(*b*) 2.5×10^5 yr, carbon dating is not useful for times on the order of 10^5 yr or longer.
73. (*a*) $\text{KE}_\alpha : \text{KE}_\beta = 5.48 \times 10^{-4}:1$.
75. (*a*) 1.6%;
(*b*) 0.67%.
77. 0.18 decays/s or 11 decays/min.

CHAPTER 31

1. ${}^{28}_{13}\text{Al}$, β^-, ${}^{28}_{14}\text{Si}$.
3. possible.
5. 5.701 MeV is released.
7. (*a*) yes;
(*b*) 19.85 MeV.
9. 4.730 MeV.
11. $\text{n} + {}^{14}_{7}\text{N} \rightarrow {}^{14}_{6}\text{C} + \text{p}$, 0.626 MeV.
13. (*a*) ${}^{3}_{2}\text{He}$ picks up a neutron;
(*b*) ${}^{11}_{6}\text{C}$;
(*c*) 1.856 MeV, exothermic.
15. 18.000953 u.
17. 173.3 MeV.

19. 1/1100.
21. 3.7×10^{-7} kg.
23. 625 kg.
25. 2.1×10^{-16} J or 1.3 keV.
29. 6.0×10^{23} MeV/g,
4.9×10^{23} MeV/g,
2.1×10^{24} MeV/g,
5.1×10^{23} MeV/g.
31. 0.38 g.
33. 5.3×10^{3} kg/h.
35. (*b*) 26.73 MeV;
(*c*) 1.94 MeV, 2.22 MeV, 7.55 MeV, 7.30 MeV, 2.75 MeV, 4.97 MeV;
(*d*) greater repulsion from higher-Z nuclei.
37. 4.0 Gy.
39. 250 rads.
41. 200 counts/s.
43. 2.5 days.
45. 8.25×10^{-7} Gy/day.
47. (*a*) ${}^{131}_{53}\mathrm{I} \rightarrow {}^{131}_{54}\mathrm{Xe} + e^- + \bar{\nu}$;
(*b*) 27 days;
(*c*) 8×10^{-12} kg.
49. (*a*) ${}^{218}_{84}\mathrm{Po}$;
(*b*) radioactive,
α decay: ${}^{218}_{84}\mathrm{Po} \rightarrow {}^{214}_{82}\mathrm{Pb} + {}^{4}_{2}\mathrm{He}$,
β decay: ${}^{218}_{84}\mathrm{Po} \rightarrow {}^{218}_{85}\mathrm{At} + e^-$,
3.1 min for both;
(*c*) chemically reacting;
(*d*) 5.7×10^{6} decays/s, 2.5×10^{4} decays/s.
51. (*a*) ${}^{12}_{6}\mathrm{C}$;
(*b*) 5.701 MeV.
53. $v_{235} : v_{238} = 1.0043 : 1$.
55. 5.1×10^{-2} rem/yr.
57. 4.7 m.
59. (*a*) 1300 kg;
(*b*) 4.1×10^{6} Ci.
61. (*a*) 4.0×10^{26} W;
(*b*) 3.7×10^{38} protons/s;
(*c*) 1.0×10^{11} yr.
63. (*a*) 3700 decays/s;
(*b*) 5.2×10^{-4} Sv/yr, about 15% of background.
65. 7.274 MeV.
67. 96 yr.
69. 3 mCi.

CHAPTER 32

1. 7.29 GeV.
3. 1.8 T.
5. 13 MHz.
7. alpha particles;
$\lambda_{\text{alpha}} = 2.2\times$ nucleon size,
$\lambda_{\text{proton}} = 4.4\times$ nucleon size.
9. 1.8×10^{-19} m.
11. 5.5 T.
15. 33.9 MeV.
17. 1879.2 MeV.
19. 67.5 MeV.
21. 2.3×10^{-18} m.
23. (*a*) charge (and strangeness);
(*b*) energy;
(*c*) baryon number (and strangeness).
27. 69.3 MeV.
29. $\mathrm{KE}_{\Lambda^0} = 8.6$ MeV, $\mathrm{KE}_{\pi} = 57.4$ MeV.
31. 52.3 MeV.
33. 7.5×10^{-21} s.
35. (*a*) 1.3 keV;
(*b*) 8.9 keV.
37. (*a*) forbidden, energy is not conserved;
(*b*) forbidden, lepton number is not conserved;
(*c*) possible.
39. (*a*) p;
(*b*) $\overline{\Sigma^-}$;
(*c*) K^-;
(*d*) π^-;
(*e*) D_S^-.
41. $D_S^+ = c\bar{s}$.
43.

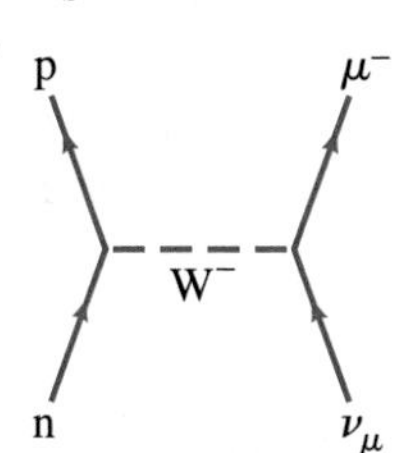

45. (*a*) 0.38 A;
(*b*) 103 m/s.
47. (*a*) 1.022 MeV;
(*b*) 1876.6 MeV.
49. (*a*) forbidden by charge conservation;
(*b*) possible by the strong interaction;
(*c*) forbidden by strangeness conservation for the strong interaction, possible by the weak interaction;
(*d*) forbidden by strangeness conservation for the strong interaction, possible by the weak interaction;
(*e*) possible by weak interaction.
53. −135.0 MeV, −140.9 MeV.
55. 64 fundamental fermions.
57. (*b*) 10^{29} K.
59. 6.59×10^{-5} m.
61.

π^- $\bar{u}$ d; π^0 $\bar{u}$ u $\bar{d}$ d
$\bar{u}$ $\bar{u}$ $\bar{d}$ $\bar{p}$; u d d n

CHAPTER 33

3. 4.8 ly.
5. 0.059″, 17 pc.
7. 110 yr.
9. 3.7×10^{-2}.
11. 1.4×10^{-4} kg/m³.
13. 1.83×10^{9} kg/m³, 3.33×10^{5} times larger.
15. −0.092 MeV, 7.366 MeV.
17. $d_1/d_2 = 0.15$.
19. 4×10^{14} m.
21. 540°.
23. 1.6×10^{8} ly.
25. (*a*) 656 nm;
(*b*) 661 nm;
(*c*) 1670 nm.
27. 0.44c.
31. 6 nucleons/m³.
33. (*a*) 10^{-5} s;
(*b*) 10^{-7} s;
(*c*) 10^{-4} s.
35. 5 ly.
37. 5×10^{9}.
39. 2×10^{28} N.
41. $d_{400\,\text{nm}}/d_{600\,\text{nm}} \approx 1.4$.
43. 0.88c.
45. 2×10^{-3}.
47. (*a*) 13.933 MeV;
(*b*) 4.7 MeV;
(*c*) 3.7×10^{10} K.
49. ≈400 ly, 2×10^{7} times bigger, 80 times smaller.

INDEX

Note: the abbreviation *defn* means the page cited gives the definition of the term; *fn* means the reference is in a footnote; *pr* means it is found in a Problem or Question; *ff* means also the following pages.

PHOTO CREDITS

Title page Art Wolfe/Getty Images, Inc. **CO-1** NOAA/Phil Degginger/Color-Pic, Inc. **1-01** Erich Lessing/Art Resource, N.Y. **1-02a, 1-02b** Franca Principe/Istituto e Museo di Storia della Scienza **1-03** Franca Principe/Istituto e Museo di Storia della Scienza **1-04a** Philip H. Coblentz/Getty Images, Inc.–Stone Allstock **1-04b** Antranig M. Ouzoonian, P.E./Weidlinger Associates, Inc. **1-05** Mary Teresa Giancoli **1-06a, 1-06b** Douglas C. Giancoli **1-07** Paul Silverman/Fundamental Photographs **1-08a** Oliver Meckes/Eye of Science/Max-Planck-Institut-Tubingen/Photo Researchers, Inc. **1-08b** Douglas C. Giancoli **1-09** Adolfo Viansson **1-10a** Douglas C. Giancoli **1-11** Larry Voight/Photo Researchers, Inc. **1-14** David Parker/Science Photo Library/Photo Researchers, Inc. **1-15** The Image Works **CO-2** George D. Lepp/CORBIS BETTMANN **2-08** John E. Gilmore, III **2-16** SuperStock, Inc. **2-17** Nimatallah/Art Resource, N.Y. **2-18** Harold E. Edgerton/Palm Press, Inc. **CO-3** Michel Hans/Vandystadt/Getty Images, Inc.–Allsport Photography **3-17** Berenice Abbott/Commerce Graphics Ltd., Inc. **3-19** Richard Megna/Fundamental Photographs **3-27a** Dave Wilhemi/Corbis/Stock Market **3-27b** Guy Vanderelst/Taxi/Getty Images, Inc.–Taxi **3-27c** Douglas C. Giancoli **CO-4** Mark Wagner/Aviation-images.com **4-01** Daly & Newton/Getty Images, Inc.–Stone Allstock **4-04** Bettmann/Corbis **4-05** Gerard Vandystadt/Agence Vandystadt/Photo Researchers, Inc. **4-07** David Jones/Photo Researchers, Inc. **4-10** Tsado/NASA/Tom Stack & Associates, Inc. **4-37** Lars Ternblad/Getty Images, Inc.–Image Bank **4-39** Kathleen Schiaparelli **4-41** AP/Wide World Photos **4-63** Tyler Stableford/The Image Bank/Getty Images, Inc. **CO-5** Earth Imaging/Getty Images Inc.–Stone Allstock **5-06c** Jay Brousseau/Getty Images, Inc.–Image Bank **5-12** Guido Alberto Rossi/Getty Images, Inc.–Image Bank **5-23** NASA/Johnson Space Center **5-27a** AP/Wide World Photos **5-27b** Mickey Pfleger/Photo 20-20/PictureQuest **5-27c** Dave Cannon/Getty Images, Inc.–Stone Allstock **5-34** C. Grzimek/Okapia **5-35** Daniel L. Feicht/Cedar Fair L.P. **CO-6** Al Bello/Getty Images, Inc.–Liaison **6-21** Harold E. Edgerton/Palm Press, Inc. **6-22** David Madison/David Madison Sports Images, Inc. **6-25** AP/Wide World Photos **6-28** Nick Rowe/Getty Images, Inc.–Photodisc **6-42** CORBIS BETTMANN **6-46** R. Maisonneuve/Publiphoto/Photo Researchers, Inc. **CO-7** Richard Megna/Fundamental Photographs **7-01** Stephen Dunn/Getty Images, Inc.–Liaison **7-08** Loren M. Winters/Visuals Unlimited **7-15** D.J. Johnson **7-18** Science Photo Library/Photo Researchers, Inc. **7-22** Berenice Abbott/Photo Researchers, Inc. **7-28** John McDermott Photography **CO-8** Tom Stewart/CORBIS BETTMANN **8-07a** Jeff Greenberg/Photo Researchers, Inc. **8-11a** Richard Megna/Fundamental Photographs **8-11b** Photoquest, Inc. **8-34** Tim Davis/Photo Researchers, Inc. **8-35** Regis Bossu/Corbis/Sygma **8-36** AP/Wide World Photos **8-38** Karl Weatherly/Getty Images, Inc.–Photodisc. **8-42** Tom Stewart/CORBIS BETTMANN **8-56a** Michael Kevin Daly/Corbis/Stock Market **CO-9** John Kelly/Getty Images, Inc.–Image Bank **9-01** AP/Wide World Photos **9-21** Douglas C. Giancoli **9-23a, 9-23b** Mary Teresa Giancoli **9-26** Fabricius & Taylor/Getty Images, Inc.–Liaison **9-28a** Douglas C. Giancoli **9-28b** Galen Rowell/Mountain Light Photography, Inc. **9-30** Douglas C. Giancoli **9-32** Giovanni Paolo Panini (Roman, 1691–1765), "Interior of the Pantheon, Rome," c. 1734. Oil on canvas, 1.280 × .990 (50 1/2 × 39); framed, 1.441 × 1.143 (56 3/4 × 45). Samuel H. Kress Collection. Photograph © 2001 Board of Trustees, National Gallery of Art, Washington. 1939.1.24.(135)/PA. Photo by Richard Carafelli **9-33** Douglas C. Giancoli **9-34a** Italian Government Tourist Board **9-56** Peter LaMastro/Getty Images, Inc. **CO-10** Verlinden, Vic/Getty Images, Inc.–Image Bank **10-10** CORBIS BETTMANN **10-19a-R, 10-19b-R,** David Hazen **10-33a** Lester V. Bergman/CORBIS BETTMANN **10-33b** Biophoto Associates/Photo Researchers, Inc. **10-34** Rod Planck/Tom Stack & Associates, Inc. **10-36** Alan Blank/Bruce Coleman, Inc. **10-46** Douglas C. Giancoli **10-48** Adam Jones/Photo Researchers, Inc. **10-52** NASA Goddard Space Flight Center/Science Source/Photo Researchers, Inc. **CO-11L** Fundamental Photographs **CO-11R** Jonathan Nourok/PhotoEdit **11-04** Robert Reiff/Getty Images, Inc.–Taxi **11-07** Gary Carter/Visuals Unlimited **11-11** Photo Researchers, Inc. **11-13** Douglas C. Giancoli **11-17** Taylor Devices, Inc. **11-19a** AP/Wide World Photos **11-19b** Corbis/Sygma **11-20** Douglas C. Giancoli **11-27** Art Wolfe/Getty Images, Inc.–Stone Allstock **11-37** Douglas C. Giancoli **11-42** Visuals Unlimited **11-48** Gallant, Andre/Getty Images, Inc.–Image Bank **11-55** Richard Megna/Fundamental Photographs **CO-12** Fra Angelico, Linaioli Altarpiece, detail. Museo di San Marco, Florence, Italy. Scala/Art Resource, N.Y. **12-04** Yoav Levy/Phototake NYC **12-09a** Andy Sacks/Getty Images, Inc.–Stone Allstock **12-09b** Getty Images, Inc.–Liaison **12-10** Bob Daemmrich/The Image Works **12-24** Norman Owen Tomalin/Bruce Coleman, Inc. **12-25b** SETTLES, GARY S./Photo Researchers, Inc. **12-28a** P. Saada/Eurelios/Science Photo Library/Photo Researchers, Inc. **12-28b** Howard Sochurek/Medical Images, Inc. **12-38** Dallas & John Heaton/CORBIS BETTMANN **CO-13** Johner/Amana America, Inc. **13-03** Bob Daemmrich/Stock Boston **13-04a, 13-04b, 13-04c** Franca Principe/Istituto e Museo di Storia della Scienza **13-06** Leonard Lessin/Peter Arnold, Inc. **13-14** Leonard Lessin/Peter Arnold, Inc. **13-15** Michael Newman/PhotoEdit **13-23** Paul Silverman/Fundamental Photographs **13-24** JACK DANIELS/Getty Images, Inc.–Taxi **13-25a, 13-25b, 13-25c** Mary Teresa Giancoli **13-28** Kennan Harvey/Getty Images, Inc.–Stone Allstock **13-29** Brian Yarvin/Photo Researchers, Inc. **CO-14** Bill Losh/Getty Images, Inc.–Taxi **14-11** Getty Images, Inc.–Hulton Archive Photos **14-14a, 14-14b** Science Photo Library/Photo Researchers, Inc. **14-16** Phil Degginger/Color-Pic, Inc. **14-22** Taxi/Getty Images, Inc. **CO-15L** David Woodfall/Getty Images, Inc.–Stone Allstock **CO-15R** AP/Wide World Photos **15-09** Will Hart **15-10a, 15-10b, 15-10c** Leonard Lessin/Peter Arnold, Inc. **15-20a** Sandia National Laboratories **15-20b** Martin Bond/Science Photo Library/Photo Researschers, Inc. **15-20c** Lionel Delevingne/Stock Boston **Table 15-4** (clockwise from top right) Ed Degginger/Color-Pic, Inc; Michael Collier; Malcolm Fife/Getty Images, Inc.–Photodisc; Inga Spence/Visuals Unlimited **15-25** Geoff Tompkinson/Science Photo Library/Photo Researchers, Inc. **15-26** Inga Spence/Visuals Unlimited **15-27** Michael Collier **CO-16** Fundamental Photographs **16-36** Michael

J. Lutch/Boston Museum of Science **16-43** Dr. Gopal Murti/Science Photo Library/Photo Researchers, Inc. **CO-17** Gene Moore/Phototake NYC **17-13c** Tom Pantages/Tom Pantages **17-17** Tom Pantages/Tom Pantages **17-18** Custom Medical Stock Photo, Inc. **17-22** Jon Feingersh/Jon Feingersh **CO-18** Mahaux Photography/Getty Images Inc.—Image Bank **18-01** J. L. Charmet/Science Photo Library/Photo Researchers **18-06b** Dave King/Dorling Kindersley Media Library **18-11** T. J. Florian/Rainbow **18-15** Richard Megna/Fundamental Photographs **18-16** Tony Freeman/PhotoEdit **18-18** Barbara Filet/Tony Stone Images **18-32** Jerry Marshall/Jerry Marshall **18-34** Scott T. Smith/Corbis/Bettmann; www.corbis.com/Scott T. Smith/Corbis Images **18-36** Jim Wehtje/Getty Images, Inc.—Photodisc. **CO-19** Courtesy of iRiver **19-24** Dept. Clinical Radiology, Salisbury District Hospital/SPL/Photo Researchers, Inc.; Department of Clinical Radiology, Salisbury District Hospital/Science Photo Library/Photo Researchers, Inc. **19-27a** Getty Images, Inc.—Photodisc. **19-27b** William E. Ferguson/William E. Ferguson **19-27c** Ed Degginger/Color-Pic, Inc. **19-29a** Paul Silverman/Fundamental Photographs **19-29b** Paul Silverman/Fundamental Photographs **CO-20** Chris Rogers/Rainbow; © 2000 Chris Rogers/Rainbow **20-01** Unidentified/Dorling Kindersley Media Library; © Dorling Kindersley **20-04a** Stephen Oliver/Dorling Kindersley Media Library; Stephen Oliver © Dorling Kindersley **20-06** Mary Teresa Giancoli **20-08a** Richard Megna/Fundamental Photographs **20-18** Pekka Parviainen/Science Photo Library/Photo Researchers, Inc. **20-50** Clive Streeter/Dorling Kindersley Media Library; Clive Streeter © Dorling Kindersley **CO-21** Richard Megna/Fundamental Photographs **21-08** Diva de Provence **21-13** Werner H. Muller/Peter Arnold, Inc. **21-21** Joe Raedle/Getty Images, Inc—Liaison **21-27b** Pete Saloutos/Corbis/Bettmann **21-30a** Richard Megna/Fundamental Photographs **CO-22** Jeremy Woodhouse/Getty Images, Inc.—Photodisc. **22-01** American Institute of Physics/Niels Bohr Library/AIP Emilio Segrè Visual Archives **22-09** Image Works **22-18** Larry Mulvehill/The Image Works; © Larry Mulvehill/The Image Works **22-21** World Perspectives/Getty Images, Inc.—Stone Allstock **CO-23** Mulvehill/The Photo Works **23-05** Douglas C. Giancoli **23-9a** Mary Teresa Giancoli/Mary Teresa Giancoli; Mary Teresa Giancoli and Suzanne Saylor **23-9b** Paul Silverman/Fundamental Photographs **23-18** Travel Pix Ltd./SuperStock, Inc. **23-21** Mary Teresa Giancoli **23-28b** S. Elleringmann/Bilderberg/Aurora & Quanta Productions, Inc. **23-30** Douglas C. Giancoli and Howard Shugat **23-46** Mary Teresa Giancoli **CO-24** Dave King/Dorling Kindersley Media Library; Dave King © Dorling Kindersley **24-04a** John M. Dunay IV/Fundamental Photographs **24-13** Science Photo Library/David Parker/Photo Researchers **24-17** George B. Diebold/Corbis/Bettmann **24-31b** Ken Kay/Fundamental Photographs **24-33** Bausch & Lomb Incorporated **24-35** Kristen Brochmann/Fundamental Photographs **24-51** Texas Instruments Incorporated **24-52b** Dan Rutter/Daniel Rutter; © Daniel Rutter/Dan's Data **24-55** Pekka Parviainen/Science Photo Library/Photo Researchers, Inc. **24-56** Spike Mafford/Getty Images, Inc.—Photodisc. **CO-25** Richard Megna/Fundamental Photographs **25-04** Mary Teresa Giancoli **25-07** Leonard Lessin/Peter Arnold **25-18** Franca Principe/Istituto e Museo di Storia della Scienza, Florence, Italy **25-20** Yerkes Observatory **25-21c** Palomar/Caltech **25-21d** Roger Ressmeyer/Starlight **25-23b** Olympus America Inc. **25-28** Reproduced by permission from M. Cagnet, M. Francon, and J. Thrier, *The Atlas of Optical Phenomena*. Berlin: Springer-Verlag, 1962. **25-31** Space Telescope Science Institute **25-32** National Astronomy & Ionosphere Center, Cornell University, Arecibo, Puerto Rico **25-36** Photo by W. Friedrich/Max von Laue. Burndy Library, Dibner Institute for the History of Science and Technology, Cambridge, Massachusetts. **25-40** Rosalind Franklin/Photo Researchers **25-44a** Martin M. Rotker/Martin M. Rotker **25-44b** Simon Fraser/Science Photo Library/Photo Researchers, Inc. **25-47** Ron Chapple/Getty Images, Inc.—Taxi **CO-26** Cambridge University Press; "The City Blocks Became Still Shorter" from page 4 of the book, "Mr Tompkins in Paperback" by George Gamow. Reprinted with the permission of Cambridge University Press. **26-01** Corbis/Bettmann **26-11** Cambridge University Press; "Unbelievably Shortened" from page 3 of "Mr Tompkins in Paperback" by George Gamow. Reprinted with the permission of Cambridge University Press. **CO-27** P. M. Motta and F. M. Magliocca/Science Photo Library/Photo Researchers, Inc. **27-11** S. A. Goudsmit/American Institute of Physics/Niels Bohr Library/AIP Emilio Segrè Visual Archives **27-12** Education Development Center, Inc. **27-15a** Lee D. Simon/Science Source/Photo Researchers, Inc. **27-15b** Oliver Meckes/Max-Planck-Institut-Tubingen/Photo Researchers, Inc. **CO-28** Richard Cummins/Corbis/Bettmann; © Richard Cummins/Corbis **28-01** American Institute of Physics/Emilio Segrè Visual Archives **28-02** ED. Rosetti/American Institute of Physics/Emilio Segrè Visual Archives **28-04** Advanced Research Lab, Hitachi, Ltd. **28-15** Mark Schneider/Visuals Unlimited **28-21** Yoav Levy/Phototake NYC **28-23** Philippe Plaily/Photo Researchers, Inc. **CO-29** Intel Corporation Pressroom Photo Archives **CO-30** Reuters Newsmedia Inc./Corbis **30-03** Center for the History of Chemistry **30-07** University of Chicago/American Institute of Physics/Niels Bohr Library/Courtesy of AIP Emilio Segrè Visual Archives **30-16** Fermilab Visual Media Services **CO-31** Peter Beck/Corbis/Bettmann; © Peter Beck/CORBIS **31-05** Gary Sheahan, "Birth of the Atomic Age," Chicago (Illinois); 1957. Chicago Historical Society, ICHi-33305. **31-08** LeRoy N. Sanchez/Los Alamos National Laboratory **31-10** Corbis **31-13** Lawrence Livermore National Laboratory/Science Source/Photo Researchers, Inc. **31-16** J. Van't Hof. **31-21** Southern Illinois University/Peter Arnold, Inc. **31-23** Mehau Kulyk/Science Photo Library/Photo Researchers, Inc. **CO-32** Fermilab/Science Photo Library/Photo Researchers, Inc. **32-01** Science Service/Watson Davis/American Institute of Physics/Niels Bohr Library/AIP Emilio Segrè Visual Archives/Physics Today Collection **32-03a** Fermilab Visual Media Services **32-03b** Fermilab Visual Media Services **32-04b** Barrie Rokeach/Aerial/Terrestrial Photography **32-05** CERN/Science Photo Library/Photo Researchers, Inc. **32-09** Science Photo Library/Photo Researchers **32-10** Brookhaven National Laboratory **CO-33** NASA/WMAP Science Team/NASA Headquarters **33-01** NASA Headquarters **33-02** Photo Researchers **33-03** U.S. Naval Observatory Photo/NASA Headquarters **33-04** National Optical Astronomy Observatories **33-09** Hubble Space Telescope/NASA/NASA Headquarters; Courtesy of NASA and the Hubble Heritage Team **33-11** National Optical Astronomy Observatories **33-22** Bell Photographers, Inc. **33-24** NASA Headquarters; GSFC/NASA **33-28** Dr. Wes Colley

Periodic Table of the Elements§

Group I	Group II	Transition Elements										Group III	Group IV	Group V	Group VI	Group VII	Group VIII
H 1 1.00794 $1s^1$																	**He** 2 4.002602 $1s^2$
Li 3 6.941 $2s^1$	**Be** 4 9.012182 $2s^2$											**B** 5 10.811 $2p^1$	**C** 6 12.0107 $2p^2$	**N** 7 14.00674 $2p^3$	**O** 8 15.9994 $2p^4$	**F** 9 18.9984032 $2p^5$	**Ne** 10 20.1797 $2p^6$
Na 11 22.989770 $3s^1$	**Mg** 12 24.3050 $3s^2$											**Al** 13 26.981538 $3p^1$	**Si** 14 28.0855 $3p^2$	**P** 15 30.973761 $3p^3$	**S** 16 32.066 $3p^4$	**Cl** 17 35.4527 $3p^5$	**Ar** 18 39.948 $3p^6$
K 19 39.0983 $4s^1$	**Ca** 20 40.078 $4s^2$	**Sc** 21 44.955910 $3d^14s^2$	**Ti** 22 47.867 $3d^24s^2$	**V** 23 50.9415 $3d^34s^2$	**Cr** 24 51.9961 $3d^54s^1$	**Mn** 25 54.938049 $3d^54s^2$	**Fe** 26 55.845 $3d^64s^2$	**Co** 27 58.933200 $3d^74s^2$	**Ni** 28 58.6934 $3d^84s^2$	**Cu** 29 63.546 $3d^{10}4s^1$	**Zn** 30 65.39 $3d^{10}4s^2$	**Ga** 31 69.723 $4p^1$	**Ge** 32 72.61 $4p^2$	**As** 33 74.92160 $4p^3$	**Se** 34 78.96 $4p^4$	**Br** 35 79.904 $4p^5$	**Kr** 36 83.80 $4p^6$
Rb 37 85.4678 $5s^1$	**Sr** 38 87.62 $5s^2$	**Y** 39 88.90585 $4d^15s^2$	**Zr** 40 91.224 $4d^25s^2$	**Nb** 41 92.90638 $4d^45s^1$	**Mo** 42 95.94 $4d^55s^1$	**Tc** 43 (98) $4d^55s^2$	**Ru** 44 101.07 $4d^75s^1$	**Rh** 45 102.90550 $4d^85s^1$	**Pd** 46 106.42 $4d^{10}5s^0$	**Ag** 47 107.8682 $4d^{10}5s^1$	**Cd** 48 112.411 $4d^{10}5s^2$	**In** 49 114.818 $5p^1$	**Sn** 50 118.710 $5p^2$	**Sb** 51 121.760 $5p^3$	**Te** 52 127.60 $5p^4$	**I** 53 126.90447 $5p^5$	**Xe** 54 131.29 $5p^6$
Cs 55 132.90545 $6s^1$	**Ba** 56 137.327 $6s^2$	57–71†	**Hf** 72 178.49 $5d^26s^2$	**Ta** 73 180.9479 $5d^36s^2$	**W** 74 183.84 $5d^46s^2$	**Re** 75 186.207 $5d^56s^2$	**Os** 76 190.23 $5d^66s^2$	**Ir** 77 192.217 $5d^76s^2$	**Pt** 78 195.078 $5d^96s^1$	**Au** 79 196.96655 $5d^{10}6s^1$	**Hg** 80 200.59 $5d^{10}6s^2$	**Tl** 81 204.3833 $6p^1$	**Pb** 82 207.2 $6p^2$	**Bi** 83 208.98038 $6p^3$	**Po** 84 (209) $6p^4$	**At** 85 (210) $6p^5$	**Rn** 86 (222) $6p^6$
Fr 87 (223) $7s^1$	**Ra** 88 (226) $7s^2$	89–103‡	**Rf** 104 (261) $6d^27s^2$	**Db** 105 (262) $6d^37s^2$	**Sg** 106 (266) $6d^47s^2$	**Bh** 107 (264) $6d^57s^2$	**Hs** 108 (269) $6d^67s^2$	**Mt** 109 (268) $6d^77s^2$	**Ds** 110 (271) $6d^97s^1$	111 (272) $6d^{10}7s^1$	112 (277) $6d^{10}7s^2$						

Symbol — **Cl** 17 — Atomic Number
Atomic Mass§ — 35.4527
$3p^5$ — Electron Configuration (outer shells only)

†Lanthanide Series	**La** 57 138.9055 $5d^16s^2$	**Ce** 58 140.115 $4f^15d^16s^2$	**Pr** 59 140.90765 $4f^35d^06s^2$	**Nd** 60 144.24 $4f^45d^06s^2$	**Pm** 61 (145) $4f^55d^06s^2$	**Sm** 62 150.36 $4f^65d^06s^2$	**Eu** 63 151.964 $4f^75d^06s^2$	**Gd** 64 157.25 $4f^75d^16s^2$	**Tb** 65 158.92534 $4f^95d^06s^2$	**Dy** 66 162.50 $4f^{10}5d^06s^2$	**Ho** 67 164.93032 $4f^{11}5d^06s^2$	**Er** 68 167.26 $4f^{12}5d^06s^2$	**Tm** 69 168.93421 $4f^{13}5d^06s^2$	**Yb** 70 173.04 $4f^{14}5d^06s^2$	**Lu** 71 174.967 $4f^{14}5d^16s^2$
‡Actinide Series	**Ac** 89 (227.02775) $6d^17s^2$	**Th** 90 232.0381 $6d^27s^2$	**Pa** 91 (231) $5f^26d^17s^2$	**U** 92 238.0289 $5f^36d^17s^2$	**Np** 93 (237) $5f^46d^17s^2$	**Pu** 94 (244) $5f^66d^07s^2$	**Am** 95 (243) $5f^76d^07s^2$	**Cm** 96 (247) $5f^76d^17s^2$	**Bk** 97 (247) $5f^96d^07s^2$	**Cf** 98 (251) $5f^{10}6d^07s^2$	**Es** 99 (252) $5f^{11}6d^07s^2$	**Fm** 100 (257) $5f^{12}6d^07s^2$	**Md** 101 (258) $5f^{13}6d^07s^2$	**No** 102 (259) $5f^{14}6d^07s^2$	**Lr** 103 (262) $5f^{14}6d^17s^2$

§ Atomic mass values averaged over isotopes in the percentages they occur on Earth's surface. For unstable elements, mass of the longest-lived known isotope is given in parentheses. 2003 revisions. (See also Appendix B.)

Useful Geometry Formulas—Areas, Volumes

Circumference of circle $C = \pi d = 2\pi r$

Area of circle $A = \pi r^2 = \dfrac{\pi d^2}{4}$

Area of rectangle $A = lw$

Area of parallelogram $A = bh$

Area of triangle $A = \frac{1}{2} hb$

Right triangle (Pythagoras) $c^2 = a^2 + b^2$

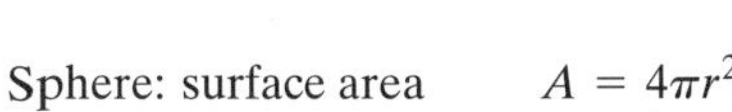

Sphere: surface area $A = 4\pi r^2$

volume $V = \frac{4}{3}\pi r^3$

Rectangular solid: volume $V = lwh$

Cylinder (right): surface area $A = 2\pi r l + 2\pi r^2$

volume $V = \pi r^2 l$

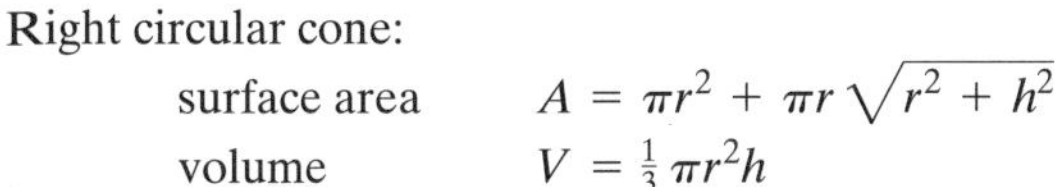

Right circular cone: surface area $A = \pi r^2 + \pi r\sqrt{r^2 + h^2}$

volume $V = \frac{1}{3}\pi r^2 h$

Exponents [See Appendix A–2 for details]

$(a^n)(a^m) = a^{n+m}$ [Example: $(a^3)(a^2) = a^5$]

$(a^n)(b^n) = (ab)^n$ [Example: $(a^3)(b^3) = (ab)^3$]

$(a^n)^m = a^{nm}$ [Example: $(a^3)^2 = a^6$; Example: $(a^{\frac{1}{4}})^4 = a$]

$a^{-1} = \dfrac{1}{a}$ $\quad a^{-n} = \dfrac{1}{a^n}$ $\quad a^0 = 1$

$a^{\frac{1}{2}} = \sqrt{a}$ $\quad a^{\frac{1}{4}} = \sqrt{\sqrt{a}}$

$(a^n)(a^{-m}) = \dfrac{a^n}{a^m} = a^{n-m}$ [Ex.: $(a^5)(a^{-2}) = a^3$]

$\dfrac{a^n}{b^n} = \left(\dfrac{a}{b}\right)^n$

Quadratic Formula [Appendix A–4]

Equation with unknown x, in the form

$$ax^2 + bx + c = 0,$$

has solutions

$$x = \frac{-b \pm \sqrt{b^2 - 4ac}}{2a}.$$

Logarithms [Appendix A–8; Table p. A–11]

If $y = 10^x$, then $x = \log_{10} y = \log y$.

If $y = e^x$, then $x = \log_e y = \ln y$.

$\log(ab) = \log a + \log b$

$\log\left(\dfrac{a}{b}\right) = \log a - \log b$

$\log a^n = n \log a$

Binomial Expansion [Appendix A–5]

$$(1 + x)^n = 1 + nx + \frac{n(n-1)}{2\cdot 1}x^2 + \frac{n(n-1)(n-2)}{3\cdot 2\cdot 1}x^3 + \cdots \quad [\text{for } x^2 < 1]$$

$\approx 1 + nx$ if $x \ll 1$

[Example: $(1 + 0.01)^3 \approx 1.03$]

[Example: $\dfrac{1}{\sqrt{0.99}} = \dfrac{1}{\sqrt{1 - 0.01}} = (1 - 0.01)^{-\frac{1}{2}} \approx 1 - \left(-\frac{1}{2}\right)(0.01) \approx 1.005$]

Fractions

$\dfrac{a}{b} = \dfrac{c}{d}$ is the same as $ad = bc$

$$\frac{\left(\frac{a}{b}\right)}{\left(\frac{c}{d}\right)} = \frac{ad}{bc}$$

Trigonometric Formulas [Appendix A–7]

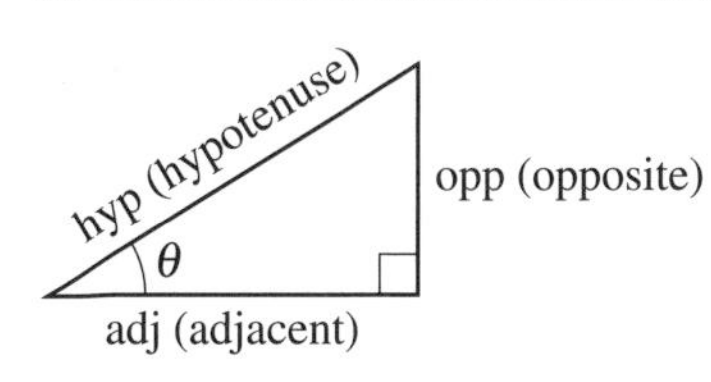

$\sin\theta = \dfrac{\text{opp}}{\text{hyp}}$

$\cos\theta = \dfrac{\text{adj}}{\text{hyp}}$

$\tan\theta = \dfrac{\text{opp}}{\text{adj}}$

$\text{adj}^2 + \text{opp}^2 = \text{hyp}^2$ (Pythagorean theorem)

$\tan\theta = \dfrac{\sin\theta}{\cos\theta}$

$\sin^2\theta + \cos^2\theta = 1$

$\sin 2\theta = 2\sin\theta\cos\theta$

$\cos 2\theta = (\cos^2\theta - \sin^2\theta) = (1 - 2\sin^2\theta) = (2\cos^2\theta - 1)$

$\sin(180° - \theta) = \sin\theta$ $\quad \cos(180° - \theta) = -\cos\theta$

$\sin(90° - \theta) = \cos\theta$, $\cos(90° - \theta) = \sin\theta$ $\quad [0 < \theta < 90°]$

$\sin\frac{1}{2}\theta = \sqrt{(1 - \cos\theta)/2}$ $\quad \cos\frac{1}{2}\theta = \sqrt{(1 + \cos\theta)/2}$

$\sin\theta \approx \theta$ [for small $\theta \lesssim 0.2$ rad]

$\cos\theta \approx 1 - \dfrac{\theta^2}{2}$ [for small $\theta \lesssim 0.2$ rad]

$\sin(A \pm B) = \sin A\cos B \pm \cos A\sin B$

$\cos(A \pm B) = \cos A\cos B \mp \sin A\sin B$

For any triangle:

$c^2 = a^2 + b^2 - 2ab\cos\gamma$ (law of cosines)

$\dfrac{\sin\alpha}{a} = \dfrac{\sin\beta}{b} = \dfrac{\sin\gamma}{c}$ (law of sines)

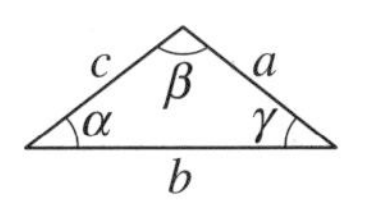